About Island Press

Since 1984, the nonprofit Island Press has been stimulating, shaping, and communicating the ideas that are essential for solving environmental problems worldwide. With more than 800 titles in print and some 40 new releases each year, we are the nation's leading publisher on environmental issues. We identify innovative thinkers and emerging trends in the environmental field. We work with world-renowned experts and authors to develop cross-disciplinary solutions to environmental challenges.

Island Press designs and implements coordinated book publication campaigns in order to communicate our critical messages in print, in person, and online using the latest technologies, programs, and the media. Our goal: to reach targeted audiences—scientists, policymakers, environmental advocates, the media, and concerned citizens—who can and will take action to protect the plants and animals that enrich our world, the ecosystems we need to survive, the water we drink, and the air we breathe.

Island Press gratefully acknowledges the support of its work by the Agua Fund, Inc., The Margaret A. Cargill Foundation, Betsy and Jesse Fink Foundation, The William and Flora Hewlett Foundation, The Kresge Foundation, The Forrest and Frances Lattner Foundation, The Andrew W. Mellon Foundation, The Curtis and Edith Munson Foundation, The Overbrook Foundation, The David and Lucile Packard Foundation, The Summit Foundation, Trust for Architectural Easements, The Winslow Foundation, and other generous donors.

The opinions expressed in this book are those of the author(s) and do not necessarily reflect the views of our donors.

Repeat Photography

Repeat Photography

Methods and Applications in the Natural Sciences

EDITED BY

Robert H. Webb, Diane E. Boyer, and Raymond M. Turner

Washington | Covelo | London

© 2010 Island Press

All rights reserved under International and Pan-American Copyright Conventions. No part of this book may be reproduced in any form or by any means without permission in writing from the publisher: Island Press, Suite 300,
1718 Connecticut Ave., NW, Washington, DC 20009

ISLAND PRESS is a trademark of the Center for Resource Economics.

Library of Congress Cataloging-in-Publication Data

Repeat photography : methods and applications in the natural sciences / edited by Robert H. Webb, Diane E. Boyer,
Raymond M. Turner.
     p. cm.
 Includes bibliographical references and index.
 ISBN-13: 978-1-59726-712-0 (cloth : alk. paper)
 ISBN-10: 1-59726-712-0 (cloth : alk. paper)
 ISBN-13: 978-1-59726-713-7 (pbk. : alk. paper)
 ISBN-10: 1-59726-713-9 (pbk. : alk. paper) 1. Landscape changes—Research. 2. Repeat photography.
3. Environmental monitoring. 4. Nature—Effect of human beings on—Research. I. Webb, Robert H. II. Boyer,
Diane E. III. Turner, R. M. (Raymond M.)
 GB405.R47 2010
 577. 072′3—dc22

                                                                                                              2010009377

Printed on recycled, acid-free paper

Manufactured in the United States of America
10 9 8 7 6 5 4 3 2 1

To Harold E. (Hal) Malde (1923–2007),
one of the pioneers of repeat photography,
who taught us the meaning and value of precision.

A

B

Figure 0.1. Mesa Encantada, Valencia County, New Mexico, USA.
A. (ca. 1899). In this view of Mesa Encantada, the mesa is surrounded by *Juniperus osteosperma* (junipers), but the foreground is barren and interrupted by gullies. That the distant horizon is visible reflects the once excellent air quality of the nineteenth century, especially considering that William Henry Jackson's glass-plate negative was orthochromatic and sensitive only to blue light. (Attributed to W. H. Jackson, 080026, courtesy of H. E. Malde).
B. (21 June 1977). Using the Sundicator (Boyer et al., chapter 2), the original view was matched with nearly the same azimuth but different inclination of the sun (Rogers et al. 1984). The once-barren foreground in Jackson's view is now crowded with *Juniperus*, and the steep banks of the gullies have eroded to more gentle slopes. (H. E. Malde, 994, USGS Photographic Library).

Contents

Foreword

George E. Gruell

The value of repeat photography to document landscape change has attracted worldwide attention, and now this book brings together recent examples of repeat photography by scientists on five continents. Representing a range of disciplines, the authors demonstrate the use of this technique to record many aspects of change over widely differing landscapes. In doing so, they underscore the many benefits of repeat photography in the earth and environmental sciences. Repeat photography is an inherently interdisciplinary endeavor, with any given set of photographs open to interpretation across a wide set of parameters. While repeat photography cannot provide the same type of quantitative data as can be obtained from satellite imagery and aerial photographs, it serves to complement them, providing not only species-specific information but also a greater time range, in many cases going back to the nineteenth century. It is an excellent technique for documenting long-term change, both by looking backward to the time of the earliest available landscape photographs and by creating a frame of reference for assessing future changes. Repeat photography has become critical in light of recent discussions of global change, be it studying the gamut from retreat and advance of glaciers, or, as I have done, documenting changes in vegetation comprising wildlife habitats.

Through my long career, I have witnessed first-hand the growth in application of this technique throughout the natural sciences. Personally, I have had the opportunity to conduct repeat photography projects in four regions of the interior western United States: northern Nevada in the Great Basin, western Wyoming's Yellowstone Ecosystem, Montana's Northern Rockies, and the California Sierra Nevada. Knowledge of vegetative conditions and trends is essential for informed management of wildlife. Lacking such knowledge, managers tend to accept current vegetative conditions as the norm, not understanding that wildlife habitats have undergone pronounced changes. Early in my career, *The Changing Mile* (1965) by J. R. Hastings and R. M. Turner—a repeat photography study depicting vegetation changes in the southwestern United States and northwestern Mexico—convinced me that repeat photography would be useful in assessing the condition and trend of wildlife habitat. As recently demonstrated by the updated version, *The Changing Mile Revisited* (2003), I also knew that my work would be part of an ongoing process of documenting constantly evolving landscapes.

As the first wildlife biologist on the Humboldt National Forest in Nevada, I sought knowledge of what wildlife habitats had looked like in the past, how they might have changed, and how such changes had affected wildlife populations. This was shortly after passage of the Multiple Use and Sustained Yield Act of 1960, which directed the Forest Service to consider the requirements of wildlife in the decision-making process. My search of Forest Service files in Nevada and Washington, DC, turned up numerous photographs of rangelands that had been grazed by livestock. Locating the same points on the ground, I repeated 121 scenes in 1965 and 1966. The original photographs, taken mostly between 1910 and 1935, showed a more open environment with perennial bunch grass predominating. As in the Hastings and Turner study, the repeat photographs showed an increase in shrubs and trees, a trend favorable to mule deer (*Odocoileus hemionus*; fig. 0.2).

At the Bridger-Teton National Forest in western Wyoming, vegetation is largely coniferous forest, with *Pinus contorta* (lodgepole pine), *Pseudotsuga*

menziesii (Douglas-fir), *Picea engelmannii* (Engelmann spruce), *Abies lasiocarpa* (subalpine fir), and *Pinus albicaulis* (white bark pine) as the primary species. Aspen communities, key wildlife habitat, are interspersed with conifers and largely found at lower and midelevations. Streams, ponds, and wet meadows support riparian vegetation, while tall forbs (herblands) characterize the plant cover on open slopes at the higher elevations. My 1967 assignment from the director of wildlife management, Intermountain Region, Forest Service, involved assessing the condition and long-term trend of wildlife habitats. Controversy had developed between agencies over the condition of elk habitat, a controversy that continues to this day. First, I conducted a study of elk summer range on Big Game Ridge in the Teton Wilderness, as its condition was thought to be deteriorated because of trampling by elk (*Cervus canadensis*).

Figure 0.2. (Foreword). Ruby Mountains, Nevada, USA.

A. (28 August–3 September 1868). In the southern Ruby Mountains of Elko County, Nevada, this east–southeast view at 2,286 meters above Flynn-Hager Spring overlooks the Ruby Marshes. The open slope in the distance is covered by herbaceous plants, *Symphoricarpos albus* (snowberry), and other deciduous shrubs. An arrow points to a *Populus tremuloides* (aspen) clone in initial stages of growth. This growth stage, scattered dead stems, and subsequent noting of charred wood in the vicinity suggest recent fire disturbance. *Cercocarpus ledifolius* (curlleaf mountain mahogany), *Pinus monophylla* (single-leaf pinyon), and *Juniperus osteosperma* (Utah juniper) are confined to steep rocky slopes that did not readily burn. A lone *Cercocarpus* is growing to the left of the arrow. (T. H. O'Sullivan, 77-KS-2-78, National Archives).

B. (31 July 1982). *Cercocarpus* and *Pinus monophylla* dominate the once open slopes, while deciduous shrubs are few. The *Populus* (arrow) have grown into a dense stand of trees. The lone *Cercocarpus* immediately to the left of the arrow has grown considerably. Woody species on rocky sites have offset losses by reproduction of seedlings. Some have carried over from the original photograph. (G. E. Gruell).

A search of various depositories produced photographs going back to 1872, taken by, among others, novelist Owen Wister and pioneer photographer W. H. Jackson. The repeat of these photographs and others in *Fire's Influence on Wildlife Habitat on the Bridger-Teton Forest, Wyoming* shows a dramatic transition in vegetation related to decades of fire suppression. The early scenes generally show conifer patches, fire-killed snags, and crown-sprouting shrubs and *Populus tremuloides* (aspen) in early stages of succession. Subsequent suppression of fires allowed development of dense forests. Trees that formerly formed a mosaic of age classes with early stages predominating grew into nearly continuous mature and overmature stands. Closure of tree canopies blocked sunlight from reaching palatable understory plants, which then died out. Displacement of *Populus* by conifer forest had been detrimental to many wildlife species. Study of repeat photography led to the conclusion that fire had been a crucial ecological agent in rejuvenating forested wildlife habitats.

At the Forest Service's Fire Sciences Laboratory in Missoula, Montana, I initiated a study of historical trends in wildlife habitats across the Northern Rockies. A search of archives including the Montana Historical Society, University of Montana Archives and Special Collections, US Geological Survey in Denver, and National Archives in Washington, DC, produced over 100 photographs taken between 1871 and 1946. Eighty-six of these were repeated, and paired comparisons were published in *Fire and Vegetative Trends in the Northern Rockies: Interpretations from 1871–1982 Photographs.*

Repeat photographs in the northern portion of Idaho and northwest Montana show a mosaic of burned and unburned landscape resulting from high-intensity fire at infrequent intervals. These landscapes had become largely covered by thick stands of *Abies grandis* (grand fir), *Tsuga heterophylla* (western hemlock), and *Thuja plicata* (western red cedar). Herbaceous plants and shrubs died or are senescent because of shading from the dense tree cover.

To the south in the Bitterroot Valley, a region of less moisture than northern Idaho or northwestern Montana, repeat photography revealed striking differences between the historic and modern vegetation. In 1909, a unique study began when a Forest Service photographer recorded the results of the 1906 Lick Creek timber sale, the first large sale in the Forest Service's Northern Region. A dozen of the 1909 photo points were relocated in the 1920s and have been rephotographed in every decade since. The early landscape supported open stands of large fire-maintained *Pinus ponderosa* thinned by low-intensity fires at intervals of 4 to 20 years. Logging of large trees, followed by the absence of fire for 70 years or more, resulted in an increase of willow and conifer saplings. *Pseudotsuga menziesii* proliferated on sites that were formerly dominated by *Pinus ponderosa*. The changes depicted in these photos allowed assessment of the effect of forest succession on timber management, wildlife habitat, livestock grazing, forest fuels, and scenic quality. These changes and initial results of restoration treatments are described in the 1999 publication *Eighty-eight Years of Changes in a Managed Ponderosa Pine Forest.*

Photographs in other regions of Montana show a more open landscape historically. Conifers were sparse and grass and shrubs were plentiful on unbroken southerly and easterly slopes of the Lewis and Clark National Forest in the Northern Rockies. By the early 1980s, *Populus* and other deciduous vegetation had been displaced by conifers. These stands were swept by wildfire in 1988 resulting in conversion to grass, shrubs, and *Populus*. Early photographs in north-central Montana show burned landscapes in early succession; fires kept pine stands open and inhibited encroachment of trees at forest–grassland ecotones. Repeat photographs show dense tree cover. In south-central Montana, the continental climate is cold and dry, and valleys have high base elevations. Here, conditions for wildfires are favorable because of dry summer periods and the interspersion of forests within grasslands. Native American ignitions were common as indicated by the reports of Lewis and Clark and others. Fire in

Pseudotsuga menziesii and *Pinus contorta* forests burning at various intensities produced a mosaic of treatments on the landscape that is apparent in early photographs. These treatments favored perennial bunch grass and suppressed sagebrush. The absence of fire resulted in a widespread increase in conifers. In the plains region of southeast Montana, *P. ponderosa* stands were kept open by frequent low-intensity surface fires. A US Army officer, Captain Raynolds, witnessed extensive fires set by aboriginals in the Little Bighorn drainage and Wolf Mountains during late August 1859. The absence of fire allowed widespread development of trees.

I contracted with the California Forestry Association to document forest succession in the Sierra Nevada in 1992. A wealth of early photographs was located, a number of which were by the first professional photographers in the West. The best of these were repeated and published in *Fire in Sierra Nevada Forests: A Photographic Interpretation of Ecological Change since 1849*. Forests of oak woodlands and pine forests at low elevations on the west slope were more open historically, having been frequently disturbed by fire. The repeat scenes show a massive increase in the density of woody vegetation. Mixed conifer growing on more productive sites at higher elevations had changed from patchy distribution to dense stands. On the drier east slope of the Sierra Nevada, open stands of *P. ponderosa* and *P. jeffreyi* (Jeffrey pine) prevailed historically. Grasses and scattered shrubs constituted the understory. Riparian zones and moist upland sites supported willows, aspens, cottonwoods, and other deciduous vegetation. In the absence of fire since the turn of the century or longer, former openings are choked by shade-tolerant, fire sensitive *Abies concolor* (white fir). At the highest elevations, *Abies magnifica* (red fir) and *P. contorta* predominated historically in a relatively open stand structure. Trees grew in patches among forest openings. Increased tree density and canopy closure resulted in displacement of *Populus*, while conifers had encroached dry meadows. This study demonstrated that Sierra Nevada vegetation has undergone major changes over the past 150 years—the forest's health has taken a change for the worse. Scenic quality is degraded, wildlife habitat has deteriorated, overstocked stands are susceptible to insects and disease, and excessive forest fuels make these stands vulnerable to catastrophic wildfire.

In 2004, I repeated the photos I had taken on the Humboldt-Toiyabe National Forest in 1965–1966. Ninety scenes repeated on five ranger districts showed a continued trend toward development of woody vegetation. *P. edulis* (pinyon pine) in pinyon–juniper woodlands had filled openings, and older trees had become more dense. *Populus* had developed into dense stands, and the encroachment of sagebrush on mountain meadows was more pronounced. This trend was particularly unfavorable for sage grouse (*Centrocercus urophasianus*) and mule deer.

My more than 40 years of experience in repeat photography have led me to the inevitable conclusion that forests of the interior West have undergone major changes because of the lack of fire disturbance. Conifers have thickened at the expense of understory plants. Woody vegetation on rangelands of northern Nevada increased in the absence of fire. On moist sites, the development of dense stands of *Populus* can be traced to marked reductions in livestock grazing, which allowed the release of suppressed suckers. Without the aid of repeat photographs, we would have little insight on the level of these changes. Also, the repeat photographs have established a benchmark against which future changes can be evaluated. What I have learned in my career mirrors what is discussed in this book, where researchers have identified many causes for increases in woody vegetation, including ones I have identified in my studies.

When I started my work in the 1960s, only a few researchers were applying repeat photography to landscape-change topics, and few with a sustained research motivation. Now, as this book documents, many researchers use this technique to address some of the same questions I have—on different continents with different ecosystems. Moreover, the applications have expanded enormously, and while most of the work published at the time I started applying this technique dealt with ecological change, now it is a more balanced application across the natural sciences, ranging from changes in geomorphology and glaciers to the long-term stability of archaeological sites. I'm honored to have had the opportunity to write the foreword for this landmark production. Its worldwide examples of vegetation and geomorphic changes should spur interest in repeat photography and motivate future studies. As the preceding review of my career indicates, I am heavily vested in this technique and wish to encourage its future use for monitoring landscape change.

Literature Cited

Gruell, G. E. 1980. *Fire's influence on wildlife habitat on the Bridger-Teton Forest, Wyoming.* USDA Forest Service Research Paper INT-252. Intermountain Forest and Range Experiment Station, Ogden, UT.

Gruell, G. E. 1983. *Fire and vegetative trends in the Northern Rockies: Interpretations from 1871–1982 photographs.* USDA Forest Service General Technical Report INT-158. Intermountain Forest and Range Experiment Station, Ogden, UT.

Gruell, G. E. 2001. *Fire in Sierra Nevada forests: A photographic interpretation of ecological change since 1849.* Missoula, MT: Mountain Press.

Hastings, J. R., and R. M. Turner. 1965. *The changing mile: An ecological study of vegetation change with time in the lower mile of an arid and semiarid region.* Tucson: University of Arizona Press.

Smith, H. Y., and S. F. Arno, editors. 1999. *Eighty-eight years of changes in a managed ponderosa pine forest.* USDA Forest Service General Technical Report RMRS-GTR-23. Rocky Mountain Research Station, Ogden, UT.

Turner, R. M., R. H. Webb, J. E. Bowers, and J. R. Hastings. 2003. *The changing mile revisited: An ecological study of vegetation change with time in the lower mile of an arid and semiarid region.* Tucson: University of Arizona Press.

Preface

Repeat photography is nearly as old as photography itself, with broad scientific, cultural, and historical applications. In a rapidly changing world, this technique graphically shows how landscapes respond to a variety of natural and anthropogenic processes. As a scientific tool, repeat photography is unique in that it can be used to both generate and test hypotheses regarding ecological and landscape changes, sometimes with the same set of images. From a cultural perspective, it provides a time capsule showing how towns, favorite places, archaeological sites, historic buildings, and even people have changed. Rephotography has long been used medically to monitor a variety of conditions, ranging from tuberculosis to retinal deterioration. Aquatic natural and cultural features are now monitored with underwater repeat photography. Because cameras were used so extensively by nineteenth- and early-twentieth-century explorers, photographs can be used to trace long-forgotten routes and allow historical reconstructions of expeditions.

This book illuminates the wide application of repeat photography in the natural sciences, specifically landscapes, emphasizing that technology, in the form of satellite remote sensing and aerial photography, has not—and will not—replace this technique, particularly not in developing countries with limited budgets for monitoring natural resources. Repeat photography complements other methods of change reconstruction for landscapes, and its best use is in concert with those techniques. Because of its long temporal reach, in some cases back to the latter third of the nineteenth century, ground photography provides an invaluable record of how landscapes appeared to the settlers of regions and continents, and evaluation of this record helps to provide an objective assessment of past landscape change as well as helping inform the potential for future change. Projects using repeat photography to document landscapes change are under way on all seven continents and in dozens of countries.

We've organized this book to emphasize the different disciplines that employ repeat photography, or as some refer to it, rephotography. We separate the chapters into groups discussing techniques, geological/geomorphic applications, applications in population ecology and landscape change, and application to societal concerns or reconstructions of changes in cultural features. Ultimately, the chapter authors have thwarted this attempt to pigeonhole their work, as many chapters span these somewhat arbitrary categorizations to present more of a holistic vision of long-term changes to landscapes. To illustrate the diversity of practitioners, we have minimized standardization of phrases or jargon used to describe the same techniques or applications; one practitioner's camera station is another applicant's photo location. This again reflects the bottom-up growth of this technique and the unique ways that it has been applied across the natural and social sciences.

Certain usages were standardized; for example, we use a rather standard citation of *Latin name* (common name) in reference to plant species throughout the book, tacitly acknowledging that the scientific names are more reliable than common names, particularly when discussing species in countries where English is not the dominant language. We try to minimize the use of acronyms throughout, defining them on first mention. However, the acronym USGS, which refers to the US Geological Survey, is ubiquitous in the book and does not require definition.

With this book, we honor and acknowledge our predecessors in this field, recognizing their accomplishments from the past as well as their contributions to this book. Prior to his untimely death in 2007, Harold E. (Hal) Malde made extremely large contributions to the use of repeat photography in documentation of geological changes to landscapes. Our dedication of this book to him is a small recognition of the large esteem with which the editors held this man and his work. Likewise, George Gruell is another researcher with sustained contributions to the field of repeat photography, particularly to its

use in documenting ecological changes in forest ecosystems in the western United States. We asked him to provide a foreword to this book to discuss his 45-plus years of experience with this technique wrapped in the context of its rapid growth in application in his lifetime, as documented in this book.

Only one discipline-spanning repeat photography publication precedes this book. In 1984, following a rapid expansion of the field, Garry Rogers, Harold Malde, and Raymond Turner published *Bibliography of Repeat Photography for Evaluating Landscape Change* with the University of Utah Press. In the more than two decades since its publication, the field of repeat photography has figuratively exploded with new methods, new areas of application, and new questions. With this book, we show the topical, temporal, and geographic range of this technique, emphasizing the wide variety of application across the natural sciences. In particular, we emphasize the growing need for and utility of this technique in a constantly changing world.

Robert H. Webb
Diane E. Boyer
Raymond M. Turner
12 February 2010

PART I

Techniques

Development of repeat photography can be compared with evolution as Charles Darwin described it in the nineteenth century. The technique evolved from a natural curiosity about the rates of environmental change, more specifically the annual change in glacier ice in the Alps. It diversified in application to include ecological and geomorphic changes, notably in arid and semiarid ecosystems. As our book documents, the technique is now applied broadly across the natural and even social sciences to address a large variety of seemingly disparate questions, adapting to the needs of diverse scientists interested in evaluating rates of change.

During the first half century of this technique's application, practitioners were often isolated from each other and addressed very specific questions. Once those questions were answered, generally the application ended, only to be initiated again by different researchers evaluating different questions in other geographic areas. Only in the middle of the twentieth century did certain practitioners of repeat photography begin to understand the broad applicability of this technique, and several interdisciplinary and sustained efforts at repeat photography began. These include the efforts of J. Rodney Hastings and Raymond Turner, who initiated the Desert Laboratory Collection of Repeat Photography (Boyer et al., chapter 2) in 1959, and George Gruell (foreword), who initiated a broad-based evaluation of long-term ecological change in forest ecosystems in the mid-

1960s. Webb et al. (chapter 1) discuss this history with a focus on the Desert Laboratory Collection, the largest of its kind in the world.

In its simplest form, repeat photography involves finding the camera station of a previous photographer, aiming a camera, and taking a photograph. Many attempts at refining this technique, either to increase the efficiency of finding a camera station or to properly position a camera, have been suggested and been abandoned in parallel with Darwin's survival of the fittest. As Boyer et al. (chapter 2) review, a number of techniques have been proposed, ranging from how to find a camera station to explicitly using parallax to adjust a camera's position. Although the latter continues to be used by some practitioners, most use relatively simple techniques of foreground–background relationships to determine where to position their cameras. In remote areas, however, use of geographic information systems (GIS) can help guide the repeat photographer to the camera station in advance (Hanks et al., chapter 3), although even this technique shows promising new evolutionary pathways with the continued development of online GIS servers, such as Google Earth (http://earth.google.com).

Some practitioners of repeat photography are exploring new ways to display imagery and the metadata associated with it (Klett, chapter 4). Particularly with the development of digital photography, as well as increasing capabilities of mobile mass-storage devices, repeat photography is moving into a new digital era where the depiction of before–after imagery, side by side in black and white, may be an evolutionary dead end as well. Books such as this one may become obsolete as the media of choice for this type of information as digital photography increases in resolution at decreased cost and as access to this type of information is more convenient with a viewer instead of on the printed page.

Finally, incorporation of GIS analytical power with repeat photography can help to quantify changes that until now have always been documented qualitatively. As Hoffman and Todd (chapter 5) show, analyses of the tonality of photographs, incorporating ground-truthing of what is actually depicted in the image, can yield an understanding of the influences of changing land uses on the ecological characteristics of landscapes. This type of analysis shows the potential for a coming together of the quantitative techniques long associated with remote sensing and the simplicity and high-resolution documentation of repeat photography.

Chapter 1

Introduction:
A Brief History of Repeat Photography

Robert H. Webb, Raymond M. Turner, and Diane E. Boyer

Repeat photography is a valuable tool for evaluating long-term ecological and/or geological change in landscapes. Its practitioners conduct repeat photography projects by acquiring historical and current images, matching those images, and then cataloging and archiving the imagery for long-term storage. The roots of repeat photography lie in the Alps of central Europe, but its uses have spread worldwide, as illustrated throughout this book. Originally developed as a means for documenting changes in glaciers, repeat photography now is used to document all manner of landscape change and ecosystem processes.

Most repeat photography projects have been ad hoc efforts to meet specific needs, such as documenting ecological change in a national park (Meagher and Houston 1998) or a larger ecoregion (Turner et al. 2003). Few efforts have attempted to collate images for collective current 'or future research use (Bierman et al. 2005). The Desert Laboratory Repeat Photography Collection, named for the Desert Laboratory research facility in Tucson, Arizona, is the largest such collection in the world (Webb et al. 2007a), with over 4,900 camera stations located throughout the southwestern United States, northern Mexico (fig. 1.1), and Kenya. We present a general history of repeat photography as well as a

more detailed one of the Desert Laboratory Collection to illustrate the development of this technique and some future trends in its application.

History of Repeat Photography

In its infancy, photography had a documentary purpose, whether it was to record news events, capture the brutality of war, or preserve images of famous people. Scientific photographic documentation in the western United States began with surveys used to determine appropriate routes for the transcontinental railways (Bell 1869), to illustrate geological features of landscapes (Powell 1895), or to show the condition of lands prior to development (e.g., George Roskruge; see Turner et al. 2003). The imagery of many of these initial photographic surveys was preserved in archives, making it available to future generations for the study of landscape change.

The Initial Science Applications

A Bavarian mathematician with a geological interest took the first steps in the development of repeat photography as a scientific tool. In 1888, Sebastian Finsterwalder began conducting photogrammetric

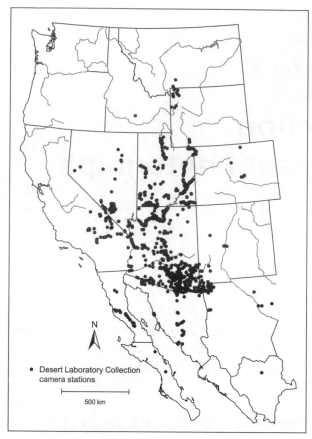

Figure 1.1. The distribution of camera stations in the southwestern United States and northern Mexico included in the Desert Laboratory Repeat Photography Collection. Not shown are camera stations in Kenya (Turner et al. 1998).

surveys of mountain glaciers in the Tyrolean Alps. He returned to his established camera stations the following year, obtaining repeated images of the same view. By comparing the old and new images, and using rudimentary photogrammetry, Finsterwalder documented change to the glaciers over time (Hattersly-Smith 1966). In doing so, he pioneered the technique of repeat photography involving deliberate and careful matching of older photographs with newer photographs from the same camera station (Rogers et al. 1984). Following Finsterwalder, early glaciologists in the Canadian Rockies (Cavell 1983), the Sierra Nevada in California (Gilbert 1904), and Alaska (Molnia, chapter 6) used repeat photography as well, establishing the groundwork for a technique that has grown in importance in

glaciology (Molnia, chapter 6; Fagre and McKeon, chapter 7).

In addition to its use in glaciology studies, repeat photography soon was used to document plant populations and landscape change. As early as 1905, ecological texts recommended using matched photographs on research quadrats as an aid to plant population measurements (Clements 1905). One of the earliest studies documented decline in rangeland species between 1903 and 1913 on the Santa Rita Experimental Range south of Tucson (Wooten 1916; see McClaran et al., chapter 12). Forrest Shreve, a pioneering desert ecologist, used repeat photography to document changes in vegetation plots at the Desert Laboratory in Tucson between 1906 and 1928 (Shreve 1929). Before the 1940s, only a few studies used repeat photography to document ecological changes in the southwestern United States (Rogers et al. 1984); most of these were narrowly focused and published in scientific journals or monographs.

In the 1940s, repeat photography became a technique frequently used to document landscape change, particularly changes induced by land use. For example, in 1946, repeat photography was used to demonstrate an increase in *Prosopis velutina* (velvet mesquite) and other woody species in Arizona grasslands (Parker and Martin 1952), a subject that continues to be of interest in the southwestern United States (McClaran et al., chapter 12; Turner et al., chapter 17). The tendency of grasslands to change into woody savannas remains a topic of research worldwide (e.g., Lewis, chapter 15).

Repeat photography also gained adherents in the geological discipline, particularly those interested in weathering and bedrock erosion. In southeastern Utah, Bryan and La Rue (1927) compared images of the Navajo Twins near Bluff for signs of weathering over about a two-decade interval. Longwell et al. (1932) illustrated a discussion of waterfall erosion using a matched pair of photographs taken of Niagara Falls, on the Niagara River on the US–Canadian border, before and after a large mass of rock at the edge of the falls fell off on 17 January 1931. Lobeck (1939) included an approximately matched pair (1886 and 1926) showing toppling of blocks from a basalt column in the Palisades of the Hudson River near Fort Lee, New Jersey.

Expeditions in the southwestern United States prior to the development of reliable photography typically included an artist to record the landscape, and expedition members often described the countryside in their journals. Lockett (1939) pioneered the combined analysis of photographic and historical records to describe anecdotally landscape change in northwestern New Mexico and northern Arizona. In 1948, Simpson (1951) used a camera to match sketches of New Mexican landscapes made in the 1870s during a paleontological expedition. Relocating and rephotographing the scenes depicted in sketches and paintings originally made by expeditionary artists are useful in determining historic routes and camps that are not known precisely (Jonas, chapter 21).

Botanist Homer Shantz of the University of Arizona in Tucson was the first person to use repeat photography to document large-scale geographic changes in plant populations. In 1956–1957, Shantz, assisted by B. L. Turner, repeated images he had taken on 1919–1920 and 1924 excursions in Africa from Cape Town to Cairo (Shantz and Turner 1958). This publication, the first book-length report of landscape change relying entirely on repeat photography, was highly influential in the fledgling field, opening up new possibilities for its application.

Shantz and Turner's work was constrained by both linear geography (roadways and railroads) and time (the dates of Shantz's original images). Many of the projects they inspired were similarly defined. In 1968–1969, Stephens and Shoemaker (1987) replicated photographs taken by John K. Hillers and E. O. Beaman during the John Wesley Powell expeditions along the Colorado and Green rivers in the American Southwest. In the early 1970s, Turner and Karpiscak (1980) also worked along the Colorado River but matched photographs from a broad variety of dates and photographers. In 1984, Robert R. Humphrey (1987) matched photographs taken in 1892–1893 along the border between Mexico and the United States. The International Boundary Commission erected 258 monuments along the bound-

ary between El Paso, Texas, and San Diego, California. One or more photographs were taken of each of these monuments. Humphrey was closely restricted geographically and temporally as he moved along this boundary line matching one photograph at each monument. The first major repeat photography project operating in a broad geographic region using historic imagery unrestricted by photographer or date was that conducted by Hastings and Turner (1965), which ultimately gave rise to the Desert Laboratory Repeat Photography Collection.

The Desert Laboratory Repeat Photography Collection

At the beginning of the twentieth century, the Carnegie Institution of Washington endowed a number of biological research stations designed to study typical ecosystems of the United States. Among the regions chosen was the Sonoran Desert, and in 1903, construction began on the Desert Laboratory on a basaltic hill west of downtown Tucson (Bowers 1990). Carnegie assigned the scientists at the Desert Laboratory the task of understanding adaptations of plants and animals to the desert environment and cataloging the species that occupied those regions, but particularly the Sonoran Desert. Carnegie's role in this research is commemorated in the monospecific genus occupied by the iconic species of the Sonoran Desert, *Carnegiea gigantea* (saguaro).

Pioneering desert ecologists flocked to the Desert Laboratory, including the highly influential Forrest Shreve (Bowers 1988) and Daniel T. MacDougal, both of whom photographed the desert environment (plate 1.1). The Desert Laboratory remained under the auspices of the Carnegie Institution until 1940, when it was transferred to the US Forest Service as headquarters for the Southwestern Forest and Range Experiment Station (Medina 1996). Forest Service scientists used repeat photography to document changes in rangeland conditions in various parts of southern Arizona (e.g., Parker and Martin 1952), but they did not formally establish a collection of repeat photographs. In the mid-1950s, the facility was transferred to the University of Arizona, where it served as the Laboratory of Geochronology for nearly four decades.

In the late 1950s, Jim McDonald, an atmospheric physicist from the University of Arizona, wanted to analyze vegetation change documented in repeat photographs as a proxy for climatological data. McDonald recruited bioclimatologist James Rodney Hastings to conduct fieldwork, creating and analyzing repeat photography of the Sonoran Desert. Hastings's project gained the attention of Raymond M. Turner, a botanist and professor at the University of Arizona. Beginning in 1959, Turner and Hastings collaborated to systematically replicate historical photographs of the region, culminating in the classic book *The Changing Mile* (Hastings and Turner 1965), which documents vegetation change from the 1880s to the 1960s in the vertical mile of elevation that encompasses the Sonoran Desert. This book was recently updated as *The Changing Mile Revisited* (Turner et al. 2003).

In 1962, Turner left the University of Arizona to join the US Geological Survey, also in Tucson. Fourteen years later, Turner moved to the US Geological Survey's satellite office at the Desert Laboratory, and with his arrival, his burgeoning photographic collection assumed its current name. The collection expanded as applications for repeat photography increased. In the early 1980s, Robert H. Webb, also with the US Geological Survey, engaged Turner to study ecosystem recovery at abandoned ghost towns in the Mojave Desert (Webb et al. 1987). In 1989, Turner retired from the US Geological Survey, and Webb assumed official guardianship of the collection. In collaboration with Turner, Webb applied the repeat photography technique to a wide variety of research questions ranging from long-term channel change of regional rivers, such as change along the Colorado River in Grand Canyon (Webb 1996, Webb et al. 1999), to documentation of changes in permanent vegetation plots in the Mojave Desert (Webb et al. 2003) and long-term change in riparian vegetation (Webb et al. 2007b). Turner continued to use repeat photography in documenting the combined effects of climate variability and land-use

practices, including an expansion of Shantz's work in Kenya (Turner et al. 1998).

As of 2010, the Desert Laboratory Collection contains approximately 35,000 negatives and transparencies taken at 4,970 camera stations (fig. 1.1). A total of 6,790 original photographs were used at these camera stations to create 9,762 matches (many images have been matched multiple times). The largest number of matches (1,586) was made in the Grand Canyon, the site of ongoing studies related to the effects of Glen Canyon Dam on the Colorado River (Turner and Karpiscak 1980, Webb 1996). The three photographers with the most original imagery incorporated into the collection are Robert Brewster Stanton (428) and Eugene C. La Rue (396), who were prominent photographers of the Colorado River, and Janice C. Beatley (402), who photographed permanent vegetation plots in southern Nevada. Most of the matches are in Arizona (4,111), followed by Utah (1,213), and Nevada (505) (table 1.1). A total of 150 camera stations have been occupied in Kenya, and 67 of these are the reoccupied Shantz and Turner (1958) camera stations. While the collection is no longer physically housed at the Desert Laboratory, the name has been kept in honor of the significance of this historic research facility.

Recent Work in Repeat Photography: Selected Studies

In the last few decades, hundreds of publications employing repeat photography have appeared in

Table 1.1. Five states of the United States and Mexico with the most photographic matches contained in the Desert Laboratory Collection of Repeat Photography

Region	Number of Matches
Arizona	4,111
Utah	1,213
Nevada	505
New Mexico	220
Sonora	190

scientific and popular literature. Few photographers or groups have conducted sustained repeat photography spanning decades; those who have include Gruell (foreword), Klett (chapter 4), Veblen (chapter 13), Hall (2002), and the authors. There are a few large-scale, systematic repeat photography projects, such as the one conducted by the US National Park Service through its Inventory and Monitoring Program (e.g., in southwest Alaska, science.nature.nps.gov/im/units/swan/). Researchers employed by the US Forest Service have conducted repeat photography projects for many years (Hall 2002). The Rocky Mountain Repeat Photography Project in Canada (bridgland.sunsite.ualberta.ca) aims to repeat topographic survey photographs taken in the early 1900s. Numerous, somewhat smaller programs are found worldwide, such as those in Vermont (Bierman, chapter 9), Mexico (Bullock and Turner, chapter 10), Kenya (Western, chapter 16), South Africa (Hoffman et al., chapter 11), and Namibia (Rohde 1997).

Many recent applications are place based, as is the case with several of the chapters within this book. Favorite locations in the United States include units administered by the National Park Service; Yosemite National Park, California, has been the location of many repeat photography efforts, resulting in numerous papers and at least two books (Vale and Vale 1994, Klett et al. 2005). Studies in Grand Canyon National Park along the corridor of the Colorado River examine changes to the river channel and riparian vegetation (Turner and Karpiscak 1980, Webb 1996). Wright and Bunting (1994) reported on environmental change in Craters of the Moon National Monument in Idaho, whereas Meagher and Houston (1998) examined changes in vegetation, wildlife habitat, and geomorphic change within Yellowstone National Park in Wyoming and Montana. Repeat photography has been used as an archaeological monitoring tool within Mesa Verde National Park, Colorado (Howard et al., chapter 22), and Chaco Culture National Historical Park in New Mexico (Malde 2000). Contemporary glaciologists continue the tradition of repeat photography within

Glacier Bay National Park, Alaska (Molnia, chapter 6), and Glacier National Park in Montana (Butler and DeChano 2001; Fagre and McKeon, chapter 7). National parks worldwide are similarly popular as subjects for repeat photography. White and Hart (2007) examined ecological changes in Banff and Jasper National Parks, Canada. Western (chapter 16) studied the ecosystem within Amboseli National Park, Kenya. Veblen (chapter 13) worked in Lanin, Nahuel Huapi, Lago Puelo, and Los Alerces national parks, Argentina. Byers (1987) reported on land-use effects in Sagamartha National Park, Nepal.

Forest ecology of mountain ranges has been another widespread application. For example, Veblen and Lorenz (1991) documented changes in the Front Ranges of Colorado, Gruell (2001) examined fire-driven ecological change in the forests of the Sierra Nevada in California and Nevada, and White and Hart (2007) examined ecological regions within the Canadian Rockies. Veblen (chapter 13) studied forests in the southern Andes of Peru and Argentina. Pioneer photographer William Henry Jackson's Colorado images are the subject of a series of three books on changes in the Rocky Mountains (Jackson et al. 1999, Noel and Fielder 2001, Fielder et al. 2005). Jackson's photographs of the northern Uinta Mountains along the Utah–Wyoming border formed the basis for a climatological study using repeat photography (Munroe 2003).

As previously discussed, grasslands and desert environments, where photography captures wide vistas generally unobstructed by trees, continue to be two of the biomes with the greatest application of repeat photography. Following the pioneering work of Hastings and Turner (1965) in southern Arizona and northwestern Mexico, Rogers (1982) studied vegetation change in the central Great Basin Desert of the American West. Klement et al. (2001) reported on vegetation and landscape change in the Northern Great Plains, and Johnson (1987) used William Henry Jackson's photographs as the basis for a study of changes in Wyoming rangeland. Gruell (foreword) monitored wildlife habitat using repeat photography in the Humboldt National Forest in

Nevada, and Kay (2003) studied vegetation change in the Fishlake National Forest in Utah.

Geomorphic and riparian change along ephemeral and perennial river channels is another area of wide application of repeat photography, particularly in the American West. Turner and Karpiscak (1980) examined changes along the Colorado River between Glen Canyon Dam and Lake Mead; Graf (1987) and Webb et al. (1991) examined channel changes in Colorado Plateau ephemeral watercourses; and Stephens and Shoemaker (1987) repeated Powell Survey photographs taken along the perennial Green and Colorado rivers. More recently, Webb et al. (2004) expanded on the work by Stephens and Shoemaker (1987) in Cataract Canyon, a reach of the Colorado River in east-central Utah.

While North America has been the continent with the most repeat photography, the technique is now used on all seven continents. Four chapters within this book (Hoffman and Todd, chapter 5; Hoffman et al., chapter 11; Nyssen et al., chapter 14; and Western, chapter 16) discuss application of repeat photography to various scientific questions in Africa, mostly related to plant demographics and land-use impacts. Europe, as previously mentioned, was the birthplace of repeat photography, and the tradition has continued (Rohde, chapter 18; Moore, chapter 19). Numerous studies have been conducted in recent years in South America (e.g., Byers 2000; Veblen, chapter 13), Asia (e.g., Byers 1987, Moseley 2006), and Australia (e.g., Pickard 2002; Start and Handasyde 2002; Lewis, chapter 15). Recently, repeat photographic studies have been initiated in Antarctica as part of the McMurdo Dry Valleys Long Term Ecological Research project (lternet.edu/sites/mcm/).

Discussion and Conclusions

Repeat photography is an excellent technique for evaluating landscape change over time, as amply demonstrated by the holdings of the Desert Laboratory Repeat Photography Collection and those of numerous other researchers. First used in the late 1800s by glaciologists as a simple method to monitor glaciers, repeat photography experienced an upswing in use, largely by American scientists and mostly after World War II. In recent years, repeat photography has become well established globally as a technique to address a vast array of research questions, such as fire effects and recovery, land-use effects, changes in archaeological features, the location of historic routes and trails, and assessing perceptions of change. It is also commonly used to do regionwide assessments of landscape change, typically with respect to general or specific land-use practices and climatic fluctuations. While photographic technology has evolved and become more accessible, the fundamental techniques of repeat photography have remained unchanged since its inception in the late nineteenth century. The increasing number, variety, and locations of repeat photography projects are directly attributable to the creative minds and needs for documenting landscape changes of those who practice this technique.

Acknowledgments

The authors acknowledge the many contributions of the late Rod Hastings and Hal Malde, and thank the dozens of amateur and professional photographers who have directly or indirectly contributed to the Desert Laboratory Repeat Photography Collection in its 50 years of existence, as well as other practitioners of repeat photography, many of whom are authors in this book. All photographs are courtesy of the Desert Laboratory Repeat Photography Collection unless otherwise noted.

Literature Cited

Bell, W. A. 1869. *New tracks in North America*. 2 vols. London: Chapman and Hall.

Bierman, P. R., J. Howe, E. Stanley-Mann, M. Peabody, J. Hilke, and C. A. Massey. 2005. Old images record landscape change through time. *GSA Today* 15:4–10.

Bowers, J. E. 1988. *A sense of place: The life and work of Forrest Shreve*. Tucson: University of Arizona Press.

Bowers, J. E. 1990. A debt to the future: Scientific achievements of the Desert Laboratory, Tumamoc Hill, Tucson, Arizona. *Desert Plants* 10:9–12, 35–47.

Bryan, K., and E. C. La Rue. 1927. Persistence of features in an arid landscape: The Navajo Twins, Utah. *Geographical Review* 17:251–257.

Butler, D. R., and L. M. DeChano. 2001. Environmental change in Glacier National Park, Montana: An assessment through repeat photography from fire lookouts. *Physical Geography* 22:291–304.

Byers, A. 1987. An assessment of landscape change in the Khumbu region of Nepal using repeat photography. *Mountain Chronicles* 7:77–81.

Byers, A. C. 2000. Contemporary landscape change in the Huascarán National Park and buffer zone, Cordillera Blanca, Peru. *Mountain Research and Development* 20:52–63.

Cavell, E. 1983. *Legacy in ice: The Vaux family and the Canadian Alps*. Banff, Alberta: The Whyte Foundation.

Clements, F. E. 1905. *Research methods in ecology*. Lincoln, NE: University Publishing Company.

Fielder, J., W. H. Jackson, and G. Klucas. 2005. *Colorado 1870–2000 II*. Englewood, CO: Westcliffe Publishers.

Gilbert, G. K. 1904. Variations of Sierra glaciers. *Sierra Club Bulletin* 5:20–25.

Graf, W. L. 1987. Late Holocene sediment storage in canyons of the Colorado Plateau. *Geological Society of America Bulletin* 99:261–271.

Gruell, G. E. 2001. *Fire in Sierra Nevada forests: A photographic interpretation of ecological change since 1849*. Missoula, MT: Mountain Press Publishing.

Hall, F. C. 2002. *Photo point monitoring handbook: Part A—field procedures*. USDA Forest Service General Technical Report PNW-GTR-526. Pacific Northwest Research Station, Portland, OR.

Hastings, J. R., and R. M. Turner. 1965. *The changing mile: An ecological study of vegetation change with time in the lower mile of an arid and semiarid region*. Tucson: University of Arizona Press.

Hattersly-Smith, G. 1966. The symposium on glacier mapping. *Canadian Journal of Earth Sciences* 3:737–743.

Humphrey, R. R. 1987. *90 years and 535 miles: Vegetation changes along the Mexican border*. Albuquerque: University of New Mexico Press.

Jackson, W. H., J. Fielder, and E. Marston. 1999. *Colorado 1870–2000*. Englewood, CO: Westcliffe Publishers.

Johnson, K. L. 1987. *Rangeland through time*. University of Wyoming, Agricultural Experiment Station, Misc. Publication 50, Laramie, WY.

Kay, C. E. 2003. *Long-term vegetation change on Utah's Fishlake National Forest: A study in repeat photography*. USDA Forest Service and Utah State University Extension. Washington, DC: US Government Printing Office.

Klement, K. D., R. K. Heitschmidt, and C. E. Kay. 2001. *Eight years of vegetation and landscape changes in the Northern Great Plains: A photographic record*. US Department of Agriculture, Agricultural Research Service, Conservation Research Report No. 45. Washington, DC: US Government Printing Office.

Klett, M., R. Solnit, and B. G. Wolfe. 2005. *Yosemite in time: Ice ages, tree clocks, ghost rivers*. San Antonio, TX: Trinity University Press.

Lobeck, A. K. 1939. *Geomorphology: An introduction to the study of landscapes*. New York: McGraw-Hill.

Lockett, H. C. 1939. *Along the Beale Trail: A photographic account of wasted range lands*. Washington, DC: Education Division, US Office of Indian Affairs.

Longwell, C. R., A. Knopf, and R. F. Flint. 1932. *A textbook of geology: Part I, physical geology*. New York: John Wiley and Sons.

Malde, H. E. 2000. *Repeat photography at Chaco Culture National Historical Park, New Mexico, based on photographs made in the 1930s, 1970s and the year 2000*. Denver, CO: Wright Paleohydrological Institute.

Meagher, M., and D. B. Houston. 1998. *Yellowstone and the biology of time: Photographs across a century*. Norman: University of Oklahoma Press.

Medina, A. L. 1996. *The Santa Rita Experimental Range: History and annotated bibliography (1903–1988)*. USDA Forest Service General Technical Report RM-GTR-276. Rocky Mountain Forest and Range Experiment Station, Ft. Collins, CO.

Moseley, R. K. 2006. Historical landscape change in northwestern Yunnan, China: Using repeat photography to assess the perceptions and realities of biodiversity loss. *Mountain Research and Development* 26:214–219.

Munroe, J. S. 2003. Estimates of Little Ice Age climate inferred through historical rephotography, northern Uinta Mountains, U.S.A. *Arctic, Antarctic, and Alpine Research* 35:489–498.

Noel, T. J., and J. Fielder. 2001. *Colorado 1870–2000 Revisited: The history behind the images.* Englewood, CO: Westcliffe Publishers.

Parker, K. W., and S. C. Martin. 1952. *The mesquite problem on the Southern Arizona ranges.* US Department of Agriculture, Circular 908. Washington, DC: US Government Printing Office.

Pickard, J. 2002. Assessing vegetation change over a century using repeat photography. *Australian Journal of Botany* 50:409–414.

Powell, J. W. 1895. *Canyons of the Colorado.* Repr., New York: Dover Publications, 1961.

Rogers, G. F. 1982. *Then and now: A photographic history of vegetation change in the Central Great Basin Desert.* Salt Lake City: University of Utah Press.

Rogers, G. F., H. E. Malde, and R. M. Turner. 1984. *Bibliography of repeat photography for evaluating landscape change.* Salt Lake City: University of Utah Press.

Rohde, R. F. 1997. Looking into the past: Interpretations of vegetation change in Western Namibia based on matched photography. *Dinteria* 25:121–149.

Shantz, H. L., and B. L. Turner. 1958. *Photographic documentation of vegetational changes in Africa over a third of a century.* University of Arizona, College of Agriculture Report 169, Tucson.

Shreve, F. 1929. Changes in desert vegetation. *Ecology* 10:364–373.

Simpson, G. G. 1951. Hayden, Cope and the Eocene of New Mexico. *Proceedings of the Academy of Natural Sciences of Philadelphia* 53:1–21.

Start, A. N., and T. Handasyde. 2002. Using photographs to document environmental change: The effects of dams on the riparian environment of the lower Ord River. *Australian Journal of Botany* 50:465–480.

Stephens, H. G., and E. M. Shoemaker. 1987. *In the footsteps of John Wesley Powell: An album of comparative photographs of the Green and Colorado rivers, 1871–72 and 1968.* Boulder, CO: Johnson Books.

Turner, R. M., and M. Karpiscak. 1980. *Recent vegetation changes along the Colorado River between Glen Canyon Dam and Lake Mead, Arizona.* US Geological Survey Professional Paper 1132. Washington, DC: US Government Printing Office.

Turner, R. M., H. A. Ochung', and J. B. Turner. 1998. *Kenya's changing landscape.* Tucson: University of Arizona Press.

Turner, R. M., R. H. Webb, J. E. Bowers, and J. R. Hastings. 2003. *The changing mile revisited: An ecological study of vegetation change with time in the lower mile of an arid and semiarid region.* Tucson: University of Arizona Press.

Vale, T. R., and G. R. Vale. 1994. *Time and the Tuolumne landscape: Continuity and change in the Yosemite high country.* Salt Lake City: University of Utah Press.

Veblen, T. T., and D. C. Lorenz. 1991. *The Colorado Front Range: A century of ecological change.* Salt Lake City: University of Utah Press.

Webb, R. H. 1996. *Grand Canyon, a century of change.* Tucson: University of Arizona Press.

Webb, R. H., J. Belnap, and J. Weisheit. 2004. *Cataract Canyon: A human and environmental history of the rivers in Canyonlands.* Salt Lake City: University of Utah Press.

Webb, R. H., D. E. Boyer, and R. M. Turner. 2007a. *The Desert Laboratory Repeat Photography Collection—an invaluable archive documenting landscape change.* US Geological Survey Fact Sheet 2007-3046. Tucson, AZ.

Webb, R. H., S. A. Leake, and R. M. Turner. 2007b. *The ribbon of green: Change in riparian vegetation in the southwestern United States.* Tucson: University of Arizona Press.

Webb, R. H., T. S. Melis, P. G. Griffiths, J. G. Elliott, T. E. Cerling, R. J. Poreda, T. W. Wise, and J. E. Pizzuto. 1999. *Lava Falls Rapid in Grand Canyon: Effects of late Holocene debris flows on the Colorado River.* US Geological Survey Professional Paper 1591. Tucson, AZ: US Government Printing Office.

Webb, R. H., M. B. Murov, T. C. Esque, D. E. Boyer, L. A. DeFalco, D. F. Haines, D. Oldershaw, S. J. Scoles, K. A. Thomas, J. B. Blainey, and P. A. Medica. 2003. *Perennial vegetation data from permanent plots on the Nevada Test Site, Nye County, Nevada.* US Geological Survey Open-File Report 03-336. Tucson, AZ.

Webb, R. H., S. S. Smith, and V. A. S. McCord. 1991. *Historic channel change of Kanab Creek, southern Utah and northern Arizona.* Grand Canyon Natural History Association Monograph Number 9. Grand Canyon, AZ.

Webb, R. H., J. W. Steiger, and R. M. Turner. 1987. Dynamics of Mojave Desert shrub vegetation in the Panamint Mountains. *Ecology* 50:478–490.

White, C., and E. J. Hart. 2007. *The lens of time: A repeat photography of landscape change in the Canadian Rockies.* Calgary, AB: University of Calgary Press.

Wooten, E. O. 1916. *Carrying capacity of grazing ranges in southern Arizona.* US Department of Agriculture Bulletin 367. Washington, DC: US Government Printing Office.

Wright, S. G., and S. C. Bunting. 1994. *The landscapes of Craters of the Moon National Monument: An evaluation of environmental changes.* Boise: University of Idaho Press.

Chapter 2

Techniques of Matching and Archiving Repeat Photography Used in the Desert Laboratory Collection

Diane E. Boyer, Robert H. Webb, and Raymond M. Turner

Although it can be used casually with little or no documentation, repeat photography is most powerful when it is employed as a reproducible scientific technique involving the establishment of permanent benchmarks, the use of high-resolution film or digital media, and an archival storage system that guarantees longevity of the data. How these tasks are performed governs, in large part, the quality of the results that can be obtained from the images, as well as their possible usefulness to future researchers.

Here we provide a general guide to procedures and techniques used over the 50-year lifespan of the Desert Laboratory Repeat Photography Collection (Webb et al. 2007a, chapter 1), drawing upon our experience as creators of and contributors to this unique collection. We focus on lessons we have learned while acknowledging that there are numerous ways that one might conduct projects in repeat photography. Finally, we discuss how imagery was acquired, replicated in the field, matched using digital or analog techniques, cataloged, and archived in the Desert Laboratory Collection.

Acquiring Images Suitable for Replicating

In undertaking a program of repeat photography, there are two ways to obtain original images. The first involves creating the original baseline photographs with or without the intention of replication by choosing a prominent camera station, making the desired images, then matching them at a future date (Hall 2002). The second is to obtain historical photographs from archives and museums, private individuals, and published sources. Repeat photography projects—the early work of the Desert Laboratory Collection is a case in point—are sometimes launched on the basis of a specific person's or expedition's collection (e.g., Homer Shantz; Shantz and Turner 1958; Turner et al. 1998).

When establishing a repeat photography program, private individuals may be identified who can provide historical materials. In such cases, one should establish a written agreement covering use, distribution, and publication. If the owner then donates the original photographs to a museum or

Table 2.1. Some online resources for historical photographs that are useful for repeat photography in the United States[1]

Archive	URL[2]
Archives with US National or International Holdings	
National Archives and Records Administration (NARA)	http://www.archives.gov/research/arc/
Library of Congress American Memory Project (searches multiple archives)	http://lcweb2.loc.gov/ammem/browse/
Smithsonian Institution	http://www.siris.si.edu/
Beinecke Library, Yale University	http://beinecke.library.yale.edu/digitallibrary/
US National Park Service	http://data2.itc.nps.gov/hafe/hfc/npsphoto2.cfm
Bureau of Land Management	http://www.blm.gov/wo/st/en/bpd.html
Harpers Ferry Center, National Park Service	http://home.nps.gov/applications/hafe/hfc/npsphoto2.cfm
US Geological Survey Photographic Library	http://libraryphoto.cr.usgs.gov/index.html
George Eastman House	http://www.geh.org/
New York Public Library	http://digitalgallery.nypl.org/nypldigital/index.cfm
Archives with US Regional Holdings	
Northern Arizona University (Cline Library)	http://www.nau.edu/library/speccoll/index.html
Southern Utah University	http://archive.li.suu.edu/archive/
University of Utah (Marriott Library)	http://www.lib.utah.edu/portal/site/marriottlibrary/
Autry National Center (includes Southwest Museum)	http://theautry.org/collections/collections-home
Denver Public Library	http://photoswest.org/presearch.html
Online Archive of California (searches multiple archives)	http://www.oac.cdlib.org/
University of California (Bancroft Library)	http://bancroft.berkeley.edu/collections/pictorial.html
Arizona Archives Online (searches multiple archives)	http://aao.lib.asu.edu/index.html
Western Waters Digital Library (searches multiple archives)	http://harvester.lib.utah.edu/wwdl/

[1]Also see Bierman et al. (2005).
[2]Accessed February 2010.

archive, the repeat photographer should establish an agreement with the repository concerning use, preferably at the time of donation.

Archives and museums hold an immense number of photographs suitable for matching. With the advent of the Internet, the task of locating such images has grown easier as increasing numbers of archives post images and descriptive guides online (table 2.1). An Internet search of the holdings of archives known to cover the study region (both national and local), as well as a broader query using standard online search engines, can yield potential photographs for a new project. Books and journals contain both valuable photographs and source references. When original photographic prints that were reproduced in a public-domain book cannot be located or were not preserved, digital scans can be made directly from the book, although quality will be inferior to that of a photographic print.

Webb et al. (2007b) provide one example of the diversity of sources available for a repeat photography study. They relied on the regional US Geological Survey office for historical photographs that documented construction and operation of long-term stream-gaging stations; these photographs were preserved, with original negatives, largely because of the diligence of past USGS employees. They also used photography from 14 public archives, including the online holdings at the National Archives and Records Administration, Northern Arizona University, the University of Utah, and the US Geological Survey Photographic Library (table 2.1) as well as more traditional archives, such as the Arizona Historical Society (Tucson).

A search for landscape images suitable for matching is an unusual one for most repositories. Archivists are accustomed to queries for portraits, buildings, and events, and their knowledge base tends to develop accordingly. Photographers rarely label landscape views with any degree of precision, and this lack of detail further confounds the archivist. In addition, photographs are commonly mislabeled; we spent years searching for the location of one 1875 image that was incorrectly identified as a place hundreds of kilometers away from the actual camera station. Few catalogers, when faced with unidentified landscape negatives, can afford research time to properly identify the images. Finally, the standard Library of Congress subject headings, used by many American catalogers, are of limited use for landscape views. Sympathetic archivists, particularly those with a good institutional memory, are invaluable in finding useful images and in promoting cataloging practices, such as geospatial referencing, that are more useful to the repeat photographer.

Standard archival practice dictates that the creator's (photographer's) original organization scheme should be maintained, and that any given collection should be kept together. This practice was not always followed by early archivists and is still not followed by some institutions. Therefore, researchers need to examine both subject files as well as individual collections for potential images. In areas with popular scenic viewpoints, one may find a series of photographs taken by a variety of photographers over time that show a similar field of view, creating an incidental time series (fig. 2.1; see also Klett, chapter 4).

Prior to the advent of digital technology, either duplicate or copy negatives were the preferred format used by the Desert Laboratory Collection for images ordered from archives. While many archives will not distribute negatives, some are willing to make an exception for researchers, including repeat photographers. When negatives are not available, 8-by 10-inch glossy prints are an acceptable substitute. Many major archives now supply photographs in electronic format or as digital prints only; some

A

B

C

Figure 2.1. Toroweap Point, Grand Canyon National Park, Arizona, USA.
A. (1872). John K. Hillers of the John Wesley Powell Expedition was the first to photograph this upstream view of the Colorado River from a height of about 1,000 meters above the river corridor at Toroweap Point. (J. K. Hillers, 66, National Archives).
B. (23 April 1934). Numerous other photographers have captured this view from slightly different positions than where Hillers stood, creating an incidental time series. This photograph was obtained a short distance away from A, but many of the same features on the background cliffs and in the river corridor are apparent. (E. D. McKee, 95.48.852, image courtesy of Cline Library, Northern Arizona University).
C. (September 1951). This view shows more of the ledges visible in A; many of the rocks are gone from left center. The beaches appear to be larger and more numerous, due in part to a lower water level but also as a result of a large flood in June 1951. (J. M. Eden, 2068, Grand Canyon National Park).

D

E

F

G

H

Figure 2.1. Continued

D. (1952). This view is aimed in a slightly different direction from A, but on the left, one can already see that some of the slab rocks on the cliffs are no longer present, likely rolled into the chasm by visitors. The size and location of the beaches along the river corridor are similar to those in 1872. (M. Litton, courtesy of the photographer).

E. (August 1959). There has been no significant change in the seven years since the previous image was taken. (N. N. Dodge, 8340, Grand Canyon National Park).

F. (10 June 1979). This first purposeful match of A shows that the beaches are smaller and fewer in number than in previous images, the result of decreased sediment load due to the presence of Glen Canyon Dam, constructed in 1963. Riparian vegetation has increased, in part because of flood control. Loose boulders have mostly disappeared. (R. M. Turner).

G. (18 August 1992). This match of A shows a model in the same spot as Hillers's model, although further back from the edge. Even more boulders have disappeared. (G. Bolton).

H. (16 May 1995). In the three years between this and the previous match, there has been no significant change to the river corridor that cannot be attributed to difference in river stage. River flow is slightly lower, giving the appearance of larger sandbars and beaches. (R. H. Webb, Stake 966).

smaller repositories, particularly those in rural areas, still use wet chemistry to produce positives. As already mentioned, many digital images can be downloaded directly from the Internet, although a significant percentage of these are low resolution and watermarked, rendering them possibly sufficient for matching but not suitable for detailed analysis or publication. Tagged Image File Format (TIFF) files, preferably at least 10 inches on the long axis and a minimum of 300 dots per inch (dpi), or a total of about 7 megabytes (MB) in size for gray scale or 18 MB for color images, are suitable for many repeat photography projects, although some types of analysis will require far greater resolution.

Working with Vintage Photographs

Proper interpretation and analysis of repeat photography require an understanding of the equipment and film that the original photographers used. Both cameras and film have evolved radically during and since the period encompassed by the original photography used in many studies. The result is a large variation in the resolution of the original images as well as differences in the captured visible spectrum. While it is unnecessary to use original equipment or have lenses with the same focal length, it is important to understand that some differences are attributable to equipment and film instead of changes in the natural environment.

Photography was developed in the late 1830s by Louis Jacques Mandé Daguerre (Newhall 1964). Daguerreotypes, as they were so named, consisted of a light-sensitive emulsion on a silver-coated copper plate, which, when exposed, produced a mirror image of the photographed subject. The finished product had a delicate surface and required storage in a sealed, glass-covered case. While Daguerre continued to perfect his technique, William Henry Fox Talbot developed photography using paper negatives, which, in turn, could be used to produce paper prints, albeit of low resolution.

Several other photographic processes were developed in the ensuing years, but all had problems—they were fragile, produced mirrored images, were cumbersome to produce, lacked detail, or made only a single image from any given exposure—that minimized commercial application. In 1851, Frederick Scott Archer discovered that collodion, a viscous mixture of guncotton, alcohol, and ether, could be used to attach light-sensitive particles to clear glass in what became known as "wet-plate" negatives. By 1860, glass-plate-negative images printed on paper had become the dominant photographic process (Newhall 1964). Ten years later, photographic technology improved dramatically with the development of gelatin-based emulsions that had much greater light sensitivity than their predecessors, making them capable of freezing motion. Of equal importance, gelatin "dry-plates" could be prepared in advance for later use, making their manufacture a commercial proposition, which in turn made photography more accessible.

While landscape images were taken using early photographic processes such as daguerreotypes, they were relatively few in number and many of them have been lost to deterioration. Hence, in many cases, the earliest images useful to repeat photographers were taken on glass plates. Pioneer landscape photographers shooting with glass plates invariably used large-format cameras (4- by 5-inch images or larger), some capable of taking images as large as 7 by 14 inches on plates designed for stereographic images on 8- by 10-inch media (Newhall 1964). Glass-plate negatives were unwieldy and fragile, but they created images with high detail and minimal distortion. Flexible, translucent roll film was developed in 1889 to lower the weight of negatives as well as provide a medium for creating enlargements (Lavédrine 2003). Whether glass plates or roll film was used, the photographic medium was orthochromatic, and these blue-sensitive emulsions, when properly exposed for typical landscapes, severely overexposed the sky (Newhall 1964). As a result, the images look significantly different from modern panchromatic black-and-white films.

Before about the turn of the twentieth century, lenses did not have shutters; to expose the negative, the lens cap was removed and then replaced after a timed period (Auer 1975). As a result, moving figures, water, or wind-blown trees appeared blurred in some images. By the early twentieth century, cameras became smaller as films gained resolution and reliability, and lenses were equipped with shutters to stop action and with better quality glass. Most landscape photographers from this time period used medium- and large-format cameras, usually with a film size no larger than 5 by 7 inches (Auer 1975).

The earliest films were cellulose-nitrate based, a flammable substance that created a fire hazard for tightly rolled films. Photographic manufacturers started to replace nitrate film with cellulose acetate "safety" film in the 1920s (Lavédrine 2003). The first color processes became available in the early 1900s (Newhall 1964) but were sufficiently cumbersome or low resolution that their popularity was limited. Color film was introduced in the 1920s, although the first color films were unstable and faded; development of Kodachrome films in the 1930s provided a stable and more archival color film, if stored properly (Reilly 1998). Polyester-based films, introduced in the 1950s, and cellulose triacetate are the predominant types in use today (Lavédrine 2003). Cameras using the 35-millimeter film format gradually gained favor with photographers in the middle of the twentieth century, although they are rapidly being replaced by digital cameras today.

Most early cameras were mounted on tripods, many of which were short and of fixed height. In later years, tripods became more adjustable; hand-held cameras, held at eye or chest level, became increasingly popular as film speeds improved. While often impossible to determine, the tripod and camera style used, along with the photographer's height, are useful in establishing the proper height of the camera when field matching photographs. Images of the photographer posing with the camera setup can be helpful in determining these parameters.

Field Matching Procedure and Techniques

Photographic matches should be as precise as possible in order to allow for the best possible interpretation of data. When matching an image established for the purpose of repeat photography, generally referred to as photo monitoring, the procedure is straightforward. By positioning the camera directly over the previously established marker and using the same camera and lens combination (or equivalent) adjusted to the same height, tilt, and azimuth, the photographer should be able to accurately replicate the image even without having a copy of the original photograph at hand. The length of time between matches varies with the purpose of the project, ranging from a few hours to several years. Hall (2002) provides detailed instructions on a method of establishing a program of repeat photography monitoring.

Repeating previously unmatched photographs entails first locating roughly where the original was taken, as few vintage photographs are labeled with any degree of precision. Local residents or guides may be of assistance, but, in the experience of the authors, they may incorrectly identify location. "Virtual" repeat photography (Hanks et al., chapter 3) using digital-elevation modeling, programs such as Google Earth, and topographic maps may aid in narrowing the possibilities. Geologic and vegetative features provide further clues, along with shadows and streamflow direction. Once the general geographic location has been identified, it is often, but not always, a simple matter to roughly locate where the original photograph was taken. In those rare instances where a member of the team happens to be a bush pilot, low-elevation flights across the potential terrain can be used to quickly locate approximate photograph positions. The position, recorded by Global Positioning System (GPS) technology in the air, is then revisited by ground vehicle and the photograph taken. Western (chapter 16) has provided this expertise for several of the Kenya photograph matches in the Desert Laboratory Collection.

Once the original camera station is determined to within a few meters, the more exact camera station used for the match is established. A variety of techniques may be used to achieve this purpose. Malde (1973) described using the principles of parallax to determine the photograph location, a technique predicated on persistent foreground and background detail, such as distinctive boulders, structures, peaks, or long-lived vegetation. The photographer must first locate an imaginary line on which foreground and background features along either the right or left side of the view fall into the same alignment as they do in the original image. By moving backward and forward along this line, other features in the center or other side of the view will come into alignment; the intersection of these two imaginary planes establishes the vertical line where the camera station is located.

The correct camera elevation is determined by comparing foreground and background features, which provide the third imaginary plane that should reveal the camera station. This can be field checked with an instant print or digital photograph. Harrison (1974) discussed a more precise but time-consuming mathematical method, which was later modified by Klett et al. (1984) to include instant prints. In this technique, a series of lines are drawn upon a copy of the original image in order to determine distance ratios among prominent, persistent features within the view. By making the same measurements on an instant print, adjustments can be made in the field to achieve a precise camera station. Should instant prints be unavailable, users of large-format cameras can make the measurements on the ground-glass viewing window on the back of the camera. Klett (chapter 4) elaborates the use of digital imagery and portable computers and printers to accurately locate camera stations within the field.

More than 100 photograph matches in the Desert Laboratory archive feature the monolithic monuments marking the international boundary between the United States and Mexico (see the Malpai Borderlands Group Web site, www.malpaiborderlands group.org/map.htm, for examples). In the field, camera placement and height can readily be determined by first measuring, on the original image, the distance between landscape features (e.g., two distant peaks) and the length of the border monument. The ratio expressed by these two values is then sought when measuring the same features on an instant print or the image on the ground-glass viewing aperture of a large-format camera. The camera is moved forward or backward until the correct ratio is achieved.

High-quality field prints that display as much detail as possible are critical to precise matching. Digital manipulation enables enhancement of details that are valuable for matching in a way that is difficult if not impossible to do with a traditional photographic enlarger. The results are often not aesthetic but are more readily matched. Nonetheless, an exact match is still sometimes impossible to obtain. Readily distinguishable features may be few, or clouds, people, animals, or other items may obscure details. We have often discovered camera stations that are no longer accessible due to erosion or deposition, construction, or vegetation growth. In these cases, a camera station is established as close to the original as possible, with the proper location described in the field notes. If vegetation or another obstruction blocks a view, the photographer may choose to take a documentary shot of the blocked view from the correct location and establish a station with an unobstructed view nearby. The likelihood of vegetation ultimately blocking the view is far greater in forested regions than in arid ones with more open vegetation.

Camera Equipment and Film

For nearly all of our fieldwork at the present time (2010), contributors to the Desert Laboratory Collection use tripod-mounted, medium-format (120 roll film) or large-format (4- by 5-inch) film cameras. These film sizes offer a reasonable balance between high resolution and portability, yielding the most possible data for scientific interpretation of change while still maintaining a practical camera gear size, weight, and cost as compared to larger film formats. Digital images are also taken at each camera station, using a digital single-lens reflex (SLR) camera, but they are considered a backup and not a substitute for film. While medium- and large-format digital cameras and camera backs that offer resolution comparable, or even superior, to that of large-format film are available, they can be prohibitively expensive for many repeat photographers. Film has better archival attributes, largely due to the transitory nature of digital storage technology (see later discussion), although archival-quality prints can serve as the archived material for digital images in lieu of film. As digital photography gains in popularity, film will become obsolete.

An assortment of fixed lenses is necessary to approximately match the varying fields of view presented in different original photographs. Wherever possible, the match is made during the same season as the original. Images made on the same date and time, under similar light/shadow conditions, are particularly revealing and are essential for some types of analysis (Hoffman and Todd, chapter 5). Rogers et al. (1984) and Malde (2000) advocate using a "Sundicator" or sun angle charts to determine time and date of a photograph based on direction and length of shadows in the original image.

For each view, our standard is to capture one color (ISO 160) and two black-and-white negatives (ISO 100), one color transparency (ISO 100), an instant print, and a digital image. Black-and-white film offers better archival properties and tonal range, while color provides additional information. Filters, typically Wratten Yellow 8, may be used with black-and-white film as a haze filter to increase contrast and bring out background scenery. Instant prints verify exposure and camera-station location, and they are also used as an aid to properly filing negatives; however, instant film is becoming both scarce and expensive, and its continued use is in doubt. Digital images may be used to field check the accuracy of a match (Klett, chapter 4) and also serve as a backup to film.

Documentation of the Camera Station and Relevant Data

Camera stations are marked using rock cairns, rebar stakes, or an "x" chiseled into rock; in some areas camera stations are not marked because of environmental concerns or the impossibility of placement, such as in streambeds or roads. In the early years of the collection, camera stations in the United States were mapped on 15- or 7.5-minute topographic maps; currently, camera station locations are documented using GPS receivers in addition to topographic maps. After taking the replicate photograph, the compass azimuth (or bearing) of the view and the camera's vertical tilt and height above ground level are recorded.

Regardless of the positioning of the original camera, the camera is leveled from side to side to allow more precise future replications. The camera models, lenses, filters, film numbers, exposures, date, time, and names of photographer and crew are recorded, as well as a potentially wide variety of view attributes, including plant species present, landforms, land-use practices, and any special considerations required to understand changes in the view. Notes identifying specific plants, rocks, or other features can be written directly upon field prints or on the instant print. Klett et al. (2004) and Klett (chapter 4) gathered a large amount of additional cultural and documentary information as part of the rephotography process.

Digital Photographic Matching on the Computer

Once a vintage image is repeated, the new view must be rotated and cropped to the original in order to show the same field of view. Before the widespread availability of digital media and workstations, this was done with a traditional photographic enlarger. Now, the images are electronically matched by overlaying digital images in different layers using appropriate computer software. With the top layer at a reduced opacity, the images are rotated, sized, and cropped until a proper alignment is achieved. Digital matching affords tighter control than conventional matching, and once completed, multiple prints or electronic copies can be generated with minimal cost and effort. In this process, digital photography offers an advantage over film because scanning negatives or prints is both time consuming and a source of potential imperfections.

When working with digital files, an unmanipulated version of each scan or original image should be saved and all work done from copies of this master. Proprietary formats should be avoided, choosing instead standard formats such as uncompressed TIFF instead of formats that use compression algorithms, such as the Joint Photographic Experts Group (JPEG). If it is likely that the collection will eventually be transferred to a specific repository, such as the National Archives in the case of federally generated records, use the standards of the repository. If hard drive space and processor speed permit, scan and archive image files at 16-bit for gray scale and 48-bit for color in order to allow for greatest tonal range and maximum flexibility in the future (Frey and Reilly 1999). Color images should be scanned in red-green-blue (RGB) rather than cyan-magenta-yellow-key black (CMYK), as the latter is a derivative color space that is specific to the device used (Puglia et al. 2004).

Cataloging Images

Maintaining intellectual control is essential in any photographic collection. In the case of the Desert Laboratory Collection, a unique number, known as a stake number, is assigned to each camera station. The number serves as the identifier for each image and is used on prints, negatives, and transparencies and in naming digital files. The year in which an image was taken serves as a suffix. For example, an 1896 image taken of the city of Tucson (fig. 2.2A) has a file name of "Stake 1061–1896," the file "Stake 1061–1981" is the match of this view made in 1981,

and the file "Stake 1061–2007" is another match taken in 2007. Alpha characters are used to differentiate views taken at the same camera location but looking in different directions. Other file-naming conventions may also be appropriate, provided they provide an unambiguous name that can be recognized by someone not intimately familiar with the collection.

While some software programs provide templates for cataloging photographs, ones more suitable to repeat photography can be created using standard spreadsheet or database programs. Fields should include a key number and camera-station number; location information, preferably in multiple fields to allow for more specific searching by descriptive location (latitude-longitude or UTM data); date; original photograph data, including source archive; and match information, including date and photographer. Other potential fields include exposure details, azimuth, subject matter, and the presence or absence of certain features, such as ephemeral or perennial stream courses, riparian vegetation, or human activity. By including drop-down lists of commonly used and standardized terms (such as index terms used by various scientific societies; e.g., the American Geophysical Union), the ability to quickly and accurately locate the desired information is greatly increased. Metadata about individual photographs can be embedded directly within digital files. When possible, metadata should conform to established conventions, such as those outlined by the Dublin Core Metadata Initiative on their Web site (www.dublincore.org).

Archival Storage of Photographs

Standard archival practices should be followed in storing collections of repeat photography, both traditional and digital. While archival-quality storage enclosures that pass the Photographic Activity Test (PAT), as standardized by the International Organization for Standardization (ISO 18916), are advisable, properly controlling temperature and relative

Figure 2.2. The city of Tucson, Arizona, USA.

A. (1896). This view from Sentinel Peak is east down Congress Street into Tucson, which at that time had 6,000 inhabitants. A small irrigation canal that flows right to left through agricultural fields in the foreground is lined with *Populus fremontii* (Frémont cottonwood). The Santa Cruz River is in the midground, and downtown is behind the line of trees. Because the orthochromatic film Roskruge used is sensitive to haze, the Rincon Mountains are fuzzy. (G. Roskruge, Arizona Historical Society/Tucson, 46397).

B. (17 December 1981). The conversion of agricultural fields to housing is not surprising given that Tucson had more than 330,000 people at this time. The tributary is no longer visible, the Santa Cruz River is stabilized by soil cement, and Interstate 10, the major thoroughfare through town, is the raised roadway in the distance. (R. M. Turner).

C. (20 June 2007). Tucson's population now is more than 800,000, and maturation of the foreground neighborhood is shown by the increase in the size of trees growing along streets and on private property. The bare area in the right midground is part of a redevelopment project for downtown Tucson. (R. M. Turner, Stake 1061).

humidity (RH) is of far greater importance (Lavédrine 2003, Adelstein 2004). High temperature and high RH enhance many types of decay, including mold (all media types), color dye fading (all forms of color media; plate 1.1B), silver oxidation of black-and-white images, bleeding of dyes in inkjet prints, and layer separation of CDs and DVDs. Large and rapid swings in humidity and temperature are particularly destructive. In addition, high ozone levels can cause color shifting in inkjet prints on porous papers, and air pollutants in general have a negative effect on most media. Light will fade all types of images, particularly color. Acids in fingerprint oils can create permanent marks, and inkjet prints are especially vulnerable. Acidic storage enclosures and adhesives also promote deterioration.

An ideal storage facility would include a fireproof vault equipped with a dry fire suppression system, and would allow for a stable RH within a range of 30 to 50 percent as well as temperature control. While black-and-white prints and CDs/DVDs are acceptably stored in cool (12 degrees Celsius) conditions, color images and acetate negatives should be placed in cold storage (4 degrees Celsius), which also benefits black-and-white prints and negatives (Reilly 1993, 1998). Good ventilation is essential for reducing contamination by off-gassing. Freezing is optimal for all media except CDs/DVDs and glass-plate negatives, dramatically increasing useful life expectancy, particularly that of color (Reilly 1998, Adelstein 2004).

Digital storage technology changes with sufficient frequency to render archiving of digital files difficult, with many storage media becoming obsolete within a few years of introduction (Rothenburg 1999). Continual migration of digital archives to the latest technology is essential, albeit time consuming and costly. In addition, all digital materials must be routinely backed up to ensure against hard drive failure. An excellent arrangement is a live, mirrored backup system. Since many backup systems are located in the same building as the master hard drive, they do not guard against disasters such as floods and fires.

Ideally, an additional set of backups is kept off site; the same is true for traditional prints and films.

Due to the transient nature of digital storage media, prints are an essential part of archiving images obtained with digital cameras or scanners (Image Permanence Institute 2003). Archivally processed wet chemistry prints generated from digital files, pigment-based inkjet prints produced on high-quality porous paper, and electrophotographic (laser) prints are all suitable choices. Dye-based inkjet and dye diffusion thermal transfer prints are somewhat less stable but also a viable option.

Discussion and Conclusions

In this chapter, we discuss nearly half a century of experience in securing, cataloging, and archiving repeat photographs. Although certain fundamental techniques apply in virtually all cases, repeat photography projects can be done quite simply or more elaborately depending upon the goals and budget of the project. While properly archiving photographs is challenging and expensive, it is an essential part of the long-term preservation of data contained in repeat photography in order to ensure continued usefulness. At the start of the twenty-first century, contributors to the Desert Laboratory Repeat Photography Collection still use film for matches, although it is almost entirely medium- or large-format film. While large-format film is still preferable to digital imagery that is the equivalent of 35 millimeter film owing to its superior resolution, large-format digital cameras now offer resolution that equals or surpasses that of large-format film cameras. In addition, digital manipulation of images is preferable to that afforded by traditional darkroom techniques, which underscores the need to digitize film media for further analysis and publication. As digital photography continues to evolve, it will likely replace film entirely, particularly if the resolution of digital images surpasses that of large-format film for a comparable cost. Whether film or digital cameras are used, an archival-quality print or negative of each photograph should be archived in order to ensure that the images will be available in perpetuity.

Acknowledgments

The authors thank Connie McCabe, Karen Underhill, Steve Tharnstrom, Dominic Oldershaw, and the many other archivists and photographers who have assisted us. All photographs are courtesy of the Desert Laboratory Repeat Photography Collection unless otherwise noted.

Literature Cited

Adelstein, P. Z. 2004. *IPI media storage quick reference.* Rochester, NY: Image Permanence Institute.

Auer, M. 1975. *The illustrated history of the camera from 1839 to the present.* Translated and adapted by D. B. Tubbs. Boston, MA: New York Graphic Society.

Bierman, P. R., J. Howe, E. Stanley-Mann, M. Peabody, J. Hilke, and C. A. Massey. 2005. Old images record landscape change through time. *GSA Today* 15:4–10.

Frey, F. S., and J. M. Reilly. 1999. *Digital imaging for photographic collections.* Rochester, NY: Image Permanence Institute.

Hall, F. C. 2002. *Photo point monitoring handbook: Part A—field procedures.* USDA Forest Service General Technical Report PNW-GTR-526. Pacific Northwest Research Station, Portland, OR.

Harrison, A. E. 1974. Reoccupying unmarked camera stations for geological observations. *Geology* 2:469–471.

Image Permanence Institute. 2003. *A consumer guide to traditional and digital print stability.* Rochester, NY: Image Permanence Institute.

Klett, M., K. Bajakian, W. L. Fox, M. Marshall, T. Ueshina, and B. Wolfe. 2004. *Third views, second sights: A rephotographic survey of the American West.* Santa Fe: Museum of New Mexico Press.

Klett, M., E. Manchester, J. Verburg, G. Bushaw, R. Dingus, and P. Berger. 1984. *Second view: The rephotographic survey project.* Albuquerque: University of New Mexico Press.

Lavédrine, B. 2003. *A guide to the preventive conservation of photograph collections.* Translated from the French by Sharon Grevet. Los Angeles, CA: Getty Conservation Institute.

Malde, H. E. 1973. Geologic bench marks by terrestrial photography. *US Geological Survey Journal of Research* 1:193–206.

Malde, H. E. 2000. *Repeat photography at Chaco Culture National Historical Park, New Mexico, based on photographs made in the 1930s, 1970s and the year 2000.* Denver, CO: Wright Paleohydrological Institute.

Newhall, B. 1964. *The history of photography: From 1839 to the present day, revised and enlarged edition.* New York: Museum of Modern Art.

Puglia, S. J. Reed, and E. Rhodes. 2004. *Technical guidelines for digitizing archival materials for electronic access: Creation of production master files—raster images.* US National Archives and Records Administration, www.archives.gov/preservation/technical/guidelines.pdf (accessed 2 February 2010).

Reilly, J. M. 1993. *IPI storage guide for acetate film.* Rochester, NY: Image Permanence Institute.

Reilly, J. M. 1998. *Storage guide for color photographic materials.* Albany: University of the State of New York.

Rogers, G. F., H. E. Malde, and R. M. Turner. 1984. *Bibliography of repeat photography for evaluating landscape change.* Salt Lake City: University of Utah Press.

Rothenburg, J. 1999. *Ensuring the longevity of digital information.* Santa Monica, CA: RAND. www.clir.org/pubs/archives/ensuring.pdf (accessed 20 July 2007).

Shantz, H. L., and B. L. Turner. 1958. *Photographic documentation of vegetational changes in Africa over a third of a century.* University of Arizona, College of Agriculture Report 169, Tucson.

Turner, R. M., H. A. Ochung', and J. B. Turner. 1998. *Kenya's changing landscape.* Tucson: University of Arizona Press.

Webb, R. H., D. E. Boyer, and R. M. Turner. 2007a. *The Desert Laboratory Repeat Photography Collection—an invaluable archive documenting landscape change.* US Geological Survey Fact Sheet 2007-3046. Tucson, AZ.

Webb, R. H., S. A. Leake, and R. M. Turner. 2007b. *The ribbon of green: Change in riparian vegetation in the southwestern United States.* Tucson: University of Arizona Press.

Chapter 3

Virtual Repeat Photography

Thomas C. Hanks, J. Luke Blair, and Robert H. Webb

In the practice of repeat photography, identifying the precise location where the original photograph was taken is crucial. The location may be readily apparent, such as at a railroad siding (Shantz and Turner 1958), along a historical track (Jonas, chapter 21), or along a river (Webb 1996), but this is not always the case. Some photographs were taken at obscure locations reached after days or weeks of rigorous travel into remote areas. Complicating the search for historical camera stations is the fact that the first detailed maps did not become available until well into the twentieth century, and most photographers left vague or nonexistent information on their travels. Here we follow the travels of the lead author's father, James J. Hanks, who accompanied two little-known expeditions as a photographer into the scarcely charted plateau lands of northern Arizona and southern Utah in 1927 and 1928 (plate 3.1). Like Jonas (chapter 21), one of the goals was to determine the routes of these expeditions.

Finding Camera Stations

Repeat photography of historical images begins with the examination of the historical view, particularly focusing on foreground–background relations that are critical to exactly reoccupying the original camera station (Rogers et al. 1984). Without background topography, or with hazy backgrounds with few if any recognizable geographic features, relocation of camera stations is difficult to impossible without additional specific information from the original photographer. None of the digital techniques discussed here can help relocate camera stations of such photographs.

Seldom does a repeat photography project begin without at least some rudimentary knowledge of where photographs were taken or with no significant background topography. In the case of the J. J. Hanks photographs, the overall trip routes—from Red Lake, Arizona, to north of Navajo Mountain (1927) or to the Kaiparowits Plateau (1928), Utah (plate 3.1)—are known. However, as is the case with most historical photographs, Hanks's captions were vague or potentially misleading; for example, the caption for plate 3.2A is "Colorado River from Gothic Point," and although the Colorado River remains a known geographic feature, Gothic Point is not a recognized name, and there are many possible overlooks in Glen Canyon where similar topography could be photographed.

The "Brute-Force Technique": Walk into the View

When only rudimentary captions or no information accompanies the photograph, the most common way of finding the original camera position is to "walk into the view." By using topographic maps and other information, including the sequence of photographs (if known) compared with the known or surmised route, one can determine an approximate view direction, and given a major geographic feature (e.g., Navajo Mountain), the photographer can determine approximately where the camera station might be within 1 to 5 kilometers. Using any number of conveyances, ranging from foot travel to helicopters, the would-be repeat photographers can access the general vicinity and find the camera station by comparing the topography and other features in the historical photograph with the current view. Boyer et al. (chapter 2), Klett (chapter 4), and Hoffman and Todd (chapter 5) discuss the mechanics of fine tuning of camera position and securing the repeat image. The brute-force approach, while effective, is not efficient in the rugged topography of the Colorado Plateau, and misjudging a ridge or access route might result in considerable time wasted in finding a camera station.

Virtual Repeat Photography

Virtual repeat photography (VRP), as we use the phrase here, is the use of geographic information system (GIS) software to display and manipulate a digital-elevation model (DEM) to locate the camera station of a historical photograph. Within a domain of 10-meter resolution, US Geological Survey 7.5-minute topographic quadrangles aggregated by the Environmental Systems Research Institute (ESRI) ArcInfo software, we use ArcScene, a three-dimensional viewer extension of ArcInfo, to view this domain, or any part of it, at any position, elevation, azimuth, and inclination we choose to find topography similar to that portrayed in a two-dimensional photograph. In addition, use of registered digital orthophotographs, widely available for the continental United States, allows illustration of the DEM with landscape elements, such as outcrops, river channels, and even trees and shrubs.

Our goal is to locate photographic sites accurately enough that finding them exactly should only take hours of on-the-ground searching, not days (see also the USDA Forest Service Repeat Photography in Southeast Alaska Web site, www.fs.fed.us/r10/spf/fhp/repeat_photo_se/index.php). Roughly speaking, this equates to identifying camera-station positions correctly to less than 0.5 kilometer, if the terrain is not difficult. For photographs featuring topographic relief of 50 meters or more, this goal is generally achievable, often very quickly and often with greater accuracy (less than 0.1 kilometer). We illustrate these capabilities by locating the camera stations of three photographs taken in the Colorado–San Juan country of southern Utah in 1928, a very remote area of very difficult terrain.

Google Earth

The ongoing proliferation of Internet map servers, led by the online product Google Earth (earth.google.com), offers an alternative to the data-intensive VRP technique. Launched as a free Web product in 2005, Google Earth has fascinated casual users and established its potential for serious geospatial reconnaissance for scientific research, including the relocation of camera stations for historical photographs. Google Earth combines a mosaic of satellite imagery and aerial photography over a wide range of resolutions with an underlying DEM to provide what would appear to be a realistic virtual landscape at any point on Earth. The imagery is frequently updated; therefore resolution is likely to continue to improve with time. As of 2006, a Google Earth image of plate 3.3 did not afford sufficient detail to be useful in locating the photograph, but by February 2008, the Google Earth view's topographic detail had greatly improved, and we could see all of the same features visible in the DEM image. Like the VRP technique, the basic software enables the user to fly around in a three-dimensional virtual space and land to inspect a 360° view. Without considering costs, the major difference between Google Earth and the VRP technique is the resolution of the underlying imagery and DEM.

The Hanks Photographs

Long before he established himself as the preeminent anthropologist of the Navajo (Kluckhohn and Leighton 1946), Clyde Kluckhohn had journeyed extensively through the American Southwest, even as a teenager (Kluckhohn 1927). Kluckhohn conceived a somewhat fanciful plan to visit the Kaiparowits Plateau in southern Utah from the Navajo Reservation in northern Arizona; he was either unaware or did not care that access would be much easier from long-established settlements in southern Utah. During the summers of 1927 and 1928, James J. Hanks accompanied Kluckhohn, as did Bill Guernon, Nel Hagen, and Lauriston Sharp, while all were undergraduates at the University of Wisconsin.

Both the 1927 and 1928 trips began at the Red Lake Trading Post in northern Arizona (plate 3.1), and both had the goal of crossing the Colorado River in the vicinity of Rainbow Bridge, the largest natural bridge in the world (Bernheimer 1929), before ascending the Kaiparowits Plateau. The 1927 trip fell short of the intended destination, whereas the 1928 trip succeeded, and both are recounted in Kluckhohn (1933). Nine of the thirteen pictures that illustrate this book were selected from the 451 photographs taken by Hanks. Hanks's images document a large variety of historical and geographical features in the region, ranging from scenic views of landscape to people and archaeological sites, and most of these photographs are available online (www6.nau.edu/library/sca/exhibits/hanks/index.cfm). Although the locations of many photographs were readily apparent from the subject matter, the captions generally were cryptic or nonexistent for many views.

Locating the camera stations for vaguely described photographs in this immense, remote, and forbidding terrain (plate 3.1), with topographic relief exceeding 2,000 meters, presents considerable challenges. Knowing how the trip traversed the landscape would help narrow the search, but the maps they used portrayed very generalized topography. Pieces of the Henry Mountains topographic quadrangle (1:250,000, edition of April 1892, reprinted 1920) and Escalante topographic quadrangle (1:250,000, presumably the edition of January 1886) survive the 1928 trip; the data used to compile these maps were collected in the 1870s and 1880s. While Kluckhohn (1933) includes sketch maps on the inside cover of his book, neither they, the text, nor Hanks's annotations of his pictures provide any real detail as to where the group was when any photograph was taken, except in the cases where well-known geographic features are given by name or are obvious (e.g., Rainbow Bridge or Hole-in-the-Rock). Many of the place names used by the Kluckhohn party were ones they established, and most of these places remain without formal names. We developed the technique of virtual repeat photography because we believed the route of the Kluckhohn party, as well as the Hanks's camera stations, were uncertain to about 5 kilometers or more, especially north of the Colorado River.

Four Examples from the Hanks Collection

The Colorado River at Oak Creek

During the 1928 trip, Kluckhohn and his group accessed the Kaiparowits Plateau and searched along its rugged southern margin for archaeological sites. Hanks took a view of the Colorado River (plate 3.2) that is clearly from this margin. Locating the camera station for this Hanks view is relatively easy—we located this station using topographic maps prior to GIS analyses—because the photographic site is high and the field of view is wide and open with significant topography, suggesting only a limited number of potential camera stations. However, it graphically provides an example of how the VRP technique works.

Plates 3.2A and 3.2B illustrate basic capabilities of VRP using ArcScene. Plate 3.2A shows the Kaiparowits Plateau as a blue-green band coming from the lower right, and the camera station appears to be near the end of that band. Plate 3.2B shows the view from that end from high above a bedrock promontory, and some landscape elements of the Hanks photograph are becoming apparent. Plate 3.2C shows the view looking south from the surface of the DEM in a position that best approximates the view from the camera station (plate 3.2D). The approximate camera position is recorded from the DEM using ArcScene, and now a helicopter would be desirable to access this camera station in extremely rugged terrain today.

Because there is little foreground in the 1928 view, a comparison of the ArcScene view with the old photograph (plates 3.2C and 3.2D) provides some general information on what has changed in about 75 years. The Colorado River, once free-flowing away from the camera position and to the right, is now impounded behind Glen Canyon Dam; Lake Powell, shown at its full-pool elevation of 1,128 meters in June 1981 (plate 3.2D), now occupies much of the canyon. The cliff on the river right near the middle of plate 3.2D is 250 meters high, but now only about 110 meters of it is exposed above lake surface. The short-wavelength features labeled N (west flank of Navajo Mountain) and C (Cummings Mesa) in plates 3.2C and 3.2D illustrate the resolution of the synthetic images.

The Colorado River at Ribbon Canyon

Greater challenges arise in reproducing photographs taken at camera stations at lower elevations and less obvious positions relative to the background topography. One example, shown in plate 3.3, was vaguely captioned "Navajo Mountain in the background. Walls of Colorado canyon in foreground." The cam-

era station could not be determined by comparing the view with topographic maps. Moreover, Kluckhohn (1933) states that the 1928 crossing of the Colorado River occurred at Hole-in-the-Rock, made famous by Mormon pioneers of the San Juan Mission (Miller 1966). Two of the Hanks photographs from 1928 document this crossing point, but the group clearly traveled along the river corridor on the north side to exit at a location other than Hole-in-the-Rock. Much of the difficulty in fixing this location involved similar river geometry and canyon walls at point 3' (plate 3.1), about 3 kilometers southwest of Hole-in-the-Rock, and at point 3.3, a site about 2 kilometers northeast of Hole-in-the-Rock; both sites have a similar perspective of Navajo Mountain, which forms the skyline of the 1928 view.

The diagnostic feature of the 1928 Hanks photograph (plate 3.3A) is the old river channel, which is left of and behind the wall of Navajo Sandstone across the river. Plate 3.3B is a DEM image looking south from point 3', showing much of the 1928 view submerged beneath Lake Powell but still indicating that the camera station was nearby. Our match (plate 3.3C), from site 3' (plate 3.1), exactly replicates the original view from a position west of Ribbon Canyon, a tributary of the Colorado River. This camera station is below the surrounding topography, which rises to about 1,340 meters to the right and behind the picture. The VRP technique allowed us to easily find this camera station, which would have been difficult if not impossible to find otherwise, and establish the Kluckhohn party's route and place of exit from the canyon of the Colorado in this difficult terrain.

In 2006, using an early version of Google Earth, we attempted to locate this camera station. Unfortunately, the resolution of the DEM at that time was insufficient to definitively determine the location of the camera station for plate 3.3A. Google Earth is periodically updated with better topographic data and photographic imagery, and by February 2008, the image rendered from Google Earth was similar to that obtained using VRP. Improvements in the online capabilities of Google Earth are now beginning to make that tool as useful as VRP in locating camera stations, as will be illustrated for the case of Piute Canyon (plate 3.5).

One interesting change in this set of images is the collapse of the buttress adjacent to the old river channel that forms the east end of the wall across the river (plate 3.3C). This failure is so massive that it is even apparent in the comparison of the historic and virtual images (plates 3.3A and 3.3B). This collapse appears to have occurred in at least two stages because the more recent collapse is evidenced by rock-fall detritus uncoated with the "bathtub ring" of calcium carbonate (white band) that is otherwise uniform along this wall at and near lake level. This event probably occurred between 1999 and 2003 when Lake Powell had been unusually low owing to drought conditions upstream in the watershed, and the earlier and larger collapse exposed a new wall on the left buttress that is coated with calcium carbonate. This event (possibly more than one) has occurred after 1928 but before the 1999 high-stand of Lake Powell.

Trail Canyon, Kaiparowits Plateau

The top of the Kaiparowits Plateau is gently undulating, and the search for camera stations here illustrates a situation for which the topography alone is not diagnostic. The 1928 photograph (plate 3.4A) shows this topography with Navajo Mountain just exposed and faint in the right background; the caption reads: "Looking down into the valley where our camp was pitched." Locating this camp was important insofar as several dozen other photographs were taken in or near this camp, where the Kluckhohn party stayed for much of August 1928. From the text and sketch maps in Kluckhohn (1933), we surmised that this picture was taken along the northeast edge of the Kaiparowits Plateau looking southeast toward Navajo Mountain. Plate 3.4B is a DEM image looking to the southeast to the head of Trail Canyon, which trends downslope from the center of the image to the right (west); this view depicts Navajo Mountain with the correct perspective and also the triple junction of valleys just left of center in plate 3.4A. Terrain of this sort is common along the 80-kilometer-long northeast edge of the Kaiparowits Plateau (plate 3.1), however, and the DEM rendering alone is not definitive for establishing the camera station of plate 3.4A.

Plate 3.4C shows an orthorectified, digital aerial photograph (USGS digital orthophoto quadrangle [DOQ], 1 meter resolution) draped over the DEM with a mapping transformation defined by the coordinate system of plate 3.4B, imparting a quasi three-dimensional character to the DOQ. This composite image provides compelling evidence that the camera station for plate 3.4A is within a short distance of this point, and we matched the 1928 view from a nearby point (3.4 in plate 3.1). Camp Kluckhohn, which the Kluckhohn expedition used as a base camp for several weeks, is at the left edge and at the base of the horizontal white ledge, just beneath the skyline at the upper left of plate 3.4D. We subsequently reoccupied 13 camera stations for 1928 photographs in the vicinity of this camp, which we would not have found without locating the camera station for the critical photograph shown in plate 3.4A.

Plates 3.4A and 3.4D are a typical repeat photography pair showing a large amount of change between 1928 and 2003. However, the combination of DEM and DOQ images (plate 3.4C) provides ancillary information at the intermediate date of 1983, when the aerial photograph was taken. Vegetation changes since 1928 are striking; on the east side of the axial valley, the *Juniperus osteosperma–Pinus edulis* (juniper–pinyon) assemblage has expanded considerably since 1928 through 1983 to 2003, at the expense of an *Artemisia tridentata* (big sagebrush) assemblage near and on the valley bottomlands. This change is typical of much of the American Southwest at these elevations and results from the combination of favorable growing conditions and fire-suppression policies of the twentieth century (e.g., Rogers 1982). Clumps of *Quercus gambelii* (Gambel oak), in the center midground, may derive from resprouting after presettlement fires. On the west side of the valley, however, the sparse hill slope in 1928 became quite vegetated by 1983 only to have thinned considerably by 2003, which we suspect is the result of a wildfire sometime between 1983 and 2003. The bare, yellowish area in the lower left of plate 3.4D is the result of a fire that occurred in 1992.

Piute Canyon

Historical photographs of Navajo Mountain from the south are more difficult to find because the terrain has much less relief, and road access is very difficult on the Navajo Reservation in northern Arizona. The VRP method, applied to views in this region, becomes much more useful in terms of saving time searching for a camera station. A Hanks photograph, taken in 1927 with the caption "Our first glimpse of Pihute [*sic*] Canyon" (plate 3.5A), offers such a challenge. Piute Canyon extends for perhaps 30 kilometers, although the perspective on Navajo Mountain narrows the range to several canyons near the upper reaches where the Kluckhohn party says it crossed in 1927 (plate 3.1). Numerous small canyons trend in a direction similar to the one shown in plate 3.5A, making location from topographic maps alone problematic.

As shown in plate 3.5B, the VRP technique shows that dominant foreground–background relations are not necessarily the most important to conclusively locating a camera station. A much smaller topographic feature, roughly resembling a saddle horn, is only apparent at one location on the rim of Piute Canyon, and we relocated the camera station for the 1927 photograph at this site (plate 3.5D). When first tested in 2006, Google Earth yielded an inconclusive rendering of the landscape from this location, and the distinctive saddle horn was not present. However, a 2008 image was much improved, and a 2010 image (plate 3.5C) substantially agrees with the VRP image and includes Navajo Mountain as well. One slight problem appears in the background, where Navajo Begay, a triangular-shaped peak between the saddle-horn mountain and Navajo Mountain, is in the Google Earth image but is not

readily apparent in the repeat photograph. Google Earth, although greatly improved, remains problematic, but its evolution suggests that it may one day be as useful as the VRP method.

In 1927, Piute Canyon had an arroyo that had recently incised into its floodplain alluvium (see Webb and Hereford, chapter 8). As shown in the Hanks photograph, the channel banks are sharp and the channel is devoid of vegetation. The repeat photograph (plate 3.5D) shows an increase in riparian vegetation in the channel that is mostly *Elaeagnus angustifolia* (Russian olive), a nonnative invasive species common in this region. *Juniperus osteosperma* (Utah juniper) and *Pinus edulis* (pinyon pine) have increased on the hillslopes in the midground as well as in the foreground.

Discussion and Conclusions

Our examples show that virtual repeat photography (VRP), a term we use to describe the use of sophisticated geographic information system (GIS) technology and high-resolution digital-elevation models (10 meter DEMs) and DOQs to digitally reference the views in old photographs, can pinpoint otherwise unknown camera stations with sufficient accuracy to not only aid in locating the original camera stations but also to aid interpretation of historical changes (Hoffman and Todd, chapter 5). We have successfully located nine camera stations originally occupied by J. J. Hanks in 1927 and 1928 using this technique, which saved us considerable field time and effort in reaching these places to capture the replicate views. Use of both DEM and DOQ data allows one to locate camera stations in terrain with just modest relief (plate 3.4). The VRP technique, however, uses expensive software to analyze large data files, and extensive training is required to apply this technique.

Google Earth, an online GIS resource, is also useful for locating camera stations, but its value depends greatly on the resolution of the underlying DEM and imagery. In our experience between 2006 and 2008, the quality of the DEMs and imagery for areas we were interested in on the Colorado Plateau improved to the point of being somewhat comparable to the VRP technique and its high-resolution DEMs and DOQs. The introductory version of Google Earth is free to download and easy to use, but the underlying data vary in quality across the Earth. As a test, we landed on camera stations for repeat photography presented in Nyssen et al. (chapter 14, plate 14.6), Lewis (chapter 15, figs. 15.2–15.5), Rohde (chapter 18, plates 18.4 and 18.5 and fig. 18.5), and Hoffman et al. (chapter 11, plate 11.2), and the DEMs at all of these sites—in Ethiopia, Australia, Scotland, and South Africa, respectively—were of insufficient resolution to allow assessment of the camera position, even though we had landed exactly where the repeat photograph was taken. In the future, Google Earth may be the tool of choice for locating camera stations in advance, and it is sufficient at some sites in the United States at the present time.

While our principal interest in replicating these photographs is to document ecological, geological, geomorphic, and hydrologic change, location of these camera stations prior to fieldwork helped enormously in finding less obvious camera stations, doing little more than "connecting the dots" and traversing these connections. In other words, the more camera stations that can be determined digitally aids in the total number of camera stations that can be reoccupied during the field phase as well as documenting expedition paths (also see Jonas, chapter 21). VRP, then, also possesses value for historians interested in the reconstruction of extended, photographically documented travels through once poorly charted terrains.

In the case of plate 3.3, the virtual photograph alone suggests the massive collapse of the eastern end of the cliff across the Colorado River. While one cannot learn much from 10-meter-resolution DEMs about geological change, except at the largest scale, the world of 1-meter-resolution DEMs is probably not far away. In the meantime, 10-meter-resolution DEM together with DOQ aerial photography offer much promise now to document ecological and geological change.

Acknowledgments

We extend our appreciation to Molly Hanks for finding J. J. Hanks's 1928 photographs; to Suki Starnes, daughter of Lauri Sharp, for making her father's collection of maps of the 1928 trip available to us; to D. E. Boyer for her skill and care in constructing digital images from the original negatives; and to her and photographer D. P. Oldershaw for their help in replicating the 1928 views. Fieldwork was funded in part by the Bureau of Land Management, Department of the Interior, with the enthusiastic support of Marietta Eaton of the Grand Staircase–Escalante National Monument. Steve Lefton and *Mooney 6709U* provided us with splendid aerial views of potential camera stations and good company. We thank R. Arrowsmith, J. W. Hillhouse, and K. A. Howard for their thoughtful reviews. All J. J. Hanks photographs are reproduced courtesy of Cline Library, Northern Arizona University; all photographic matches are courtesy of the USGS Desert Laboratory Repeat Photography Collection.

Literature Cited

Bernheimer, C. L. 1929. *Rainbow Bridge*. Garden City, NY: Doubleday, Doran and Company.

Kluckhohn, C. 1927. *To the foot of the Rainbow*. New York: Century Company.

Kluckhohn, C. 1933. *Beyond the Rainbow*. Boston, MA: Christopher Publishing House.

Kluckhohn, C., and D. C. Leighton. 1946. *The Navaho*. Cambridge, MA: Harvard University Press.

Miller, D. E. 1966. *Hole in the Rock*. Salt Lake City, UT: University of Utah Press.

Rogers, G. F. 1982. *Then and now: A photographic history of vegetation change in the central Great Basin*. Salt Lake City: University of Utah Press.

Rogers, G. F., H. E. Malde, and R. M. Turner. 1984. *Bibliography of repeat photography for evaluating landscape change.* Salt Lake City: University of Utah Press.

Shantz, H. L., and B. L. Turner. 1958. *Photographic documentation of vegetational changes in Africa over a third of a century.* University of Arizona, College of Agriculture Report 169, Tucson.

Webb, R. H. 1996. *Grand Canyon, a century of change.* Tucson: University of Arizona Press.

Chapter 4

Three Methods of Presenting Repeat Photographs

Mark Klett

Many different techniques have been used in repeat photography since its development in the late nineteenth century (Rogers et al. 1984; Boyer et al., chapter 2; Hoffman and Todd, chapter 5). This chapter addresses alternative methodologies for the practice of making repeat photographs, or rephotographs, of landscapes. Making rephotographs in both remote and urban settings has drawn interest from a variety of disciplines from the natural and social sciences to history and the arts, so it is necessary to acknowledge that the work is driven by divergent interests. At the same time, as amply demonstrated throughout the chapter, there are also photographic methods common to all fields, and these form the basis of shared goals.

The most important shared goal for rephotography has been the visualization of change as represented through photographs made of the same subject taken at different times.

A comparison of original and repeated photographs may indicate that change occurred in the intervening time interval, but since the photographs alone cannot explain the reasons for change, additional research or other materials collected on-site may be necessary to understanding the paired images. Thus, including contextual material may become an essential part of a project's methodology. In some studies, what is surrounding or outside the frame of the photograph may be as important as what was originally included in the view.

Although the methods used to make rephotographs vary, there are two major physical variables in the making of every landscape photograph that dictate a common basis for a shared rephotographic methodology. The first is based on the idea that every photograph has a unique physical point of origin, or vantage point, and results from the way camera lenses record real space. The vantage point corresponds to the center of the camera's lens, and, if this point can be reoccupied, a new photograph made from the same position will duplicate the former photograph's information.

In the sciences, the location of the vantage point is often critical because accurate measurements of change over time cannot take place unless the variable of the observer's position has been eliminated. In this work, reoccupying an original vantage point within centimeters can make a difference in interpreting the data (Hoffman and Todd, chapter 5). And even in studies where objective measurement is not a concern, reoccupying an existing vantage point with accuracy leads to a more convincing rephotograph. Therefore this technique has been adopted by most careful rephotographic workers in the fields of both art and science. Several publications describe methods for accurately reoccupying an existing vantage point (Malde 1973, Harrison 1974, Klett et al. 1984, Rogers et al. 1984). Vantage points are typically found by carefully realigning the remaining objects that appear in the original photograph until they reappear in the same configuration. To do the work successfully, a copy of the original photograph is carried in the field, and measurements taken on the print surface determine numeric spatial relationships between objects in the scene. These measurements are then compared to test prints made while on location (if possible), and adjustments are made in the vantage point until the measurements line up with the original. In the world of digital photography, it is also possible to overlay first and second images onto the screen of a laptop computer and compare vantage points while on-site. Whatever the techniques used to verify them, relocation of an original vantage point is the most common basis for rephotography in every discipline.

The second characteristic of landscape photographs is that they are made at unique times. The most common indication of the original photograph's moment in time is the lighting, and this can also be duplicated by matching both the time of day and the time of year in which the photograph was made (Malde 1973; Hoffman and Todd, chapter 5). The accuracy of duplicating the original photograph's lighting can greatly enhance the look of a rephotograph when compared to the original. Lighting defines the shapes of three-dimensional objects depicted in a flat picture plane, so that choosing not to duplicate lighting may limit how a viewer interprets space in the scene.

Together, repetition of vantage point and duplication of lighting correspond to a basic two-part rephotographic methodology that transcends disciplinary boundaries. When done carefully, the resulting work may be useful to researchers in the sciences, the humanities, and the arts as well as for general viewers. But beyond this basic methodology there are several other choices available to rephotographers. I will describe three projects as examples of extended rephotographic methodologies: the Rephotographic Survey and Third View projects (Klett et al. 1984, 2004), Yosemite in Time (Klett et al. 2005), and After the Ruins (Klett et al. 2006). Each project begins with the basic two-part rephotographic methodology and then alters those initial methods from the experience gained from each project.

Using Interactive Technologies in Rephotography

The Rephotographic Survey Project (RSP) began in 1977 to repeat the western survey photographs originally made for the King, Wheeler, Hayden, and Powell expeditions of the 1860s and 1870s (Bartlett 1980). These photographs were originally made by William Henry Jackson, Timothy O'Sullivan, William Bell, John K. Hillers, and others. The survey photographs were first housed in science and government archives such as the US Geological Survey and the National Archives; copies have since become collected by art museums. The photographers themselves have become the subjects of much historical research (Naef and Wood 1975, Osteroff 1981, Snyder 1981, Hales 1988, Nickel 1999).

Because the project was designed by artists, many of the questions behind the RSP's work took aim at the visual effects of pairing two similar but different photos that were taken at two distinct times. We found that any two images placed together form a new whole that creates a unique context in which neither photo exists in its time alone, and the images sometimes defy expectations as to which image came first. Further, the changes evident do not always conform to linear expectations.

The limitations of this work were also revealing. We found that rephotographs can imply the passage of time but not the causes or even existence of external forces that cause change. Rephotographs cannot show what is excluded by the view on either side of a picture's frame, and the photographs alone cannot explain the influences or context behind the making of the images.

There were three observations about landscape photographs that resulted from the RSP's fieldwork. First, landscape photographs result from an interaction of the photographer's personal experience and cultural influences. The photographs reflect an individual response to a particular moment in time and space and are also shaped by cultural forces that influence the vision of their makers; in a similar way, many forces shape viewers' interpretation of photographs. Second, photographers are always participants in the making of a landscape image, not simply impartial witnesses, and their photographs are never neutral visions of the world. Third, landscape photographs are timely even when they seem timeless, and the moment of each image is unique and can never be exactly repeated.

The resulting work of 122 rephotographs was published by Klett et al. (1984) using the conventional technique of pairing images either on the same page or on opposing pages (fig. 4.1). Side-by-side comparison of images has been the standard method of displaying the results of repeat photographs (Rogers et al. 1984) but has certain limita-

tions. Two horizontal landscape images displayed together require large and expensive amounts of page space when published, and if small, the details are often not reproduced well. This is especially true if there are more than two images constituting a rephotographic group, for example, when a site has been rephotographed multiple times. In such cases display can become a greater design challenge, requiring that images be scaled smaller to fit pages or that more pages be added when the images are published.

This was one of the problems faced by the project Third View as it attempted to repeat, once again in the late 1990s, the original survey photographs and the first rephotographs made by the RSP 20 years earlier (Klett et al. 2004). The Third View project began in 1997 on the twentieth anniversary of the RSP and was designed to address questions raised by the limitations of the first project.

The new project's field team explored technological advances that made it possible to collect new data in the field as well as provide a solution to the problem of displaying the work of multiple photographs.

Figure 4.1. Comstock Mine, Virginia City, Nevada, USA.
A. (1868). Photographer Timothy O'Sullivan captured this image of the Comstock Mine in Virginia City, Nevada, for the Geological Exploration of the Fortieth Parallel. (T. H. O'Sullivan, University of New Mexico).
B. (11 July 1979). When the Rephotographic Survey Project (RSP) returned to Virginia City, the individual mining claims had been replaced by a single strip mine as shown by this image, entitled "Strip Mine at the site of Comstock Mines, Virginia City, Nevada." This is a typical example of rephotography involving an original and a single match, readily displayed on opposing pages (Klett et al. 1984). (M. Klett for the RSP, 79-6-8).

This was important since most of the rephotographic sets now had three photos for comparison, making the earlier side-by-side approach impractical.

Third View's methodology began with the conventional two-part approach to repeating vantage point and lighting, adding a second rephotograph to continue the earlier series. The RSP's methods were used to make these rephotographs, and the only differences were that location coordinates were newly marked by Global Positioning System (GPS). Exposure data and camera information were also recorded for each exposure. A 4- by 5-inch view camera was used for the photography, and test shots were made onto instant film. Measurements were taken between copy prints and the instant print, then compared to calculate the most accurate placement of the camera (Klett et al. 1984). In the case of discrepancies between first- and second-view vantage points, the Third View images were based on the vantage points of the second views made by the RSP.

But after the rephotographs were made, other data were collected that could not have been used in the first attempts at rephotography. New photographs of the landscape were taken surrounding the original vantage point; video footage was recorded of areas around or near the site or from the journey to the location. Sound recordings and oral history interviews were recorded with people connected to sites. Other made-for-computer images such as panoramas were created to place the scenes in a contemporary context and show what is behind the camera or outside of the picture frame. In addition, artifacts (contemporary and not of antique or archaeological value) were collected at significant locations. The project was an updated version of the traditional western geographical survey, only rather than exploring unknown territory the field team reexplored the once open spaces of the West as it had been transformed into home to millions of inhabitants. Thus, besides physical change, the project was concerned with the evolving perception of place, the shifting mythologies of the West of the imagination, and the many concerns of documentary photographic practice.

Fieldwork collected a large array of visual and aural documents, and later these components were edited and combined into interactive presentations. The project launched a Web site (thirdview.org) containing many of the less memory intensive works. The fully edited materials were presented in an interactive DVD that was published and included in the back jacket of Klett et al. (2004). In the DVD, viewers may access the video, sound files, and other images that connect places to events, people, and ideas. In addition, there were several significant changes to the techniques used to view rephotographs themselves. The software used to create this disk changed over the period of several years, and as this chapter is written some programs used are no longer available. But the concept of the work is not software dependent, and new or different software may be used to accomplish the same results both in stand-alone disk form and on the Web. The purpose here is not to describe how such work is technically created because the process is constantly changing, but rather to outline the range of possibilities that are available with increasing capacity within the digital arena.

Perhaps the most effective electronic technique used by Klett et al. (2004) was the simple overlay and dissolve of one image into the next. Images at each time interval were sized and aligned using photo editing software, and then using the interactive software, the images were placed into a single window and made to dissolve one into the other by clicking a switch with the computer mouse. By constantly changing the view from one to the next and back, specific places in each photograph could be compared. While we found this technique was effective, we also found that it required a greater degree of precision in making the repeat views than was required by side-by-side comparisons. If the rephotographs were not exactly remade from the original vantage points, or if the lighting was significantly different between photos, the changes between the dissolves became glaring, immediately recognizable, and distracting.

In addition to the simple dissolve between photographs, project programmer and designer Byron

Wolfe created another way to compare and view selected parts of the overlapped images of the rephotographic sets. By creating a mask that acts like a moveable window, the overlapped rephotographs also offered a "Time Reveal" that placed a view from one time period on top of another (plate 4.1). A viewer is able to choose the time period revealed in this window (from first, second, or third views) as well as that of the underlying photograph. The window appears to float on top of the photograph beneath it and is moved by grabbing the window and dragging it across the screen with the computer mouse. The Time Reveal window has proved the project's most dramatic way to illustrate the changes evident at any single location.

Besides the Time Reveal window, each rephoto-graphic set offers other options for viewers, including a magnifying glass to enlarge details in each photo (fig. 4.2). The magnifier helps alleviate one problem with viewing photographs on computer screens, their relative coarseness when compared to printed forms, and lack of fine detail. In order to enable the magnifying glass option, we scanned images at twice the normal resolution for screen viewing, and when deployed the magnifier acts as another window mask, calling forward the enlarged version of the photograph.

Other buttons with each rephotographic set allow viewers to turn color versions of the photographs on or off (figure 4.1). The photos may be seen in monochrome or as full color versions of third views. Other buttons turn sound recordings on and off; sound

Figure 4.2. From the Third View DVD (Klett et al. 2004), the magnifying glass used to examine details in photographs when viewed on the computer screen. The magnifier offers a 2× increase in scale. Another button on the side of the viewing window allows users to examine images in monochrome or full color. (B. Wolfe).

that was often recorded at sites ranged from ambient noise to interviews with people associated with the photographs.

Each site rephotographed by Third View included a panorama made at the original vantage point. The panoramas are overlaid with the original images that can be turned off and on using a mouse-controlled switch. The panorama can also be enlarged or reduced in scale, and navigation is possible in the up/down and left/right screen.

Geography was used as the primary navigation interface for the project. A map of the western states contains colored dots indicating rephotographic locations (plate 4.2). When rolled over with the computer's mouse, these sites become active and once clicked bring viewers to the specific site locations. Some locations, such as the Green River, Wyoming, site (plate 4.3) contain multiple rephotographs

along with other materials. The site locations are displayed as photographs that act as "hot" buttons or icons to take viewers further into the program. The icon buttons are aligned in linear fashion with a slider used as the navigation device to scroll down the choices. Together the interface allows viewers access to all the materials the project offers as part of the "Journey" section of the DVD. One such choice is an interactive tutorial that explains how to verify the accuracy of a chosen vantage point through measurements (plate 4.4).

The Third View DVD (Klett et al. 2004) also contains an "Archive" feature. The database is a selection available upon first opening the disk on the computer. If the viewer only wishes to research the project's photographs, this part of the disk enables searches based on location, title, keywords, photographer, and other parameters (fig. 4.3).

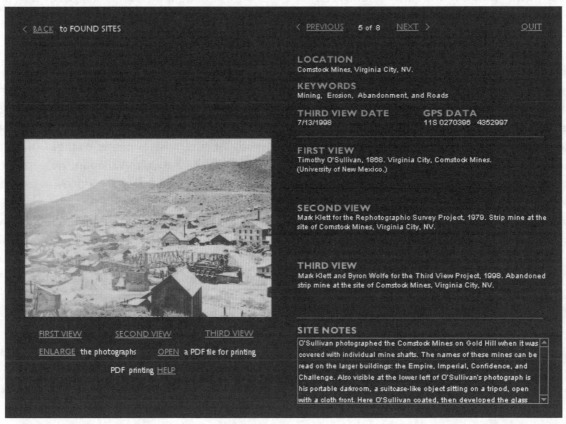

Figure 4.3. The "Archive" section of the Third View DVD (Klett et al. 2004) enables searches by location, title, photographer, and keywords.

Third View was working with what was cutting edge technology in the late 1990s, but 10 years later, new digital technologies offer even greater options, particularly that of using digital cameras to make rephotographs. Third View used conventional films with large-format view cameras and then scanned the film to create a combined analog/digital product (see also Boyer et al., chapter 2).

More recently, medium- and high-resolution digital sensors offer many advantages to rephotographers, including instant feedback at the site. However, the small screens used to view images on the backs of most digital cameras will be of little use in judging accurate rephotography. In order to take full advantage for precise rephotographic work, the digital image must be compared to the original in detail to determine whether or not the chosen vantage point is accurate. There are two ways to do this. First, the digital image may be printed on-site and the proof print compared to the original. Small battery-powered printers make this option possible. In this case, measurements between prints are compared, and calculations are made using conventional methods.

Second, the digital image may be transferred to a portable computer and overlaid on top of a digital copy of the original photograph. The two must then be scaled to match, and, using photo editing software, the images can be transformed, made semi-transparent, and adjusted in position to provide a quick visual check of the chosen vantage point. The drawback is that some computer monitors are difficult to see in bright daylight, and extra time is required for the image transfers and software manipulations. Digital cameras and backs, computers, and printers also require maintaining battery power for their operation, and the combination can add up to significant weight and bulk. Altogether, the advantages may still outweigh the disadvantages at readily accessed sites, especially in knowing that the site has been properly photographed before one leaves the field.

These are only two potential methods for making accurate rephotographs through digital capture.

Since digital technologies have only recently overtaken film as the dominant photographic material of our era, using digital cameras or digital camera backs to make rephotographs will undoubtedly require other modifications in the future.

Yosemite in Time: Photographs as Stratigraphic Layers

The Yosemite in Time project (Klett et al. 2005) chose Yosemite National Park in California's Sierra Nevada mountains as its subject partly because of its cultural status as a leading symbol of the American wilderness. For our purposes, the park was also important because a tremendous number of important historic photographs were made there over two centuries (Naef and Wood 1975, Osteroff 1981, Scott 2006). Many of the park's photographs are among the most famous landscape images of the American West (Scott 2006).

The project began with the standard two-part methodology used by both the RSP and the Third View projects. Initial work concentrated on dozens of 1873 mammoth plate (approximately 51 by 61 centimeters) photographs made by photographer Eadweard Muybridge. These rephotographs showed many of the most dramatic changes that have occurred in the Yosemite Valley in the past 130 years, including infill of valley meadows by trees and the meandering of the Merced River (fig. 4.4).

The project quickly expanded to include work by other photographers in both the nineteenth and twentieth centuries. Our interests in the history of fine art photography led us to limit this group to four key photographers: Muybridge, Carleton Watkins, Ansel Adams, and Edward Weston, who worked from the 1860s through the 1960s and made many of the most widely exhibited and published photographs of Yosemite.

We quickly found that while Yosemite National Park is itself very large, the photographs that helped make the park famous tend to be clustered closely together. The photographers we followed were mak-

Figure 4.4. Bridal Veil Falls, Yosemite National Park, California, USA.
A. (1872). From the Yosemite in Time project (Klett et al. 2005), Eadweard Muybridge titled this image "Pohona (Bridal Veil Falls), Valley of the Yosemite. No 6." (E. Muybridge, 6, Bancroft Library).
B. (2002). The match made 130 years later had changed sufficiently to justify an entirely different caption: "View from the old road on the valley's north side across an overgrown meadow." (M. Klett and B. Wolfe).

ing views of the park's scenery, and they often chose the same features as subjects; as a result, their vantage points were often quite close to one another. If one were to plot the location of vantage points of the best known scenic photographs on a map of Yosemite, there would be small clusters of great image density in certain locations and vast open areas without representation throughout most of the park.

This discovery led to an adaptation of the standard rephotographic methodology. While discovering the vantage point used by one photographer to photograph a selected landscape subject, we sometimes found another vantage point nearby that was used by a second photographer from another era. Sometimes these separate photographers were photographing the same subjects, sometimes their focus was on slightly different subjects in different directions, but their framing of the views might overlap in space. Sometimes we could actually connect the view of one photographer to that of another by walking the space between vantage points.

This was the case in the panorama we titled "Above Lake Tenaya, 2002" (plate 4.5). The panorama connects Edward Weston's photograph of a

juniper tree in 1936 to an Eadweard Muybridge photograph of glacier-polished granites made in 1872. The two vantage points were approximately 300 meters apart, and originally the cameras were pointed almost 180 degrees in opposite directions. Byron Wolfe and I were able to rephotograph each original using conventional techniques. We then walked the distance between vantage points, and using instant film positives on a 4- by 5-inch view camera, we were able to plot a path and make a series of intervening photographs that connected the two vantage points. Each panel was made at a separate point dozens of meters from the last, and this distance was not kept constant; instead, it varied according to the space and objects in the scene. The individual panels of the panorama were not meant to be stitched together using panoramic software, but instead retain their individuality while allowing the background of the scene to flow continuously from panel to panel. Common graphic elements in the foreground, rocks, cracks, and trees, were used to connect unrelated bits of foreground.

The result is a variant of the normal rephotographic process and a direct offshoot of the methods used to make location panoramas at sites visited for the Third View project (Klett et al. 2004). Only in this case, the camera does not rotate on a stationary tripod but moves through the scene. The panorama does not replace the usefulness of the two rephotographs made at each original vantage point but it does inform how the two vantage points are related. The final image composites all photographs, including copies of the original images, onto one digital file. It is also very easy to make a variation of this image by removing the two original photographs from the panorama to show the areas now covered by the overlays. A rectangular window may be drawn into the panorama to indicate the location of the earlier image in the larger contemporary space.

A similar overlapping panorama from Lake Tenaya illustrates a concept that became important to the Yosemite project (plate 4.6). In this panorama, titled "Four views from four times and one shoreline, Lake Tenaya, 2002," three historic images are combined and overlaid onto a wider panorama made at the scene. The larger photograph on the left, made by Eadweard Muybridge in 1872, is perhaps the first photograph ever made at the lake. In the center of the panorama an Ansel Adams, ca. 1942, of the rocks and dramatic clouds across the lake lays partially over an earlier view made by Edward Weston in 1937. Each photographer, in separate trips to the lake, managed to choose vantage points that were within 6 meters of one another. Muybridge and Weston, while traveling to the lake some 65 years apart, were standing less than a meter from one another!

This is an example of a high image density within a limited spatial area. The individual photographs may be seen as the equivalent of visual stratigraphic layers and can be peeled away to reveal separate periods of time. Thus one strategy for rephotography in areas of sufficient density is to map photographs that may not appear at first to be related into spatially extended time layers. The combined and layered panorama is similar to the "Time Reveal" window of the Third View project, except in this case, the historic views do not line up as precise rephotographs but rather as overlapping images assembled from different time periods and from different vantage points.

After the Ruins: Rephotography in a Rapidly Changing Urban Environment

Urban environments present different challenges for rephotography than remote locations. Cities change quickly, and contrary to work in wilderness settings where time intervals of 100 years may evidence little natural change between photographs, the same time span in the life of a city can render a historic view unrecognizable. This was the case in San Francisco, California, and the project After the Ruins (Klett et al. 2006). Unlike the Third View or Yosemite projects, the rephotographs do not reveal slight differences between first and second photographs; instead the scenes in downtown San Francisco have changed so dramatically in 100 years that the comparison between photographs is distinguished more by landmarks that have remained the same than by those that have changed.

The 18 April 1906 earthquake in San Francisco has been considered one of the most significant earthquakes of all time not simply because of its magnitude but because of the scientific data realized as a result of studying the event (Ellsworth 1990). Similarly, the social implications of the earthquake have also been extensively studied (Fradkin 2005). The earthquake caused severe damage to buildings, but more importantly the fire that followed came close to obliterating the entire city and added to a death toll numbering between 3,000 and 5,000 people (Fradkin 2005).

One unique characteristic of the 1906 San Francisco earthquake is that it is perhaps the first well-photographed natural disaster. The majority of the photographs were made by amateurs with cameras of unknown characteristics. Small portable roll-film cameras had recently been introduced, and as a result there were thousands of photographs made during and after the fire that destroyed over half the city.

Researching the photographs on several online archives led to some immediate observations about these disaster photographs that were later verified by fieldwork. Many of the pictures were taken from vantage points inside the ruins of buildings, and new buildings have been built at the same locations, meaning the original positions were now inside new structures and unusable. The remaining photographs tended to be clustered around buildings that were landmarks at the time, are still standing today, and can be identified. Finally, because scores of different photographers used a wide variety of cameras, it is not possible to draw conclusions about how the photographs were made from knowledge of the photographers themselves. It is often possible, for example, that when working with the photographs of a known photographer, we can predict the use of certain lens focal lengths, a consistent choice of vantage points relative to subjects, or a favored time of day in which to make photographs (Klett et al. 1984, Webb 1996). Furthermore, the work of professional photographers usually has a technical consistency that makes repeating the work easier. This was not the case with the 1906 photographs of San Francisco.

Once the landmarks that remain standing from the disaster were found, roughly relocating vantage points in the city became easy. Exact repetition of these earlier vantage points posed greater problems, however, since much of the information that could be used to verify the camera position has been altered or is missing.

Here it must be stated that every attempt to relocate a preexisting vantage point has limitations in accuracy. The degree of accuracy becomes greater when more information can be compared between the original and a rephotograph. This is especially true when an original contains information at a variety of distances from the camera. Photographic vantage points that provide clearly identifiable objects or markers in the foreground, middle ground, and background of the photograph are the most ac-

curately reoccupied. Foreground in this case can mean anything close enough to the camera for the photographer to touch, which is especially true when wide-angle lenses were used. Under these circumstances, when information at multiple distances from the camera is reliable, it is often possible to position another lens within centimeters of the original. But if the information is limited or the information is limited to similar distances from the camera, the degree of accuracy is significantly reduced.

Reduction of information was often at work in San Francisco, where only small amounts of what was left standing in 1906 remain to be identified from the original photographs. In many street scenes in the heart of the city, I estimated the position error of relocating some vantage points to be in the range of 5 to 10 meters. It is simply not possible to verify the accuracy of one vantage point over another within a circle of that diameter.

While limited information could not verify the accuracy of some vantage points, the circumstances in the city did suggest a new approach to making and pairing original and rephotographed images together. Typically, accurate rephotographs are made to repeat both the vantage points and the cropping of the original image. The framing window of the rephotograph is meant to duplicate the exact information of the original. But in 100 years of rebuilding San Francisco, what were once the tallest buildings in the city are now dwarfed by those structures that have been rebuilt around them. Consequently tightly cropped rephotographs of the same scenes leave out much of the city as it exists today. A new approach was used to show as much of this growth as possible while preserving the ability to directly compare information between images.

The original vantage point was relocated as carefully as possible, but then a lens much wider than used to make the original photograph was chosen to repeat the photograph. The effect was to show more of the area around the image while still preserving the accuracy of the rephotograph (fig. 4.5). This is possible because accurately repeating a vantage

A

B

Figure 4.5. Mills Building, San Francisco, California, USA.
A. (2004). This view, entitled "Mills Building, Montgomery and Bush Streets," was taken in downtown San Francisco, California, USA. (M. Klett with M. Lundgren).
B. (1906). The set of photographs shows the relation of the rephotograph (left) to the original (right). The typical left–right order has been changed from historic, left to contemporary, right to contemporary, left to historic, right. The change reflects the
work's message that the earthquake event is not a one-time occurrence. The original vantage point was reoccupied as closely as
possible, but the contemporary photo was made with a wider-angle lens. The two images appear at the same scale, and the pairs
are composed so that the spaces both occupy the same position on each page. (Photographer unknown, Bancroft Library).

point does not rely on matching the focal length of
an original lens, only the position of that lens in
space. The choice of lens is only a matter of decid-
ing how much space to include in the photograph.
Put another way, the vantage point is independent
of the lens focal length as it is a position in real
space, and the choice of this position establishes all
the relationships between objects in the resulting
photo. The lens focal length is a choice in the angle
of the view to be contained by the photograph, and
in the case of the San Francisco rephotographs, the
use of a wider angle lens simply provides more information around what has come to occupy the
original scene.

The San Francisco rephotographs were printed as
diptychs on the same sheet of paper. Each half of the
paper represented one time period, and the rephotographs were sized so that both the rephotograph
and the original appear at the same scale. That
meant the rephotograph as a wider view appeared
larger than the original image placed beside it, and

that the earlier image could be envisioned to fit
within the larger, more contemporary space. Viewing the pair required knowledge of this strategy. In
cases where that position is not centered, the original
appears in a position relative to the same space on
the other side of the diptych.

Another change to the typical method of presenting rephotographs came in the reversal of the normal horizontal order of the older photo on the left,
the newer image on the right. In most Western cultures a sequence in time is usually read from left
(earlier) to right (later), and most rephotographic
pairs are composed with the first image on the left as
the first in time. In the San Francisco rephotographs,
this order was intentionally reversed, and the newer
image appears on the left. In one sense this combination looks as expected since the left to right views express the natural expectation of entropic change
from structure (on the left) to chaos (on the right).
But on further inspection, we know the order that
appears on the left has not broken down into chaos,

A

B

Figure 4.6. Towne Mansion Portico, San Francisco, California, USA.
A. (2003). In this photograph, entitled "Portico of the Towne Mansion, moved to Lloyd Lake, Golden Gate Park," the portico was treated as the subject of the rephotograph and the basis for selecting the vantage point, not the original space. (M. Klett and M. Lundgren).
B. (ca. 1905). "A. N. Towne Residence, view from the Chas. Crocker Residence—California Street." Since the portico was moved to a new location after the earthquake, it necessitated the selection of a new vantage point from which it was rephotographed from the same relative lens-to-subject distance and angle as the original. (Photographer unknown, Bancroft Library).

at least not yet. Of course that is one message of the work, that earthquakes are not one-time events but instead are cyclical occurrences. The display of the work is therefore meant to make use of and enhance that understanding.

The San Francisco work provided other challenges and opportunities. Figure 4.6 (right side) shows the Towne Mansion on Nob Hill in 1905 before the earthquake. Notably, the portico of the mansion survived the fire that destroyed the rest of the structure, and later it was moved to a new location in Golden Gate Park. The portico sits there today at the edge of Lloyd Lake as the only surviving ruin of the old city and is now called "Portals to the Past." The question relative to making a rephotograph was whether it was more important to rephotograph the site of the mansion or the subject of the mansion's portico. The site itself on Nob Hill has changed completely, and there is virtually no information to be found that connects the past and present scenes. All buildings visible in the 1905 photo have been replaced, including the steps of the Crocker residence in the foreground.

Ultimately the portico became the subject of the rephotograph, and this necessitated moving to the park and re-creating an entirely new vantage point from elements of the old. The portico was rephotographed from the same relative subject-to-lens distance and angle of view as it appeared in the original photograph. This choice is not typical for rephotography of landscapes that are otherwise known for their spatial stability, but the selection of a moveable object or person changing over time rather than a fixed physical position in space may be the proper subject of rephotographs for practitioners in several disciplines, particularly the social sciences. An extreme example is to be found in Ganzel (1984). Ganzel traced the photographs made famous by the Farm Security Administration photographers of the 1930s and concentrated on the people in relocating the people in the photographs over their environments.

Discussion and Conclusions

Rephotography has been practiced by a wide range of disciplines but common to all is the need to coordinate the relationship between space and time in ways that enable viewers to visualize change. The methods used by most practitioners employ a two-part technique of finding the original photograph's vantage point and matching the lighting of the original. However, depending on the challenges of the location and the desired outcome, the work may require modifications to this standard methodology.

The use of electronic media may satisfy the need to expand the context of the view both inside and outside of the picture's frame. Interactive media may also be used to visualize the changes between photographs. Digital cameras are quickly becoming the tool of choice and will require further adaptation of standard rephotographic methods.

Another modification of rephotographic methodology takes advantage of sites that have a high density of photographs related by proximity of vantage points. By treating the combined space as a visual stratigraphic record, image composites map the relationships of multiple vantage points within a larger area.

Finally, strategies that focus on presenting rephotographs may compel viewers to interpret the images in new ways. Information about the scene being rephotographed can be expanded or reduced by selecting lenses with focal lengths appropriate to the desired angle of coverage.

Many alternative methods are available to enhance standard rephotographic methodologies, and the projects described here are only three examples. The expansion of new methods will depend on the needs of the study, the ability to adapt to new settings, and the willingness to explore creative solutions.

Acknowledgments

The author thanks the Anderson Ranch Arts Center, Snowmass, Colorado, for its financial support of the Third View Project; and Byron Wolfe, Rebecca Solnit, Philip Fradkin, and Michael Lundgren for their collaboration.

Literature Cited

Bartlett, R. A. 1980. *Great surveys of the American West*. Norman: University of Oklahoma Press.

Ellsworth, W. L. 1990. Earthquake history, 1769–1989. In *The San Andreas fault system, California*, ed. R. E. Wallace, 152–187. US Geological Survey Professional Paper 1515. Washington, DC: US Government Printing Office.

Fradkin, P. 2005. *The great earthquake and firestorms of 1906*. Berkeley and Los Angeles: University of California Press.

Ganzel, B. 1984. *Dust Bowl descent*. Lincoln: University of Nebraska Press.

Hales, P. B. 1988. *William Henry Jackson and the transformation of the American landscape*. Philadelphia, PA: Temple University Press.

Harrison, A. E. 1974. Reoccupying unmarked camera stations for geological observations. *Geology* 2:469–471.

Klett, M., K. Bajakian, W. L. Fox, M. Marshall, T. Ueshina, and B. Wolfe. 2004. *Third views, second sights: A rephotographic survey of the American West*. Santa Fe: Museum of New Mexico Press.

Klett, M., M. Lundgren, P. L. Fradkin, R. Solnit, and K. Breuer. 2006. *After the ruins, 1906 and 2006: Rephotographing the San Francisco earthquake and fire*. Berkeley: University of California Press.

Klett, M., E. Manchester, J. Verburg, G. Bushaw, R. Dingus, and P. Berger. 1984. *Second view: The Rephotographic Survey Project*. Albuquerque: University of New Mexico Press.

Klett, M., R. Solnit, and B. G. Wolfe. 2005. *Yosemite in time: Ice ages, tree clocks, ghost rivers*. San Antonio, TX: Trinity University Press.

Malde, H. E. 1973. Geologic bench marks by terrestrial photography. *US Geological Survey Journal of Research* 1:193–206.

Naef, W., and J. Wood. 1975. *Era of exploration: The rise of landscape photography in the American West, 1860–1885*. Boston, MA: Albright-Knox Gallery and the Metropolitan Museum of Art, distributed by the New York Graphic Society.

Nickel, D. 1999. *Carleton Watkins: The art of perception.* New York: Harry N. Abrams.

Osteroff, E. 1981. *Western views and Eastern visions.* Washington, DC: Smithsonian Institution Traveling Exhibition Service.

Rogers, G. F., H. E. Malde, and R. M. Turner. 1984. *Bibliography of repeat photography for evaluating landscape change.* Salt Lake City: University of Utah Press.

Scott, A. 2006. *Yosemite: Art of an American icon.* Berkeley and Los Angeles: Autry National Center in association with the University of California Press.

Snyder, J. 1981. *American frontiers: The photographs of Timothy H. O'Sullivan, 1867–1874.* New York: Aperture.

Webb, R. H. 1996. *Grand Canyon, a century of change.* Tucson: University of Arizona Press.

Chapter 5

Using Fixed-Point Photography, Field Surveys, and GIS to Monitor Environmental Change: An Example from Riemvasmaak, South Africa

M. Timm Hoffman and Simon W. Todd

Repeat photography provides an important tool for environmental monitoring, particularly since photographs are so visually appealing and easily interpreted. Little technical skill is required for a basic examination of a photograph, and they have wide appeal for anyone interested in the environment. There are also several manuals available that provide detailed instructions on fixed-point photographic monitoring (e.g., Elzinga et al. 1998, Howery and Sundt 1998, Hall 2001, Rasmussen and Voth 2001). However, the approach has not generally been promoted as a useful quantitative tool for assessing landscape change and is rarely linked to additional ecological measures. In addition, the significant advances that have been made in the development of change detection techniques in the remote-sensing arena, particularly with regard to the analysis of satellite imagery (Coppin et al. 2004), have rarely been employed by those using ground-based imagery such as repeat photography.

In this chapter, we show how an analysis of repeat photography can be used in conjunction with detailed ecological surveys to provide a robust and accessible qualitative and quantitative measure of environmental change. We also detail a novel geographic information system (GIS) approach to repeat photographic analysis and demonstrate its usefulness in monitoring the impact of livestock on different components of the landscape.

Background

The monitoring program from which this analysis is drawn in Riemvasmaak, a 75,000 hectare arid rangeland (125 millimeter mean-annual rainfall) in the Northern Cape Province of South Africa (fig. 5.1), was initiated in 1995 and repeated in 2005 (see Hoffman et al. 1995, 2005). The region, which lies just north of the Orange River, has a long history of human settlement. However, it was always sparsely occupied until the start of the twentieth century when local seminomadic farmers moved into the region. Most of the households owned domestic animals and lived peacefully together in Riemvasmaak until 1974. In that year, under the South African apartheid government's plan of "separate development," the approximately 1,500 people of different race groups

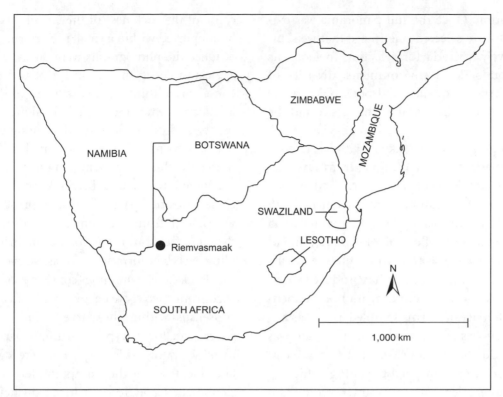

Figure 5.1. Location of Riemvasmaak in a southern African context.

were forcibly removed to distant regions in southern Africa.

As one of the first land restitution cases in postapartheid South Africa, those inhabitants who were forcibly removed in 1974 were granted the right to return to their ancestral land in 1995. However, their homes and agricultural infrastructure (e.g., dug wells) had been destroyed, and returnees appealed for assistance in rebuilding their economies, which were largely centered on their agricultural activities. FARM-Africa, a nongovernmental organization (NGO) with experience in African agricultural systems, assisted in this process and commissioned a baseline survey and inventory so as to compile a "state of the environment" report for the region upon which a monitoring program could be based. The lead author was involved in the initial 14-day survey (Hoffman et al. 1995) and developed the approach to monitoring vegetation change described in this chapter, although the GIS-based

matched photograph comparison was developed later. We returned to the area in 2005 for 10 days to repeat the study (Hoffman et al. 2005). Here we present an outline of the general approach we adopted for repeat photography monitoring as applied to the Riemvasmaak region.

Methods

Field Survey

We adopted a landscape approach and sampled the vegetation within different landforms at 29 localities (camera stations) spread out across the 75,000 hectare study area (Hoffman et al. 1995). We focused our sites around major historical settlements but also selected sites based on their aesthetic appeal and panoramic views. A high position on a hill or rocky slope overlooking the locality was usually sought for a camera station. Once the full panorama was evident before us, we selected one image that we sampled intensively. Criteria for selection included its representation of local environments, diversity of landforms, aesthetic quality, and potential as a site from which future landscape changes could be assessed.

All photographs were taken with a tripod-mounted camera at heights ranging from 1.39 meters to 1.6 meters. Four cameras were used at each photo station. A medium-format camera with black-and-white film was used together with two 35 millimeter single-lens reflex (SLR) cameras, one of which contained black-and-white film, and the other color slide film. A camera that produced an instant print was also used and drawn upon to document the different landforms sampled as well as any interesting features in the landscape. Due care was taken to record the location (using a Global Positioning System), camera height, photographic details (i.e., camera lens, f-stop, shutter speed, film type, focal length), exact time of photograph, and weather conditions at the time of exposure. The camera station location and number were marked on a 1:50,000 topographic map and the direction of the field of view of the main, sampled image recorded by means of a standard compass.

After photographing the main image, the field of view was swiveled at 30° increments to the left until that part of the panorama to the left of the main image was photographed. Then the field of view was swiveled at 30° increments to the right to obtain a panorama to the right of the main image. A 360° view was rarely possible because of the presence of tall vegetation or a hill slope that obscured a portion of the panorama. However, the five to eight overlapping images photographed by this approach provide for an extensive coverage of the landscape. With the aid of a built-in tripod spirit level, appropriate care was taken at each position to ensure that the camera remained horizontal at all times. Finally, the camera station itself was photographed from about 50 meters away, and the spot directly beneath the camera station was marked with a metal dropper, which, be-

cause of the rockiness of the terrain, usually had to be supported within a small rock cairn.

Once the photographs were taken, the main image captured on the instant print was classified by consensus within the survey team into its different landforms, which included rocky slopes, sandy pediments, and dry river channels, among other potential landforms. These different landforms were marked with a permanent pen on the instant print, numbered (a–f), and briefly described from the camera position. With the instant print as our guide we walked from the camera position "into" and through the central part of the main image, sampling each landform in detail as we moved through the landscape. Sampling consisted of identifying each plant species we encountered and assigning, by consensus within the survey team, a percentage cover value for that species within a particular landform. Unknown plant species were collected and later identified at the Compton Herbarium at Kirstenbosch, Cape Town, where voucher specimens were deposited. The different landforms effectively formed our plots and ranged in size from less than 1 hectare to hundreds of hectares; generally, they were about 10 to 100 hectares. A two-way matrix of species by sites was constructed and formed the basic data set for understanding floristic–landform relationships within the region.

The first field survey was carried out from 19 to 26 January 1995. We returned to Riemvasmaak 10 years later and repeated the photographs from 9 to 17 January 2005. We used the same tripod, cameras, and film, and we tried to match the photographs as close to the same time of day as possible.

Geographic Information System Analysis of Repeat Photographs

Hoffman et al. (2005) give a complete account of the repeat photographs from the 29 camera stations used to document the impact of livestock on the Riemvasmaak environment after a decade of renewed grazing. In addition to presenting the photographic pairs for qualitative examination, we also developed a method for matching the photographs using geographic information system (GIS) software (plate 5.1). In this technique, two or more images of the same site were overlaid onto one another, and differences between them were identified digitally. This allows for the quantification of major changes in particular areas of the landscape. However, to use this method, the matched photographs had to be taken from precisely the same location, because even a few meters' difference would cause large errors. Differences caused by different cameras or lenses can to some extent be compensated for during the photo-matching process that will be described here. It is important that the photographs are taken at the same time of day and year to preserve the location of the sun in its azimuth and the effect this has on the size and location of shadows (see Boyer et al., chapter 2, for information on the Sundicator approach).

The first step in matching repeat photographs using GIS software is to overlay one image on the other by using common points of reference (fig. 5.2). In cases where there are several images in a series, one of the images is simply nominated as the base image and the other images are matched to this base image. Common points of reference should be distributed evenly across the image and should consist of easily recognizable and immovable objects such as cracks and points of large boulders and rocks. In some cases, such objects are not available and more changeable objects such as the bases of trees may have to be used.

Once a suitable number of common points have been identified, usually in the order of 30 points, the images can be overlaid onto one another. Slight differences between the images are usually corrected by the software during this process. However, if the photographs are inadequately matched, this will quickly become apparent as a large amount of distortion in the processed image will occur. The image-matching software is usually able to provide goodness-of-fit or error terms for each point that can be used to identify incorrectly placed common points.

Image subtraction is an obvious next step in which two photographs in a time series are matched

to a common coordinate system. In a manner similar to that often utilized with remote-sensing data, images of the same scene are subtracted from one another to identify areas where changes have taken place. However, as with remote-sensing data, this is not a straightforward matter of subtracting one image from the other due to subtle differences in the exposure and range of pixel values present in each image. Even with repeat landscape photographs taken at the same time of year and under similar weather conditions, there are subtle differences in the images due to small differences in the camera, ambient lighting conditions, film, and developing. These differences can usually be minimized by comparing the pixel values within selected areas in one image to the pixel values in the same selected areas in the other image. The areas selected should be of invariant objects that are not likely to change texture or color over time. Appropriate examples include rocks and bare sand, although care should be taken with the latter to avoid places where animal activity, such as hoof action, may have changed the texture of the soil surface, resulting in different reflectance values. Objects such as vegetation and the sky should be avoided as these are variable and do not provide stable reference points through time. Also, this is where the significance of taking photographs at the same time of year and day becomes important as this ensures that the incident angle of light is the same and shadows are in the same places.

In practice, image subtraction can be accomplished by making small polygons over the selected objects and then using the software to extract the pixel values for these areas from both images. Regression analysis is then used to compare the pixel values from each image for the corresponding locations. In our experience, the two images are usually related to one another in a linear manner, and the tonal range of one image can be corrected to match the other by transforming it using the regression equation so obtained. After correction, the images can be subtracted from one another to obtain a difference image. This image consists of a range of negative and positive values. As there is usually some

Figure 5.2. Lower Kourop Valley, Riemvasmaak, South Africa.

A. (21 January 1995). A view of camera station 11 in the Lower Kourop Valley (28.46465° S; 20.14012° E), taken at 16:03 hrs, looking 134° SE.

B. (21 January 1995). A detail view from A showing the abundance and cover of different grass species and a few shrubs within the gently sloping river wash.

C. (15 January 2005). The same view 10 years later, at 16:11 hrs.

D. (15 January 2005). A detail view from C showing the decline in grass and shrub cover as a result of heavy grazing in the area. This pair illustrates the amount of detail and the accuracy of matching possible with this technique. Due to the high level of precision in the matching, it is possible to follow the fate of individual shrubs even when a large amount of change has taken place because their locations are exactly matched at the same scale.

degree of uncertainty involved in the regression, the 95 percent confidence limits of the regression can also be used to remove those areas that fall within these limits. The GIS or other image-processing software can be used to make those areas that fall within the confidence limits transparent, leaving only those areas where a significant difference has been detected.

The brightness of these areas can also be scaled according to their pixel values, thereby highlighting the areas where the difference between the images is greatest. It is useful to overlay this image on either of the original images to highlight those areas where changes have taken place. The difference image is also able to differentiate areas where vegetation has appeared from those where vegetation has disap-

Figure 5.3. Near Bok se Puts, Riemvasmaak, South Africa.
A. (24 January 1995). Camera station 18, near Bok se Puts (28.32617° S; 20.03810° E), looking 258° WSW. The image was taken at 10:19 hrs in bright sunlight.
B. (24 January 1995). A detail view of A showing the abundance and cover of the main grass and shrub species in the view.
C. (12 January 2005). The same view, photographed at 09:11 hrs in overcast conditions.
D. (12 January 2005). A detail of C. Although the main shrubs are still present, the number of separate grass tufts has clearly increased, as evidenced in a comparison of the detailed views.

peared. We emphasize that such change detection relies on the fact that real differences in the vegetation result in corresponding changes in the pixel values of the image. Situations where one type of vegetation replaces another of the same tone do not show up well in such an analysis, nor do cases where the vegetation such as grass tussocks are a similar tone to the soil surface, although such cases are, in our experience, quite rare. Increases in the size of plants due to growth usually result in an increase in the size of their shadows. To date we have used this type of analysis only on black-and-white images, although the technique applies equally to color images where each color band can be corrected independently.

Color imagery also presents the possibility that different species can be identified by their spectral signatures and quantified through time.

Results

We present the results primarily from photo station 22 to illustrate the approach used in our study. However, we also include the paired images from photo stations 11 and 18 (figs. 5.2 and 5.3, respectively), which show the level of detail that can be derived from well-matched repeat photographs. In this case, the braided stream network that was present in 1995 at site 11 has been replaced by a homogenized sandy pediment with a significant loss of grass cover. At photo station 18, the fate of individual grass tussocks and the dynamics of populations can be traced over time.

Detailed changes in the cover of species within different landforms (fig. 5.4) at photo station 22 are shown in table 5.1, whereas more general changes in the cover of growth forms are discussed later (see table 5.2). Despite the significant change in species composition, few changes in the cover of different growth forms between the two dates are evident. The repeat photograph pairs for photo station 22 are shown in figure 5.4. While clear differences between the two photographs are discernible, they are better illustrated by the image in plate 5.2, which shows a substantial decrease in coarse grass in the river channel and an increase in tree biomass, but not density, in all of the landforms where they occur. The change in the number of pixels in each change class (table 5.3, fig. 5.4, and plate 5.2) shows that it has been the increase in tree volume that has contributed mostly to landforms showing an increase in pixel density (e.g., c and d). Conversely the loss of grass cover has

Table 5.1. List of species and growth forms and their cover values (percent) associated with different landforms at photo station 22 in 1995 and 2005

Species		Rocky Slopes and Pediments				Sandy Pediments				River Channel	
		22a		22e		22c		22d		22b	
		'95	'05	'95	'05	'95	'05	'95	'05	'95	'05
Cleome oxyphilla	A		+	r	r	+	r	+	r		+
Codon royeni	A			+	+	r	r		r	+	+
Forsskaolea candida	A	r		r						+	
Osteospermum microcarpum	A		r							+	
Pergularia daemia	A										r
Sesamum capense	A	r				+					
Trichodesma africanum	A									+	r
Zygophyllum simplex	A			+	+			+			
Enneapogon cenchroides	G	r									
Enneapogon scaber	G	2	+								
Leucophrys mesocoma	G		r								
Schmidtia kalahariensis	G			r							
Setaria appendiculata	G	r									
Stipagrostis ciliata	G					r	r				r
Stipagrostis hochstetteriana	G			r	r	+	+				
Stipagrostis namaquensis	G					+	10			12	10
Stipagrostis uniplumis	G	2			1	7	+	+	r	+	
Triraphis ramossisima	G					+					
Aptosimum spinescens	LS			+	+		r				
Giseckia pharnaceoides	LS			+		+		+			
Hermannia spinosa	LS	+	+								
Limeum aethiopicum	LS										r
Dyerophytum africanum	MS	r	r			r					r
Galenia africana	MS					r					
Hibiscus elliotiae	MS	+	+								
Indigofera heterotricha	MS	+	+					r			
Indigofera spinescens	MS	12	10								
Lebeckia sericea	MS								r		
Monechma spartioides	MS	+	+	1	+	1	1			r	+
Peliostemon leucorhizum	MS	1	1								
Petalidium lucens	MS					r					

Table 5.1. Continued

Species		Rocky Slopes and Pediments				Sandy Pediments				River Channel	
		22a		22e		22c		22d		22b	
		'95	'05	'95	'05	'95	'05	'95	'05	'95	'05
Plexipus gariepensis	MS					+					
Sericocoma avolans	MS		r								
Zygophyllum microcarpum	MS			2	2	4	5	1	2		1
Zygophyllum suffruticosum	MS							+			
Cadaba aphylla	TS				r						
Lycium prunus-spinosa	TS						r		r		
Rhus populifolia	TS		r								
Sisyndite spartea	TS		r								
Commiphora gracillifrondosa	SS		r								
Euphorbia gregaria	SS	10	10	+	5						
Kleinia longiflora	SS	r									
Psilocaulon absimile	SS							r			
Acacia erioloba	T			r	r	1	1	8	8	2	2
Acacia mellifera	T			4	r	5	5			r	1
Aloe dichotoma	T	r									
Boscia albitrunca	T	r	r	r	r				r		
Boscia foetida	T	r	r			+	+		r		
Parkinsonia africana	T			r	r				r		
Prosopis glandulosus	T							1	1		
Schotia afra	T									r	r
Ziziphus mucronata	T									r	r
Total cover (%)		28	22	9	8	19	22	11	11	15	14

A = annuals, G = grasses, LS = low shrubs (less than 0.25 meter), MS = medium shrubs (0.25–1.5 meters), TS = tall shrubs (greater than 1.5 meters), SS = stem succulents, T = trees. A symbol 'r' indicates that the species was rare in the landform (one or two individuals only) while '+' indicates a cover value of below 1 percent in the landform.

Table 5.2. Changes in the estimated cover (percent) of different growth forms at photo station 22 in Riemvasmaak between 1995 and 2005

Growth Form	Rocky Slopes and Pediments				Sandy Pediments				River Channel	
	22a 1995	22a 2005	22e 1995	22e 2005	22c 1995	22c 2005	22d 1995	22d 2005	22b 1995	22b 2005
Annuals/forbs	0	0.1	0.2	0.2	0.2	0	0.2	0	0.4	0.2
Grasses	4.1	0.1	1.0	0	7.3	10.2	0.1	0	12.1	10.0
Low shrubs (<0.25 m)	0.1	0.1	0.2	0.1	0.1	0.01	0.1	0	0	0.01
Medium shrubs (0.25–1.5 m)	13.3	11.3	3.0	2.1	5.1	6.0	1.1	2.0	0	1.0
Tall shrubs (>1.5 m)	0	0.02	0	0.01	0	0.01	0	0.01	0	0
Stem succulents	10.0	10.0	0.1	5.0	0	0	0.01	0	0	0
Trees	0	0.02	4.0	0.1	6.1	6.1	9.0	9.0	2.0	3.0
Total cover (%)	27.5	21.6	8.5	7.5	18.8	22.3	10.5	11.0	14.5	14.2
Number of species	18	18	16	14	17	13	10	11	11	14

Rocky pediments are located in the foreground (22a) and left midground (22e) of the photograph at photo station 22 (see fig. 5.4), the sandy pediments to the right (22c) and center (22d) midground while the wide river channel (22b) runs directly through the image.

Table 5.3. The percent change between 1995 and 2005 in the number of pixels in each change class for the landforms in figure 5.4 at photo station 22 as derived from an analysis of the closely matched photograph pairs

| Zone | Landform | Cover Change Class | | |
		Unchanged (percent)	Increase in Cover (percent)	Decrease in Cover (percent)
a	Rocky foot slope	49	17	34
b	River channel	61	13	26
c	Sandy pediment	35	26	39
d	Sandy pediment	51	24	25
e	Rocky pediment	73	12	15

contributed most to the decrease in pixel density in landforms a–d.

Discussion and Conclusions

The value of repeat photography for the historical assessment of landscape change is well established (e.g., Turner et al., chapter 17). However, in our view, its potential for environmental monitoring purposes has yet to be fully realized. In particular, closer linkages between repeat fixed-point photography, on-

Figure 5.4. Droeputs, Riemvasmaak, South Africa.
A. (25 January 1995). Camera station 22, Droeputs (28.44608° S; 20.24869° E), looking 317° NW, taken at 11:55 hrs. a = rocky foot slope, b = river channel, c and d = sandy pediments, e = rocky pediment.
B. (15 January 2005). The same view, taken at 11:35 hrs. This pair shows the landform units identified and sampled to produce the results in tables 5.1 and 5.2. These images have been accurately matched for both location and tonal quality using GIS software and the difference between them is represented in plate 5.2.

the-ground ecological measurements, and quantitative image analysis need to be made. Our study in Riemvasmaak is an attempt to combine all three elements in one approach. Its novelty lies in the emphasis we have placed on the detailed ecological information accompanying the fixed-point photographs at a particular level of spatial scale, as well as the postsurvey GIS analysis of the images. We believe that it is in this latter area that greater advances need to be made if the full value of the technique is to be realized. Much can be learned from the broad field of remote sensing, where techniques for detecting change in spectral reflectance patterns, particularly from satellite imagery, are well advanced (Coppin et al. 2004). Oblique ground photographs pose their own unique challenges for image analysis, but with some care and precision, these are not insurmountable problems.

Green et al. (2005) suggest that an evaluation of measurements made within a particular monitoring program should be done in terms of (1) their relevance to interested parties, (2) their precision and accuracy, and (3) the availability of expertise, equipment, and money. In terms of the first criterion, our approach was well received by local land users and community leaders, the NGO that commissioned the study, and the research community with whom we have shared our information. Postsurvey community meetings were held and the results presented to people from Riemvasmaak. Photographs were used to engage with local herdsmen who were surprised at the degree of detail that could be seen in the images and the information that they conveyed. The choice of a landform approach was also of particular relevance as this is the level of spatial scale understood by local farmers, who talk of range condition associated with plains, hills, and rivers. By scaling our unit of analysis at this level, we were able to share our findings more readily with end users at an appropriate level of spatial scale. While we did not pursue this approach, it is also possible to incorporate local oral testimony to help explain the changes evident in the photo pairs.

In terms of the need for precision and accuracy in our approach, we noted several advantages but also a number of important limitations. The advantages were that we incorporated measures of species diversity and abundance as well as growth-form composition and cover during the field surveys. We were also able to relate these to quantitative changes over a 10-year period within specific landforms. This allowed us to answer questions such as what has changed, where it has changed, and by how much. By incorporating information about the long-term rainfall record and stocking rates since 1994 (Hoffman et al. 2005), we were able to interpret these changes in the context of arid rangeland dynamics and the impact of livestock on forage production (Gillson and Hoffman 2007). Once the photographic images were accurately matched and overlain on each other, it was also possible to pan around the site and zoom in and out to interrogate the image at the desired level of detail. On print images, only large or isolated objects can be readily matched. However, on well-matched digital images, objects as small as 100 millimeters can be matched and followed through time. As a result, the histories of scores of individual plants, including trees, shrubs, and even grass tussocks, can be reliably traced through time. Also, it is possible to more accurately measure growth rates and relative sizes of plants over the survey period. Furthermore, by using this approach we were able to sample a significant proportion of the area of Riemvasmaak. Since a single panorama can cover up to 1,000 hectares, most of the major catchment areas were covered by our survey.

The limitations relate both to the field measurements as well as to the problems of matching images accurately. The difficulty of assigning an accurate percentage cover score in a relatively unbounded landform that varied considerably in area from location to location was a weakness of the survey. A subjective assessment of cover is always problematic as is shown in the discrepancy between the cover values for plants in the river channel at photo station 22 (compare tables 5.1 and 5.2; fig. 5.4). Our score of 12 percent for *Stipagrostis namaquensis* in the river channel in 1995 is clearly an underestimate as is borne out by the change-detection analysis in fig. 5.4. The use of a modified Whittaker plot of 0.1

hectare (Stohlgren et al. 1998), centered in the middle of the image, might provide a more bounded plot-based approach, although the time taken to complete such a plot would have added considerably to our effort. There are also significant problems with matching fixed-point photographs accurately. The focal length of the lenses used, as well as camera position, are critical in this regard. While it is essential that the repeat photographs are taken at the same time of day and year to minimize the influence of shadow on the quantitative analysis, the vagaries of the weather can play havoc with the best efforts to match fixed-point photographs.

Finally, in terms of the last criterion posed by Green et al. (2005) concerning the availability of expertise, equipment, and money, our approach shares problems with many monitoring programs. The continuity of personnel and equipment was a significant factor in the success of the repeated survey in 2005. Knowledge of the flora was also important and could be a serious limiting factor if the survey team is unfamiliar with the dominant species at a location. Postsurvey analysis of the images also demands considerable GIS skill, although most research teams today have this expertise on hand. However, when assessing the cost-effectiveness of the survey, we believe our approach reflects a relatively rapid, robust, quantitative method for monitoring environmental change that is both scientifically defensible and easily interpreted by land users and decision makers alike.

Acknowledgments

We thank FARM-Africa for financial support. The Mazda Wildlife Fund is thanked for the use of a courtesy vehicle. All photographs were taken by M. T. Hoffman.

Literature Cited

Coppin, P. R., I. Johnkheere, K. Nackaerts, B. Muys, and E. Lambin. 2004. Digital change detection methods in ecosystem monitoring: A review. *International Journal of Remote Sensing* 25:1565–1596.

Elzinga, C. L., D. W. Salzer, and J. W. Willoughby. 1998. *Measuring and monitoring plant populations*. Bureau of Land Management, Technical Report BLM/RS/ST-98/005+1730. Denver, CO.

Gillson, L., and M. T. Hoffman. 2007. Rangeland ecology in a changing world. *Science* 315:53–54.

Green, R. E., A. Balmford, P. R. Crane, G. M. Mace, J. D. Reynolds, and R. K. Turner. 2005. A framework for improved monitoring of biodiversity: Responses to the World Summit of Sustainable Development. *Conservation Biology* 19:56–65.

Hall, F. 2001. *Ground-based photographic monitoring*. USDA Forest Service General Technical Report PNW-GTR-503. Pacific Northwest Research Station, Portland, OR.

Hoffman, M. T., D. Sonnenberg, J. Hurford, and B. W. Jagger. 1995. *The ecology and management of Riemvasmaak's natural resources*. Cape Town: National Botanical Institute and FARM-Africa. www.pcu.uct.ac.za/resources.html (accessed 3 June 2010).

Hoffman, M. T., S. W. Todd, and J. Duncan. 2005. *Environmental change in Riemvasmaak 10 years after resettlement*. Cape Town: Leslie Hill Institute for Plant Conservation and FARM-Africa. www.pcu.uct.ac.za/resources.html (accessed 3 June 2010).

Howery, L. D., and P. C. Sundt. 1998. *Using repeat color photography as a tool to monitor rangelands*. University of Arizona, College of Agriculture, Cooperative Extension AZ 1024, Tucson. ag.arizona.edu/pubs/natresources/az1024.pdf (accessed 3 June 2010).

Rasmussen, G. A., and K. Voth. 2001. *Repeat photography monitoring made easy*. Utah State University, Cooperative Extension NR504, Logan.

Stohlgren, T. J., K. A. Bull, and Y. Otsuki. 1998. Comparison of rangeland vegetation sampling techniques in central grasslands. *Journal of Range Management* 51:164–172.

PART II

Applications in the Geosciences

As discussed in Webb et al. (chapter 1), repeat photography began as a tool for monitoring glacier advances and retreats in the Alps of central Europe. As we move through the twenty-first century, monitoring of changes in glaciers continues to dominate application of repeat photography in the geological sciences (Molnia, chapter 6; Fagre and McKeon, chapter 7). One cornerstone of global-change monitoring, glacial retreat—and the occasional glacial advance—contributes to the ongoing collation of environmental data that speak to questions of the extent to which humans are modifying climate and their environment.

Long-term changes in fluvial systems have been an intensely studied topic in geomorphology worldwide, with approaches ranging from morphological analyses to frequency analysis of flow and sediment transport to interdisciplinary analyses of hydrological and biological interactions. Changes to the fluvial system, when viewed at the landscape scale through repeat photography, are not easily resolvable as to their cause, but their timing can be resolved with photographic evidence. Both Webb and Hereford (chapter 8) and Bierman (chapter 9) address landscape changes and their effects on the fluvial system from two different perspectives in two different environments, yet both have common ground in documenting channel narrowing in two parts of North America. Webb and Hereford (chapter 8) use repeat photography to take a time series approach to documenting channel change in a group of heavily studied but poorly understood channels locally referred to as arroyos, while Bierman (chapter 9) discusses how watershed-scale land-use practices potentially affect large perennial rivers with impacts on flood frequency. In both cases, repeat photography increases our understanding of geomorphic change by documenting its pace through dates of photographs documenting the status of alluvial systems and their attendant watersheds.

Chapter 6

Repeat Photography of Alaskan Glaciers and Landscapes from Ground-Based Photo Stations and Airborne Platforms

Bruce F. Molnia

Repeat photography is a technique in which a historical photograph and a modern photograph, both having the same field of view, are compared and contrasted to quantitatively and qualitatively determine their similarities and differences (Molnia et al. 2007; Boyer et al., chapter 2; Klett, chapter 4). I have used this technique at a number of locations in Alaska, including Glacier Bay National Park and Preserve, Kenai Fjords National Park, and northwestern Prince William Sound (plate 6.1), to document and understand changes to glaciers and landscapes as a result of changing climate.

The use of repeat photography to document temporal change is not new. It originated as a glacier-monitoring technique in the Alps about 150 years ago (Rogers et al. 1984). What is unique in this Alaskan application of repeat photography is the systematic approach used to obtain photographic documentation of glacier and landscape change for every fiord in the study areas. Through analysis and interpretation of these photographs, both quantitative and qualitative information is extracted to document Alaskan landscape evolution and glacier dynamics for the last century and a quarter on local and regional scales and the landscape response to retreating glaciers.

The purpose of this chapter is threefold: first, to describe how repeat photography is being used to study Alaskan glaciers and landscapes; second, to summarize the history of Alaskan glacier photography; and third, to show examples of the usefulness of repeat photography in understanding the complex behavior of Alaskan glaciers.

Photography of Alaskan Glaciers

Considering Alaska's remoteness, its early photographic record is surprisingly rich. I have located more than 500 pre-twentieth-century historical photographs of Alaskan glaciers and adjacent landscapes. All postdate the purchase of Alaska from Russia in 1867.

What may be the earliest Alaskan glacier photograph, a photograph titled "*Steamer Queen* amongst the ice," depicts a two-masted steamship in front of part of the terminus of an unidentified tidewater calving glacier. Unfortunately, no permanent landscape features are displayed on the photograph that can be used to determine the exact location. The photograph, held in the collections of the Kingston Museum, Kingston upon Thames, United Kingdom, is attributed to Eadweard J. Muybridge (born Edward James Muggeridge), the father of motion picture photography (for more of Muybridge's legacy, see Klett, chapter 4; Rohde, chapter 18; Moore, chapter 19). As Muybridge's only known trip to Alaska was made in August 1868, the photograph would have been made less than a year after the 18 October 1867 ceremony transferring Alaska to the United States. Muybridge, an accomplished San Francisco landscape photographer, accompanied Major General Henry Wager Halleck, commander of the US Army's Military Department of the Pacific, to photograph military ports and harbors of the newly acquired Alaska. Halleck's survey visited and photographed Fort Tongass on Tongass Island, Fort Wrangel (now Wrangell), and Sitka. No information exists as to whether the survey actually visited any glaciers. The closest would have been along the Stikine River or in LeConte Bay, both north of Fort Wrangel.

Identifying this photograph as the "first" photograph of an Alaskan glacier is problematic. First, there is no mention of an *S.S. Queen* operating in Alaska prior to the 1880s (McDonald 1984). Second, this image is not one of the 29 Alaska images attributed to Muybridge in the "Lone Mountain College Collection of Stereographs by Eadweard Muybridge," housed at the Bancroft Library at the University of California at Berkeley, or the 37 Alaska images contained in the "Eadweard James Muybridge Photograph Collection, 1868–1869," housed at the Alaska State Library in Juneau, Alaska. Consequently, the validity of this photograph as the first Alaskan glacier photograph is unconfirmed.

The earliest photograph I have located that conclusively depicts part of the terminus of an identifiable Alaskan glacier was made nearly 15 years later. It is a 13 July 1883 photograph of the terminus of Muir Glacier made by an unidentified photographer working for "Isaac G. Davidson Photo," a commercial photographic studio based in Portland, Oregon (fig. 6.1). The ridge on the left side of the photograph appears in more than 50 other nineteenth-

Figure 6.1. Muir Glacier and Muir Inlet, Alaska, USA.
(13 July 1883). The earliest documented photograph of an Alaskan glacier depicts the retreating terminus of Muir Glacier. It is a northwest-looking view, taken from the lower East Arm of Glacier Bay, south of Point Muir. The ridge to the west (left) of the glacier was covered by Muir Glacier through the middle nineteenth century. Known as "The Rat," it appears in many late-nineteenth-century photographs, including plate 6.5. The ship is the *S.S. Idaho*. (Isaac G. Davidson, NSIDC).

century photographs and is a distinctive feature used to identify many late-nineteenth- and twentieth-century camera stations. The Davidson photograph and many other early photographs that depict glacier termini and lateral margins permit some degree of accuracy not only in identifying fields of view, but also in mapping glacier positions, features, and extent.

Aerial photography of Alaskan glaciers began in the 1920s (see later section). Plate 6.2A shows one of the best known Alaskan glacier aerial photographs, a 1937 view by Bradford Washburn of Barnard Glacier. It has appeared in hundreds of publications. Plate 6.2B is a 2006 oblique aerial photograph that replicates the field of view of the 1937 photograph.

Many historical photographs record glacier terminus positions that are at or near Little Ice Age (LIA) maxima. Others document retreat histories under way for more than a century. All provide conclusive visual evidence of former glacier positions and vertical extent.

Change in Alaskan Glaciers during the Photography Era

Surveys of recent Alaskan glacier behavior (Molnia 2007, 2008) confirmed that more than 99 percent of all of the valley glaciers in Alaska are currently retreating. Therefore, it is not surprising that all the landscapes observed in Kenai Fjords National Park with repeat photography are characterized by long-term glacier retreat. In Glacier Bay National Park and Preserve and the fiords of northwestern Prince William Sound, the picture is more complicated. In Glacier Bay, all East Arm glaciers have experienced retreat while several West Arm glaciers have undergone periods of advance during the twentieth and early twenty-first centuries. In northwestern Prince William Sound, several glaciers are currently advancing. Two, Harvard and Meares glaciers, are located on either side of the rapidly retreating Yale Glacier. With these few exceptions, all regions have experienced significant post-LIA ice loss.

The driver for recent landscape and glacier change is an Alaska-wide increase in air temperature, accompanied by changes in precipitation. Since 1949, Alaska temperatures have increased by about 1.9 degrees Celsius (Alaska Climate Research Center 2007). Similarly, between 1968 and 1990, precipitation increased an average of 30 percent over all of Alaska except the southeastern Alaska panhandle (Groisman and Easterling 1994).

Methods

Ground-Based Repeat Photography

I have examined more than 15,000 historical ground-based photographs depicting Alaskan glaciers and have selected more than 1,000 for repeat photography. Scans of these photographs have been compiled into a digital database. To date, more than 200 sites have been revisited and the fields of view replicated. The pre-1920 photographic record, which exceeds 2,000 images, provides visual information about more than 200 Alaskan glaciers. For Glacier Bay alone, more than 1,400 late-nineteenth- and early-twentieth-century photographs, taken either from land or from boats, have been found. More than half of these depict glacier termini and related features. About 300 were taken during the nineteenth century, with the earliest dating from 1883.

Sources of these photographs include the USGS Photographic Library in Denver, Colorado; the National Archives in College Park, Maryland, and Anchorage, Alaska; the Alaska State Library in Juneau, Alaska; the National Snow and Ice Data Center (NSIDC) in Boulder, Colorado; the University of Alaska (UAF), in Fairbanks, Alaska; the Glacier Bay National Park and Preserve archive in Bartlett Cove, Alaska; travel narratives; scientific publications; Internet sites; and antique dealers.

Most historical photographs lack metadata, most significantly the location of the photo station, camera specifics, lens information, and film and exposure data. In many cases, only the photographer's name or date of acquisition is known. Some do not even have this information. Very few photographs have the latitude and longitude of the collection site. Most photographs have never been published.

Thousands of later Alaskan ground-based and aerial photographs also exist. Many of these were systematically collected to document temporal changes in Alaskan glacier-covered environments. I have duplicated many of these as well. At some locations, these later photographs represent the earliest photographic documentation of a region that has just emerged from under a rapidly retreating glacier. For example, upper Muir Inlet in the East Arm of Glacier Bay and Northwestern Fiord in the Kenai Mountains only became ice free during the later twentieth century. Photographs taken since 1975 are the earliest photographs of these ice-free regions that exist. Plate 6.3 shows photographs taken in 1976 and 2003 from the same location on the southwest shoreline of upper Muir Inlet that document rapid changes in this area.

Information derived from a photograph can sometimes be used to determine when the photograph was made. Comparing glacier positions with other photographs of known age also helps to identify the age of undated photographs.

Since 2000, I have revisited more than 200 sites of historical photographs and systematically duplicated the original fields of view. At each site, metadata, including date and time of visit, latitude, longitude, site elevation, and bearing to the center of each photographic target, were recorded. Locations were determined with a Global Positioning System (GPS) receiver and compass. At each location digital images and/or color film photographs were made to capture the same field of view as that of the historical photograph. Often, different focal length lenses were used to maximize later image manipulation. Typically, a larger field of view is imaged so that the resulting images can be cropped to match the area of the historic image. Many historical photographs were made with rotating lens, panoramic, or mapping cameras, typically with fields of view that exceed those of most modern "normal" or even "wide-angle" lenses. Consequently, at some locations, overlapping, sequential photographs are obtained that can be digitally mosaicked. Following fieldwork, new photographs are compared and contrasted with corresponding historical photographs to determine changes, and to understand rates, timing, and mechanics of landscape evolution. The resulting photographic pairs provide striking visual documentation of the dynamic landscape evolution occurring in southern Alaska in response to changing climate.

Since 2000, about 75 locations in northwestern Prince William Sound were revisited. Prior to fieldwork, "working" locations were identified by comparing historical photographs with USGS topographic maps and recent aerial photographs. This same technique was also employed for Glacier Bay and the Kenai Mountains. Actual locations for most sites are only found through a trial-and-error field process in which features on historical photographs are matched by comparing the spatial relationships between mountain peaks and foreground features. As a result it was determined that many historical photographs were made from boats or ships. Several were also made from the surface of no longer existing glaciers. Approximately 10 percent of the land-based locations had cairns placed by the original photographers.

From 2003 to 2006, about 125 locations in Glacier Bay were revisited. Additionally, in upper Muir Inlet, several locations from which I had photographed McBride, Riggs, and Muir glaciers between 1976 and 1980 were also revisited. Nearly all of these locations were under glacial ice prior to the 1970s. Hence, the 1976–1980 photographs are the "historical" photographs in this area. Approximately 30 percent of the land-based locations had cairns.

For Kenai Fjords National Park, very few historical photographs exist. Less than 50 early-twentieth-century land- and sea-surface-based photographs have been located. Almost all depict glacier termini and related features. Except for a handful of 1920s–1940s photographs and postcards, most date from 1909 and were made by U. S. Grant, D. F. Higgins, or

Sidney Paige. These photographs are archived at the USGS Photographic Library. About 20 of their photographs were previously published with sketch maps that identified photo sites and glacier termini (Grant and Higgins 1911, 1913). During the summers of 2004–2006, about 40 locations were identified and revisited using the same methodologies as in Prince William Sound and Glacier Bay. Grant and Higgins's (1913) text and sketch maps were useful in identifying many of the locations. At many Kenai Fjords National Park sites, twentieth-century glacier retreat exceeded 15–20 kilometers. As a result, many glaciers in the 1909 photographs are not visible from the original photo locations. Therefore, new photo station sites were established to provide for future repeat photography. Cairns were not found at any location.

Repeat Aerial Photography

Since 2004, repeat photography efforts have been expanded to include historical oblique aerial photographs depicting glacier termini. To date, of the more than 300 pre-1940 aerial photographs selected, about 35 have been duplicated. Flight line information and precise location information are lacking for all nonvertical historical aerial photographic images.

Trial and error has produced the following methodology for duplicating historical oblique aerial photographs: (1) Preflight comparison with topographic maps and triangulation of features on the images to be duplicated often narrows down the location of the photo point; (2) The line of flight of the airplane needs to be parallel to the historical field of view; (3) If altitude information is not part of historical metadata, then several passes are usually needed to zero in on and identify the altitude from which the historical image was made; (4) Both the pilot and the photographer should have copies of the photograph to be duplicated so that both can focus on the desired field of view and mutually identify how to approach the location from which to duplicate the image; (5) The photographer needs to be seated directly behind the pilot so that both

have the same perspective; (6) A third person is very helpful to record location, altitude, and any other comments about the photographic process; and (7) Exposing multiple frames as the airplane flies parallel to the field of view often results in one or more photographs that capture the original field of view.

History of Ground-Based Alaskan Glacier Photography

An 1883 US military expedition, led by Lt. Frederick Schwatka, was probably the first land-based military expedition to transport a camera and to expose glass photographic plates in Alaska. Beginning on 10 June 1883, this expedition explored the Yukon River and captured images of several glaciers located north of Skagway. Similar US military expeditions to the Copper River, led by Lt. William Abercrombie in 1884, and to the Copper, Tanana, and Koyukuk rivers, led by Lt. Henry T. Allen in 1885, also photographed Alaskan glaciers. Charles A. Homan of the US Engineers, a member of the Schwatka expedition, is described as a topographer and photographer in the expedition report. Two woodcut illustrations in the report (Schwatka 1885) are described as having been "taken from photographs by Mr. Homan." They show the termini of small valley glaciers. One, labeled "Finger of Saussure Glacier" (Schwatka 1885: 77), is probably a photograph of an unnamed glacier on the east side of Mount Hoffman. Like essentially all other pre-mid-twentieth-century photographs of Alaskan glaciers, Schwatka's photographs lack any geographical information. Efforts to relocate these photo sites have been unsuccessful, partially due to the development of dense vegetation and possibly because of the artistic license taken by the artist who made the woodcuts from the original photographs (see Jonas, chapter 21).

In his expedition report, Schwatka comments on difficulties inherent in making photographs in Alaska. In a caption, he wrote:

A glimpse of Baird Glacier covered with fog is given. The mountains holding the glacier being twice as high as the one shown on the left, their crests, if they had been visible, would not have been shown in the photograph from which this illustration is made, being above the line where it is cut off. It is only at night that the fog-banks lift, when it is too late to take photographs. (Schwatka 1885: 73)

Schwatka's comment clearly identifies that the long interval of time needed to expose an image and the limitations of glass plate sensitivity, lens aperture, and focal length were significant problems that early, camera-equipped explorers encountered.

Allen's 1885 expedition report (Allen 1887) contains lithographs of the termini of the Childs and Miles glaciers that were based on photographs. I have been unsuccessful in locating the photo points from which Allen's photographs were made. Here, stylized renditions of the glaciers and mountains were the reason for lack of success. Allen's report describes a unique problem that he encountered that drastically affected his photographs.

The photography of the Copper River ... expresses in a poor manner the results of much patience and perseverance under the most trying circumstances. The plates were necessarily entrusted to natives to be carried to the mouth of the river. Recent developments show that their curiosity led them to open the box containing them, thus exposing the plates to the light, and totally injuring all but the few we had developed. (Allen 1887: 12)

Early Visits to Glacier Bay and Vicinity

Concurrent with the military expeditions were numerous visits to glaciers in southeastern Alaska by commercial steamships. These voyages attracted adventures and photographers and produced photographs of the termini of many southeastern Alaska glaciers, including Muir, Taku, Norris, and Davidson

glaciers. The first visit to Muir Glacier was by the steamship *S.S. Idaho*, which arrived on 13 July 1883 (fig. 6.1). Eliza Ruhamah Scidmore, one of the first women to explore remote glacier-covered Alaska, was a passenger on the *S.S. Idaho*. She obtained a copy of the Davidson photograph of the glacier, which she later gave to Lawrence Martin, who gave it to William O. Field, who placed it in the World Data Center for Glaciology archive, which eventually became part of the NSIDC collection.

A month later, Frank LaRoche, a commercial photographer from Seattle, Washington, arrived at Muir Glacier on the *S.S. Queen*. A LaRoche photograph of Muir and Morse glaciers (fig. 6.2) is one of the most famous photographs ever made of an Alaskan glacier. Due to dense vegetation that has developed in the subsequent 127 years, I have not been able to gain access to the location of the photograph, which was taken from a ledge about 600 meters up the east flank of Mount Wright.

Harry Fielding Reid

Expeditions headed by Harry Fielding Reid in 1890 and 1892 (Reid 1892, 1895) made more than 200 photographs of the glaciers and landscapes of Glacier Bay. Reid (1859–1944) was the first to conduct comprehensive investigations of the glaciers of Glacier Bay. More than 500 of Reid's Glacier Bay and European Alps photographs are archived at the NSIDC. I have revisited and rephotographed five of Reid's 1890s photographic sites in lower Muir Inlet (an example appears in plate 6.4).

Reid presented a summary of how to make "Observations of Glaciers." In a section titled "Future Changes," he described how to photographically document the location of a glacier's terminus. Some of his comments are more applicable to Europe than to Alaska:

(1) All photographs at the end of a glacier are useful; particularly if the magnetic bearings of the camera, and the approximate distance from the glacier are given. (2) Select two stations, one on

Figure 6.2. Muir and Morse glaciers, Alaska, USA.
(August 1883). West-looking photograph of the retreating termini of tidewater Muir and terrestrial Morse glaciers, from an elevation of about 600 meters on the flank of Mount Wright. The glaciers separated prior to the 1880s. Pictured are Frank LaRoche, the photographer (standing), and Captain Richardson, the master of the *S.S. Queen*. Note the absence of vegetation. Compare this to a recent view of the same area (plate 6.4B). Barren zones around Morse Glacier's perimeter and along Muir Glacier's lateral margins document significant nineteenth-century thinning and narrowing of both glaciers. (Frank LaRoche, NSIDC).

each side of the valley, commanding a view of the glacier's end. Photograph the end from these two stations; two photographs from each station may be required to show all the end. Mark the station, describe them carefully and leave an account at the ranch or hotel from which the glacier is usually visited so that they can easily be found by later observers, who should take photographs from these points in preference to all others. This will be the beginning of a systematic record of the glacier. From these photographs it will be possible to make a map of the glacier's end if we know: the distance between the stations; the angle at each station between the other and some points in the photograph and the focal length of the lens. . . . In taking the photograph have the camera as level as possible, with the plate vertical; do not use the swing back, and do not alter the focus in taking two or more pictures from the same station. Hand cameras should rest on some support; a rock answers well. The longer the focal length of the lens, the better; negatives or positives on glass yield more accurate results than prints, as the paper may become distorted in the manipulation. (Reid 1896: 286–287)

Tourists, Climbers, Travel Writers, and Postcards

In 1883, George Eastman invented photographic film in rolls. Five years later, he patented the first handheld camera. It was sold by the Eastman Kodak

Company of Rochester, New York, and came loaded with enough film to make 100 photographs. After the film was exposed, the camera was returned to the factory, where the film was developed and prints were made.

By the late 1880s, Glacier Bay, Taku Inlet, and Lynn Canal had become popular destinations for thousands of visitors, many equipped with Kodak cameras. Several steamships that had previously been making scheduled trips to southeast Alaskan towns and forts from San Francisco, Port Townsend, and Seattle began to visit Muir Glacier. The voyages attracted many photographers, both amateur and professional. I have located several hundred late-nineteenth-century photographs that depict the terminus of Muir Glacier. William H. Partridge, a popular Boston photographer, visited Glacier Bay's East Arm in 1886. Several of his photographs were published in Wright (1887, 1889) and were seen around the world. Widely distributed photographs of Muir Glacier attributed to photographers Frank LaRoche (fig. 6.2), George M. Weister, and Isaiah W. Taber also elevated public interest in visiting Alaska. I have revisited and rephotographed more than 25 sites where pre-1900 photographs were taken in the Muir Glacier area. Historical and modern pairs of photographs from several of these sites are presented in plates 6.4–6.6.

Elsewhere, scientific, climbing, and exploratory expeditions were also using cameras. In 1890 and 1891, two USGS expeditions explored Yakutat Bay and the Malaspina Glacier area as they attempted to climb Mount St. Elias. These expeditions, led by Israel C. Russell, created more than 150 photographs, many of which captured the positions of the termini of Turner, Hubbard, Lucia, and Malaspina glaciers. Many were published in illustrated reports of Russell's expeditions (Russell 1891, 1892, 1893). Some were used by other investigators, including Tarr and Martin (1914), to assist them in determining subsequent change.

An 1897 Italian climbing expedition was the first to successfully climb Mount St. Elias. Led by Prince Luigi Amedeo Di Savoia, Duke of the Abruzzi, it transported five cameras and enough film to make more than 1,300 photographs. The expedition report (de Fillippi 1899) describes the photographic equipment that they brought with them as follows:

(1) one camera obscura of 30 by 40 centimeters with 4 double frames for negatives and 1 rapid rectilinear lens. This camera had 60 "London Plates," each measuring 30 by 40 centimeters; (2) one camera of 10 by 8 inches supplied by Ross and Company of London with 12 double negative frames for film, one double anastigmat Ross-Goetz lens with an equivalent focus of 12 inches, one $6^1/_2$ inch telephoto attachment made by Dallmeyer, one ordinary three-legged stand, one low stand specially adapted for telephotographs, one deep yellow glass screen, 180 medium isochromatic films, 60 instantaneous isochromatic films, and an additional 360 exposures of unspecified type; (3) one folding Kodak number five camera of 7 by 5 inches from Eastman & Co, London, with a rapid lens and stereoscopic lenses, 20 sensitive "Kodak spools" of 32 exposures each (640 exposures); (4) one pocket red lantern; (5) one black tent for changing negatives; and (6) two personal cameras, one carried by the Duke and one by another party member. (de Fillippi 1899: 187–188)

The most significant products of the Italian expedition were a number of foldout panoramic photographs of the surrounding region. To date, none of the expedition's panoramic photographs have been duplicated. These photographs were among the first photographs of the St. Elias Mountains and the first of the large valley glaciers at the base of Mount Logan and the adjacent eastern Chugach Mountains.

Postcards

The printing of picture postcards, many featuring glacier termini, began in the 1890s. Many postcard photographs included enough of the geography around the glacier termini to roughly determine the photo point and the location of the glacier at the time of the photograph. Copyright dates, information in messages, specifics of the style on the address side of the card, and postmarks on the cards help in determining the date of the photograph. The glaciers of Glacier Bay, the lower Copper River area (especially Miles and Childs glaciers), and the Hubbard, Mendenhall, Davidson, Taku, and Spenser glaciers were frequently depicted.

Between 1893 and 1943, Juneau commercial photographers Lloyd Winter and E. Percy Pond produced more than 200 photographs of Alaskan glaciers, many of which were made into postcards. Several of their Glacier Bay postcards have been rephotographed; an example is presented in plate 6.6.

Alaska–Canada Boundary Surveys

Hundreds of photographs were made between the early 1890s and 1920 by surveyors of the International Boundary Commission and the Alaska Boundary Commission during their attempts to document the position of specific landscape features necessary to determine the location of the Alaska–Canada boundary. Routinely, these surveyors would transport their photographic equipment to the top of a previously unsurveyed boundary peak. There they would "shoot" complete panoramas of the territory around each peak, some using a "photo-topographic camera." In subsequent years, the growth of dense vegetation at lower elevations has made many of these sites inaccessible.

Otto Klotz of the Canadian Topographic Survey was among the first to recognize that photography was a useful tool for surveying and documenting the position of glaciers and for determining changes in glacier position. In 1894, he conducted a "photo-topographic survey" of the terminus of Baird Glacier (Klotz 1895), taking photographs on four dates—15 May, 19 May, 13 July, and 11 August—from a 260-meter-long baseline about 520 meters from the glacier margin. From the 13 July and 11 August photographs, Klotz determined that the terminus was lowered by about 0.6 meters and the average motion

of the ice was about 0.3 meters per day. Klotz (1899: 534) wrote that "It is desireable [*sic*] that future investigators leave readily recognizable marks near the ice-front (as was done by the writer in the survey of the Baird in 1894, with white lead on the adjoining bare rock wall), as such are preferable for the determination of the smaller fluctuations of the glacier."

It is also apparent that Klotz had a thorough working knowledge of cameras and optics as he instructed his peers to develop consistency in future photographic surveys: "Whatever methods of measurement and survey are used, it cannot be too strongly recommended that photographs be taken with a camera of fixed and known focal length from a properly oriented base-line. The study of the motion of glaciers will then be reduced to an exact science."

The 1899 Harriman Alaska Expedition

The 1899 Harriman Alaska Expedition photographed glaciers in Lynn Canal (6 and 8 June 1899), Glacier Bay (9–14 June 1899), Lituya Bay and La Perouse Glacier (18 June 1899), Yakutat Bay and Malaspina Glacier (18–23 June 1899), and Prince William Sound (25–29 June 1899). Several thousand photographs, some depicting glacial features, were made by the expedition's official photographer, Edward S. Curtis, later known for his photographs of Native Americans.

Expedition scientists included William H. Brewer, John Burroughs, Frederick Colville, William H. Dall, Benjamin K. Emerson, Bernhard E. Fernow, Grove Karl Gilbert, George B. Grinnell, Charles A. Keeler, Trevor Kincaid, C. Hart Merriam, John Muir, Charles Palache, William F. Ritter, and William Trelease. Gilbert and Merriam took photographs of many glaciers that they visited. In particular, Gilbert photographed the termini of Favorite, Grand Pacific, Reid, Charpentier, Hugh Miller, and Muir glaciers in Glacier Bay; and Hubbard, Turner, Nunatak, Cascading, and Hidden glaciers in Yakutat Bay. At many locations, Gilbert recorded multiphotograph panoramas. Glacier terminus photographs were published in the geology and glacier volumes of the expedition (Burroughs et al. 1902, Gilbert 1904). About 10 of Gilbert's Glacier Bay photographic sites and several of Merriam's sites have been revisited and rephotographed. An example is presented in plate 6.7.

Late-Nineteenth- and Early-Twentieth-Century USGS Photography

Cameras were a standard part of the field equipment of many USGS geologists who came to Alaska during the early twentieth century. Many of their photographs of glaciers and glacierized landscapes serve to document the face of Alaska soon after the end of the LIA. Thousands of these USGS Alaska photographs are archived at the USGS Photographic Library in Denver, Colorado.

Between 1899 and 1924, Alfred H. Brooks took more than 1,300 photographs, including Tanana Glacier in 1899; the Kigluaik Mountains in 1900; the Tordrillo, Kichatna, and Talkeetna mountains in 1902; Miles and Childs glaciers and the Chugach Mountains in 1909 and 1910; Davidson Glacier in 1912; and the glaciers of Glacier Bay and Prince William Sound in 1924. Several of Brooks's Glacier Bay photo sites have been revisited and rephotographed.

Between 1904 and 1939, Fred H. Moffitt took nearly 2,000 Alaskan photographs. He photographed Resurrection Bay and its glaciers, and Turnagain Arm with Portage and Twentymile Glacier in 1904; glaciers of the Wrangell, Skolai, and Chugach mountains, including Kluvesna, Kennicott, Childs, Miles, Russell, Frederica, and Valdez glaciers in 1905; Gulkana Glacier in 1910; Kuskulana, Kennicott, and Chitina glaciers in 1919; Kennicott Glacier in 1922; glaciers of Prince William Sound including McCarty, Chenega, and Columbia glaciers in 1924; glaciers of Prince William Sound including Port Nellie Juan, Taylor, Cottrell, Columbia, Kings, Falling, and Contact glaciers in 1925; Chitistone, Russell, Frederica, Nizina, and Rohn glaciers in 1927; Kennicott Glacier in 1928; Muldrow Glacier in 1930; Black Rapids Glacier in 1937; and Black Rapids, Jarvis, and Gerstle glaciers in 1939.

Between 1907 and 1916, James W. Bagley, a USGS topographer, experimented with the use of panoramic cameras for mapping (Bagley 1917). He used several in his Alaskan topographic surveys to photograph locations with adjacent glaciers. Prior to World War I, Bagley was commissioned by the US Army and assigned the task of developing a multiple-lens camera for map surveys from aircraft. This type of multiple-lens camera "is regarded primarily as an instrument which makes use of the principles of plane-table surveying for constructing a map rather than an instrument for picturing the surface of the earth" (Sargent and Moffitt 1929). It was this type of camera, referred to as a T-1 camera, which would be used in 1926, for the first systematic aerial photographic survey of Alaska. According to Sargent (1930) the camera is actually "three cameras in one large case about 28 inches high, 20 inches wide, and 7 inches thick; and weighs approximately 75 pounds [34 kilograms]."

Other USGS geologists who photographed Alaskan glaciers included Stephen Capps (1908–1936), Ernest Leffingwell (1906–1914), Alfred G. Maddren (1906–1917), Walter C. Mendenhall (1898–1902), John B. Mertie (1911–1942, including Glacier Bay in 1919), Sidney Paige (1905 to about 1908, including Prince William Sound in 1905), and Charles Will Wright (Glacier Bay in 1906 and 1931). I have revisited and rephotographed about a dozen of Wright's photographic sites and several of Mertie's sites. Historical and modern pairs of photographs from two sites in Queen Inlet, originally photographed by Wright in 1906, are presented in plates 6.8 and 6.9. These photographs demonstrate how much additional information can be derived when multiple historical pairs of the same glacier are analyzed and compared.

U. S. Grant III and D. F. Higgins

As contractors for the USGS, U. S. Grant III and D. F. Higgins spent the 1908 and 1909 field seasons photographing and investigating many of the glaciers of Prince William Sound and the Kenai Peninsula. This was a continuation of work started by Grant and

Sidney Paige in 1905. In all, more than 235 photographs were made, more than 60 of which depict glaciers. Many were published in a well-illustrated summary of their collective observations (Grant and Higgins 1913). Unlike most of their contemporaries, Grant and Higgins realized the significance of systematic photography. They state: "In any study of the positions of glacier fronts dated photographs are of prime importance, for they furnish accurate records and can be obtained when there is no time for detailed observation. If the photographs are taken from easily recognized stations which can be occupied in later years their value is still greater" (Grant and Higgins 1913). Realizing that documented photographs "will be of so great value in the study of future fluctuations of these ice streams," they present comprehensive lists of glacier photographs available from other sources. Grant and Higgins examined and analyzed photographs taken by many other geologists in the preparation of their study (Grant and Higgins 1913). In addition to their photographs, they describe other Prince William Sound and Kenai Mountain glacier photographs by F. C. Schrader in 1898; by Gilbert and W. C. Mendenhall in 1899; by A. C. Spencer in 1900; and by Paige in 1905. I have revisited and rephotographed about 50 of the Grant and Higgins photographic sites and several others that they describe. Plate 6.10 presents a historical and modern photographic pair from a site photographed by Grant, whereas plate 6.11 presents a set from a site photographed by Paige.

Ralph Tarr and Lawrence Martin

Expeditions in 1905, 1906, 1909, 1910, 1911, and 1913, led by Ralph Stockman Tarr of Cornell University and/or Lawrence Martin, also of Cornell and later of the University of Wisconsin, produced hundreds of glacier photographs. Although Tarr was a contract geologist working for the USGS, funding was also provided by the National Geographic Society and the American Geographical Society (AGS). Tarr and Martin were very innovative in their use of repeat photography to document glacier changes. They drew successive years' terminus positions on earlier or later photographs to depict amounts of change. They would also superimpose correctly scaled pictures of well-known architectural features, such as the Washington Monument or the US Capitol, on photographs of glaciers to indicate the size of glacier features. Their work (Tarr and Martin 1914) is the finest summary of Alaskan glacier behavior published during the first 75 years of the twentieth century. It contains more than 100 well-captioned, often annotated photographs of Alaskan glaciers.

In 1906 and 1909, most of the photography was done by Oscar D. von Engeln, a professional photographer who later studied with Tarr and Martin (Cornell University BA in 1908 and PhD in 1911). Following Tarr's death in 1913, von Engeln became Cornell's professor of geomorphology. In 1910, he published an article to provide readers with tips on equipment, subjects, and locations to maximize their success in obtaining photographs of Alaskan glaciers (von Engeln 1910). Von Engeln wrote that "it is the rain and the consequent humidity which makes photographic processes so difficult. On the other hand, the same humid conditions provide the snowfall on the higher ranges, which in turn, gives rise to the glaciers, on whose presence is dependant much of the pictorial interest of the region."

Bill Field

In 1925, William Osgood "Bill" Field Jr. (1904–1994) made his first visit to Alaska, visiting Glacier Bay, one of many visits that would span seven decades and produce more than 10,000 photographs. Its purpose was to relocate and occupy stations established by earlier scientists, particularly those of Harry Fielding Reid (Field 1926). In 1931, Field visited Prince William Sound to reoccupy locations established by the 1899 Harriman Alaska Expedition. His photographic observations continued into the mid-1980s and resulted in numerous publications documenting glacier variations (e.g., Field 1932, 1937, 1942, 1948, 1975).

According to Morrison (1995: 10), Field's ultimate goal was to "continue the work of Tarr and Martin." He not only accomplished this goal, he took the science of glacier monitoring to new levels. His photographs and observations constitute the largest historic photographic collection and database on Alaskan glaciers in existence. Field established numerous photographic stations that were reoccupied by investigators, often in successive years. Field used these sequential photographs to produce maps that systematically depicted termini changes of individual glaciers through the mid-1960s. I have updated many of Field's maps to reflect continued glacier change through the beginning of the twenty-first century (Molnia 2008). More than a dozen of Field's photographic sites have been revisited and rephotographed; examples are presented in plates 6.12 and 6.13.

In his own words, Field describes how, following his first trip to Alaska in 1925, he became involved in repeat photography:

I went to the science library at Harvard and asked about books on glaciers in Alaska. This is probably when I first saw Reid's report on "Glacier Bay and its Glaciers," the report of his 1890 and 1892 expeditions to Glacier Bay. I also ran into the report of the Harriman Expedition . . . In due course, I caught up with Tarr and Martin's *Alaskan Glacier Studies.* These books are classics now. They opened my eyes to what had been done in the 1890s and the early 1900s. I found that quite a lot of what these people had done had not been repeated. And the men themselves hadn't been able to go back, although they would have liked to return. I saw that somebody had done some work, and I saw that there was a possibility of continuing it. I wanted to go back and find out exactly what happened. But at that time my plan was by no means to undertake a life's work. It was simply to try to carry out studies on the changes of the glaciers at intervals of every five years or so. When I met these people later, they were so enthusiastic about somebody continuing their observations, and their encouragement greatly influenced me to keep going back for the next sixty years. (Field 2004: 28)

In 1940, Field began working for the AGS, eventually establishing a World Data Center for Glaciology site at its New York City headquarters. More than 8,000 glacier photographs taken by or collected by Field or his assistants, a small part of its extensive collection, are archived at the World Data Center for Glaciology, now at the NSIDC in Boulder, Colorado. The William O. Field Collection, consisting of an extensive collection of books and more than 10,000 photographs, is archived at the UAF.

At the AGS, Field began a project titled "Observations of Glacier Behavior in Southern Alaska." In 1975, the Cold Regions Research and Engineering Laboratory of the US Army Corps of Engineers published Field's *Mountain Glaciers of the Northern Hemisphere* (Field 1975), a two-volume set complete with atlas that was an outgrowth of the glacier behavior investigation. Although it lacks photographs, this publication is the most complete summary of the glaciers of Alaska and the Northern Hemisphere published through the end of the twentieth century.

Bruce F. Molnia

As a graduate student in 1968, I began to photograph glaciers of the Coast, St. Elias, and Chugach mountains. In 1974, as part of a USGS assessment of the impact of oil and gas activities in the Gulf of Alaska region, I began an ongoing aerial and ground-based photography effort that has produced about 100,000 photographs of Alaskan glaciers. Areas of emphasis include Bering and Malaspina glaciers, Icy and Glacier bays, the fiords of the Kenai Peninsula, and the Juneau Ice Field. Some of these photographs are used as the modern half of each of the repeat photography pairs presented in this chapter. Several of my publications (Molnia 1982, 2000, 2007, 2008) summarize the distribution, behavior, and history of investigation of many of Alaska's glaciers. Collectively, they contain hundreds of my Alaskan glacier photographs. I have revisited several of my early Glacier Bay photographic sites and rephotographed the fields of view. Examples are presented in plates 6.3 and 6.14.

History of Aerial Alaskan Glacier Photography

1926 and 1929 USGS and US Navy Expeditions

In 1926 and 1929, the USGS and the US Navy conducted aerial photographic expeditions that produced the first known aerial photographs of Alaskan glaciers. USGS expedition leader, topographic engineer Rufus H. Sargent, wrote: "It is doubtful whether many exploratory expeditions of late years have contributed so much of financial and scientific value as these two aerial surveys" (Sargent 1930). During fiscal year 1924–1925, funds were identified to pay for a 1926 effort to obtain aerial photographs of 26,000 square kilometers of Alaska. This would "test this new method in regions where large areas are particularly adapted to its success and the common methods are difficult to apply because the country is remote and many parts of it are, at present, almost inaccessible" (Sargent and Moffitt 1929). The surveys wanted to compare conventional plane-table mapping or modified plane table–panoramic camera mapping to the aerophotographic method. Sargent and Moffitt wrote:

> This method (plane-table) is undoubtedly the best that was applicable, but the field work is slow in comparison with the aerophotographic method, it involved a great amount of physical labor expended in overcoming the natural obstacles to travel through the country and a consequent loss of time, and because of its relative slowness it is unable to take as great advantage of favorable weather. On the other hand, the time required to photograph the surface of the land is limited only by the speed of the plane in traversing the chosen area; the resulting photographs contain an embarrassing amount of detail, so that it becomes necessary to choose what shall be used: and most of the physical labor of climbing and getting about is done by the machine. (Sargent and Moffitt 1929: 144)

In response to a request from the USGS, the Navy Department, in 1926, organized the Alaskan Aerial Survey Expedition. Under the command of Lt. Ben Wyatt, the 112-man expedition consisted of the tender *Gannet*; three open-cockpit Loening amphibian biplanes; and a 40-meter-long barge equipped to perform all necessary photographic and film processing operations. Sargent and Moffitt (1929: 146–147) state that two of the planes were "fitted up for photographic work by providing them with the proper mountings for holding the cameras and by cutting a small observation hatch in the bottom of the fuselage." Sargent was responsible for selecting the specific areas to be flown, preparing flight lines, inspecting the film, and transmitting the film to the USGS in Washington, DC. Photographs, at a scale of approximately 1:20,000, were obtained from an altitude of 10,000 ft [about 3,050 m] on flight lines of $3^{1}/_{2}$ miles [about 6 kilometers]. This resulted in 60 percent overlap and 25 percent sidelap on all lines. In all, 5,760 three-image aerial photographs, a total of 17,280 negatives, were exposed with the Bagley Trilens T-1 cameras. According to Sargent and Moffitt, each triplet included:

> a central picture which represents the ground directly under the airplane and two side pictures which represent adjoining areas on each side of the central picture. The central picture is taken with the camera pointed vertically downward, and the two side pictures are made at the same moment by two supplemental cameras directed obliquely to each side and fixed at a definite angle to the vertical. A set of three pictures thus taken represents an area of about 11 square miles [28.5 square kilometers] when the plane flies at the preferred elevation of 10,000 feet [about 3,050 meters]. (Sargent and Moffitt 1929: 160)

Oblique photographs and rolls of moving-picture film were also made. Photographed were Admiralty, Annette, Dall, Duke, Etolin, Gravina, Heceta, Kosciusko, Kuiu, Kupreanof, Long, Mitkof, Prince of Wales, Revillagigedo, Sikkwan, Tuxekan,

Woronkofski, Wrangell, and Zarembo islands; the Cleveland and Lindenberg peninsulas; and the area of the Chickamin River. Glaciers are present on Admiralty and Kupreanof islands and in the headwaters of the Chickamin River.

The 1926 expedition was considered to be so successful that a similar mission was conducted in 1929. Photographing glaciers was one of the highest priorities of this expedition. Sargent wrote that:

> It was my purpose to have the fronts of all large glaciers in the region photographed, both by the mapping and the oblique cameras, so as to record the positions of the fronts of the glaciers in 1929. The oblique photographs taken from the side of the plane are marvels of grandeur and beauty, and the mapping photographs present wonderful views of the glaciers from directly overhead. These vertical photographs reveal the phenomenon of glacier flow in a manner never before recorded in the United States, so far as I know. It is believed that there is a wealth of scientific information for the glaciologists in these pictures. (Sargent 1930: 131–132)

The 1929 expedition produced about 7,600 three-image photographs, covering "approximately 12,750 square miles [about 33,000 square kilometers] of country, much of which had never been seen by the human eye." Photographed were Baranof and Chichagof Island; parts of the Coast Mountains; Lynn Canal; the glaciers of Glacier Bay; Icy Strait; and the coastline of the Fairweather Range to Fairweather Glacier. Glaciers exist in all of these areas. An additional 692 oblique photographs were also taken. As was the case in 1926, the 112-man expedition consisted of the aeroplane tender *Gannet* and a 40-m-long barge completely equipped to perform all necessary photographic and film processing operations. However, this time the expedition supported four open-cockpit Loening amphibian biplanes, each equipped with a camera hatch in the bottom of the fuselage with waterproof hatch cover.

Film was exposed "at an altitude of 10,900 ft [about 3,320 m] producing a picture more than 6 miles [9.5 kilometers] wide and about $2^1/_2$ miles [4 kilometers] in the direction of the flight."

Additional Navy Alaskan Survey Expedition Missions acquired photography of Alaskan glaciers in 1932 and 1934, with the 1934 mission photographing Finger Glacier, La Perouse Glacier, Malaspina Glacier, Icy Bay, and Bering Glacier. Systematic oblique aerial photographic missions flown between 1960 and 1993 by Austin Post and Robert Krimmel, and by the author between 1974 and 2010, have revisited a number of the glaciers photographed by the 1926 and 1929 expeditions.

Bradford Washburn

In 1933, Bradford Washburn began photographic reconnaissance of Alaska as a prelude to his climbing expeditions. His photographs include the first aerial observations of many glaciers in the Fairweather, St. Elias, Chugach, and Alaska ranges and the Wrangell Mountains. In 1933 and 1934, Washburn photographed the glaciers of the Fairweather Range between Finger and Malaspina glaciers; in 1937, he photographed Lituya Bay and Fairweather Glacier; and in 1938 he photographed the area between Yakutat Bay and Bering Glacier. Much of his 1935 aerial photography was of Canada's Yukon Territory. A dog-team survey of Alaska's Art Lewis and Nunatak glaciers was also conducted. Support for these photographic expeditions was provided by the NGS. Washburn returned to Alaska many times, photographing glaciers at many locations. In 1966, he conducted a photographic survey of the Mount Hubbard–Mount Kennedy area for the NGS (Washburn 1972). More than 5,000 of Washburn's Alaskan aerial photographs are archived at the UAF. With the support of Michael Sfraga and Susan Hazlett of UAF, in 2005, 2006, and 2009, I successfully repeated more than 30 of Washburn's aerial glacier photographs. Two of these repeat aerial photography pairs are presented in plates 6.2 and 6.15.

Austin Post

From 1960 through 1993, Austin Post, first with the University of Washington and later with the USGS, conducted annual photographic missions to document changes in Alaskan glaciers. In 1955, Richard C. Hubley of the University of Washington introduced Post to "organized aerial photography of the glaciers in western North America" (Post and LaChapelle 1971: I), when Post assisted Hubley in a photoreconnaissance of the North Cascade Mountains. During the International Geophysical Year of 1957, Post was involved in planning the logistics of the AGS Glacier Mapping Project, which included obtaining aerial photography of the glaciers to be mapped.

In 1960, Post began a systematic aerial photographic effort titled the "Program for Aerial Photographic Surveying of Glaciers in Western North America." The program's first three years were funded by the National Science Foundation. The program's primary foci were glaciers in southeastern and south central Alaska, although, during the life of the program, glaciers in the Alaska Range, the Alaska Peninsula, and the Wrangell Mountains were also photographed. Not every glacier was photographed annually. However, by 1993, some glaciers were photographed more than a dozen times.

After the end of the 1963 collection year, the program was on the verge of ending. However, the 27 March 1964 Alaskan earthquake played a major role in the continuation of the program. Post wrote the following:

> After the quake, Mark Meier, Director of the US Geological Survey's Glaciology Project Office in Tacoma, Washington wanted me to continue the photography and record changes resulting from the shaking. Again, so many interesting things were observed that Mark continued this program the next year and two years later, after a monumental effort, actually obtained for me a professional position with the Survey. (Post 1995: 19)

Initially, Post used aircraft of opportunity and flew with William R. Fairchild and many other well-known Alaskan bush pilots. Beginning in the early 1970s, many surveys were conducted with Robert Krimmel of the USGS. Krimmel frequently piloted the aircraft. Typically, both vertical and oblique photographs were made on each flight. Between 1960 and 1993, more than 100,000 negatives were exposed. All are archived at the UAF Geophysical Institute in Fairbanks, Alaska.

Vertical Aerial Photography Programs

Between 1941 and the early 1950s, several US government aerial photographic programs systematically obtained vertical aerial photographic coverage of many Alaskan areas. Gulf of Alaska region glaciers were photographed with a nine-lens mapping camera in 1941 and with a "Trimetragon" camera system in 1948. Between 1948 and the mid-1950s, many southern Alaskan glaciers were photographed with a single-lens vertical camera by several US Air Force photographic missions designated by the prefix "SEA." Since the early 1950s, many vertical photographic missions conducted by federal and State of Alaska agencies and private companies have imaged glaciers. For example, during the 1993–1995 surge of Bering Glacier, USGS, the Bureau of Land Management, and the Department of the Interior's Office of Aircraft Services flew nearly a dozen missions to monitor the changes in the terminus region of the glacier.

In 1978, federal and Alaskan state agencies pooled their resources and initiated an integrated and standardized program, the Alaska High-Altitude Aerial Photography (AHAP) Program, to develop a uniform aerial photographic database of Alaska. The purpose of the program was to document Alaska's landscape with "a set of unified and coordinated photographs" (Brooks 1988: 3). By 1986, the last year of data acquisition, 90 percent of Alaska had been imaged with both black-and-white and color-infrared photography.

These pictures were collected by NASA high-altitude aircraft from 65,000 ft [about 19,810 meters]. Black-and-white photography is at a scale of approximately 1:125,000, while color-infrared photography is at a scale of approximately 1:63,360. A black-and-white image covers an area of about 250 square miles [about 650 square kilometers], while a color-infrared photograph covers an area of about 64 square miles [about 166 square kilometers]. (Brooks 1988: 3)

Lessons Learned from Repeat Photography

Kenai Fjords National Park

Repeat photography provides significant insights into the post-LIA evolution of the landscapes of the fiords of the Kenai Mountains. Photographic evidence documents more than 80 percent of the post-LIA period from 1850 to the present. Information derived from the repeat photography pairs has documented (1) rapid vegetative succession and the transformation from glacier till and bare bedrock to forest at many locations; (2) post-1909 retreat of the largest fiord glaciers, some more than 20 kilometers; (3) Bear Glacier's late-twentieth-century and early-twenty-first-century retreat and thinning, resulting in about 3 kilometers of retreat between 2002 and 2005 (plates 6.16 and 6.17); (4) Pedersen Glacier's substantial retreat and the subsequent development of an extensive wetland (plate 6.18); (5) Aialik Glacier's relatively small changes during the past century; (6) Northwestern Glacier's substantial post-1909 retreat, resulting in the opening of Harris Bay and Northwestern Lagoon; (7) McCarty Glacier's substantial post-1909 retreat, resulting in the opening of McCarty Fiord; (8) Holgate Glacier and Little Holgate Glacier's gradual thinning and retreat; and (9) the transition from tidewater to land-based, stagnant or retreating, glacier termini, including Yalik and Petrof glaciers.

Glacier Bay National Park and Preserve

For Glacier Bay, photographic evidence documents approximately half of the 250 years of the post-LIA period. Information derived from repeat photography has been useful in documenting the following: (1) rapid influx of vegetation and the transformation and progression from barren glacier till and bedrock to forest, such as in upper Muir Inlet (plate 6.12); (2) the late-nineteenth-century timing and magnitude of glacier retreat in East Arm, a trend continuing to the present (figs. 6.1 and 6.2; plates 6.3–6.6, 6.12, and 6.14); (3) retreat of the glaciers in the Geikie and Hugh Miller inlet areas of West Arm; (4) early-twentieth-century retreat and the subsequent variability of Reid and Lamplugh glaciers (plate 6.7); (5) Johns Hopkins and Grand Pacific glaciers' early-twentieth-century readvances, followed by the continued advance of Johns Hopkins Glacier and the retreat and thinning of Grand Pacific Glacier; (6) decadal-scale fluctuations of smaller glaciers, such as the hanging glaciers of Johns Hopkins Inlet, including Hoonah and Toyatte glaciers; (7) transitions from tidewater termini to land-based, stagnant, or retreating, debris-covered, glacier termini; (8) filling of upper Queen Inlet with more than 125 meters of sediment (plates 6.8 and 6.9); (9) rapid erosion of fiord-wall lateral moraines following ice retreat; and (10) development of outwash and talus features at many locations.

Northwestern Prince William Sound

Repeat photography provides significant insights into the post-LIA evolution of the landscapes of northwestern Prince William Sound. For College Fiord, Barry Arm, Harriman Fiord, Blackstone Bay, Unakwik Inlet, Kings Bay, and Port Nellie Juan, photographic evidence documents from 50 percent to more than 80 percent of the post-LIA period. Information derived from repeat photography has been useful in documenting (1) rapid influx of vegetation and the transformation from barren glacier till and bedrock to forest, especially in lower College Fiord; (2) the post-1950 advance of Harvard Glacier (plate 6.10), while its immediate eastern neighbor Yale Glacier has retreated nearly 6 kilometers (plate 6.15); (3) advances of Harriman and Meares glaciers; (4) the substantial retreat and/or disappearance of many low-elevation glaciers, such as Baltimore Glacier (plate 6.10); (5) pre-twentieth-century shrinkage of Toboggan Glacier and its continued retreat during the twentieth and twenty-first centuries (plate 6.11); (6) continuing retreat of Nellie Juan Glacier; (7) significant retreat of Tebenkof Glacier with the transitions of its terminus from tidewater to land based; and (8) retreat of every glacier in Kings and Chenega bays.

Discussion and Conclusions

This chapter describes the use of repeat photography at Glacier Bay and Kenai Fjords national parks and northwestern Prince William Sound, Alaska, to document and understand changes to glaciers and the adjacent landscape as a result of changing climate. Repeat photography is the core of a systematic approach used to obtain photographic documentation of glacier and landscape change for every fiord in these areas. Through analysis and interpretation of these photographs, both quantitative and qualitative information is extracted to document Alaskan landscape evolution and glacier dynamics for the last century and a quarter on local and regional scales.

Literature Cited

Alaska Climate Research Center. 2007. *Temperature change in Alaska, 1949–2006.* climate.gi.alaska.edu/ClimTrends/Change/TempChange.html (accessed 9 March 2009).

Allen, H. T. 1887. *Report of an expedition to the Copper, Tanana, and Koyukuk rivers, in the Territory of Alaska, in the year 1885, for the purpose of obtaining all infor-* *mation which will be valuable and important, especially to the military branch of the government.* Washington, DC: US Government Printing Office.

Bagley, J. W. 1917. *The use of the panoramic camera in topographic surveying.* US Geological Survey Bulletin 657. Washington, DC: US Government Printing Office.

Brooks, P. D. 1988. *The Alaska High-Altitude Aerial Photography (AHAP) Program: A state/federal cooperative program.* Anchorage, AK: AHAP Program Office.

Burroughs, J., J. Muir, and G. B. Grinnell. 1902. *Alaska: Narrative, glaciers, natives; volume I of the Harriman Alaska Expedition with cooperation of the Washington Academy of Sciences.* New York: Doubleday, Page, and Company.

de Fillippi, F. 1899. *The ascent of Mount St. Elias [Alaska] by H.R.H. Prince Luigi Amedeo Di Savoia, Duke of the Abruzzi.* New York: Frederick A. Stokes.

Field, W. O. 1926. The Fairweather Range: Mountaineering and glacier studies. *Appalachia* 16:460–472.

Field, W. O. 1932. The glaciers of the northern part of Prince William Sound, Alaska. *Geographical Review* 22:361–388.

Field, W. O. 1937. Observations of Alaskan coastal glaciers in 1935. *Geographical Review* 27:63–81.

Field, W. O. 1942. Glacier studies in Alaska. *Geographical Review* 32:154–155.

Field, W. O. 1948. The variation of Alaskan glaciers, 1935–1947. *International Geodetic and Geophysical Union, Association of Scientific Hydrology, Assemble Generale d'Oslo* 2:277–282.

Field, W. O., editor. 1975. *Mountain glaciers of the Northern Hemisphere.* Volume 2. Hanover, NH: US Army Corps of Engineers, Cold Regions Research and Engineering Laboratory.

Field, W. O. 2004. *With a camera in my hands: A life history as told to C. Suzanne Brown.* Fairbanks: University of Alaska Press.

Gilbert, G. K. 1904. *Alaska—glaciers and glaciation: Volume III of the Harriman Alaska Expedition with cooperation of the Washington Academy of Sciences.* New York: Doubleday, Page, and Company.

Grant, U. S., and D. F. Higgins. 1911. Glaciers of Prince William Sound and the southern shore of the Kenai Peninsula, Alaska. *Bulletin of the American Geographical Society* 43:401–417.

Grant, U. S., and D. F. Higgins. 1913. *Coastal glaciers of Prince William Sound and the Kenai Peninsula, Alaska.* US Geological Survey Bulletin 526. Washington, DC: US Government Printing Office.

Groisman, P. Y., and D. A. Easterling. 1994. Variability and trends of precipitation and snowfall over the United States and Canada. *Journal of Climate* 7:184–205.

Klotz, O. T. 1895. Experimental application of the photo-topographical method of surveying to the Baird Glacier, Alaska. *Journal of Geology* 3:512–518.

Klotz, O. T. 1899. Notes on the glaciers of Southeastern Alaska and adjoining territory. *Geographical Journal* 14:524–534.

McDonald, L. S. (in cooperation with the Puget Sound Maritime Historical Society). 1984. Alaska steam: A pictorial history of the Alaska Steamship Company. *Alaska Geographic* 11(4): 88 pp.

Molnia, B. F. 1982 (rev. 1993). Alaska's glaciers. *Alaska Geographic* 9(1): 144 pp.

Molnia, B. F. 2000. Glaciers of Alaska. *Alaska Geographic* 28(2): 112 pp.

Molnia, B. F. 2007. Late nineteenth to early twenty-first century behavior of Alaskan glaciers as indicators of changing regional climate. *Global and Planetary Change* 56:23–56.

Molnia, B. F. 2008. Glaciers of Alaska (glaciers of North America). In *Satellite image atlas of glaciers of the world*, ed. R. S. Williams Jr. and J. G. Ferrigno, chapter 8. US Geological Survey Professional Paper 1386-K.

Molnia, B. F., R. D. Karpilo Jr., J. Pfeiffenberger, and D. Capra. 2007. Visualizing climate change—using repeat photography to document the impacts of changing climate on glaciers and landscapes. *Alaska Park Science* 6:42–47.

Morrison, M. 1995. William O. Field and the American Geographical Society: The early years. *Physical Geography* 16:9–14.

Post, A. 1995. Annual aerial photography of glaciers in northwest North America: How it all began and its golden age. *Physical Geography* 16:15–26.

Post, A., and E. R. LaChapelle. 1971. *Glacier ice.* Seattle: University of Washington Press and the Mountaineers.

Reid, H. F. 1892. Studies of Muir Glacier, Alaska. *National Geographic Magazine* 4:19–84.

Reid, H. F. 1895. The variations of glaciers. *Journal of Geology* 3:278–288.

Reid, H. F. 1896. Glacier Bay and its glaciers. In *US Geological Survey 16th annual report, 1894–1895, Pt. 1*, 415–461.

Rogers, G. F., H. E. Malde, and R. M. Turner. 1984. *Bibliography of repeat photography for evaluating landscape change.* Salt Lake City: University of Utah Press.

Russell, I. C. 1891. An expedition to Mount St. Elias, Alaska. *National Geographic Magazine* 3:53–204.

Russell, I. C. 1892. Mt. St. Elias and its glaciers. *American Journal of Science* 43:169–182.

Russell, I. C. 1893. Second expedition to Mount Saint Elias. In *Thirteenth annual report of the United States Geological Survey to the secretary of interior, 1891–1892; part II—Geology*, ed. J. W. Powell, 7–98. Washington, DC: US Government Printing Office.

Sargent, R. H. 1930. Photographing Alaska from the air. *Military Engineer* 22:131–138.

Sargent, R. H., and F. H. Moffitt. 1929. Aerial photographic surveys in southeastern Alaska. In *Mineral industry in Alaska in 1926*, ed. P. S. Smith, 143–160. US Geological Survey Bulletin 797. Washington, DC: US Government Printing Office.

Schwatka, F. 1885. *Along Alaska's great river.* New York: Cassell and Company.

Tarr, R. S., and L. Martin. 1914. *Alaskan glacier studies of the National Geographic Society in the Yakutat Bay, Prince William Sound and Lower Copper River regions.* Washington, DC: National Geographic Society.

von Engeln, O. D. 1910. Photography in glacial Alaska. *National Geographic Magazine* 21:56–62.

Washburn, B. 1972. Oblique aerial photography of the Mount Hubbard–Mount Kennedy area on the Alaska–Yukon Border. In *National Geographic Society research reports 1966*, ed. P. H. Oehser, 283–297. Washington, DC: National Geographic Society.

Wright, G. F. 1887. Muir Glacier. *American Journal of Science* 33:1–18.

Wright, G. F. 1889. *The ice age in North America and its bearings upon the antiquity of man.* New York: D. Appleton.

Chapter 7

Documenting Disappearing Glaciers: Repeat Photography at Glacier National Park, Montana

Daniel B. Fagre and Lisa A. McKeon

In 1997, the USGS Global Change Research Program at Glacier National Park (GNP) initiated the Repeat Photography Project with photographs repeated from historic images of Grinnell and Boulder glaciers. The paired images revealed dramatic glacial recession and became, for many people, some of the first visual representations of the effects of climate change. Since then, repeat photography has proved to be a critically important tool for documenting and analyzing the retreat and disappearance of glaciers at GNP. Of equal importance, it has served as a compelling communication medium for educating the public and policymakers about the dramatic transformation of the park during the last century of warming temperatures. Because humans are predisposed toward visual information, photographic evidence often trumps other types of data in convincing people that fundamental changes have occurred (also see Molnia, chapter 6). In this chapter, we describe the legacy of glacier photography and the use to which it has been put in a systematic effort to chronicle the response of glaciers to climate change.

Glacier National Park and Its Early History

GNP consists of 4,082 square kilometers of mountain wilderness in northwestern Montana, adjoining its sister park, Waterton Lakes National Park in Canada (fig. 7.1). It has an imposing topography with over 175 named summits along 171 kilometers of continental divide, 762 lakes up to 2,761 hectares in size (Lake McDonald), and 906 kilometers of streams and rivers (NPS 2002). GNP and Waterton Lakes National Park were jointly designated the Waterton–Glacier International Peace Park in 1932, a United Nations Biosphere Reserve in 1976, and a World Heritage Site in 1995.

Composed of sedimentary rock up to 1.3 billion years old (Rockwell 2002), the landscape of the park has been sculpted into a diversity of landforms by repeated past glaciations. This has resulted in features such as arêtes, horns, cirques, and one of the highest headwalls in North America, the north face of Mt. Cleveland (1,400 meters). GNP's landscape is dominated by conifer forests that cover approximately 70

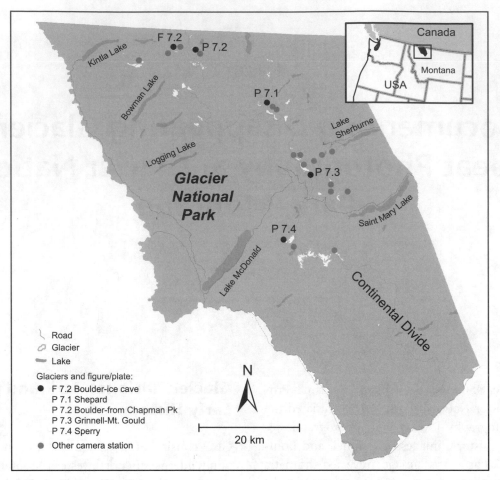

Figure 7.1. Map of Glacier National Park's location and camera stations for repeat photography of glaciers in Montana, USA.

percent of the park. Alpine tundra, rock, and ice cover 30 percent of the park above the treeline at around 2,000 meters, although the elevation of tree-line is quite variable. Home to numerous charismatic fauna such as grizzly and black bears, wolves, bighorn sheep, mountain goats, and wolverines, this mountain ecosystem is subjected to periodic, extensive forest fires during dry summers and as much as 9-meter-deep snow packs during long winters.

The area that was to become GNP was explored by early fur trappers in the 1800s and was regularly used by Native Americans, especially during summer months. George Bird Grinnell, an influential author, became an early advocate of setting this area aside as a preserve after visits in the 1880s. He first explored the glacier that was to bear his name, Grin-

nell Glacier, in 1887 and described it "as 1,000 feet [300 meters] high and several miles across" (Diettert 1992). His fascination with glaciers was apparent when he described them as the "jewels in the Crown of the Continent," the inspirational name he gave to this region during his campaign to create GNP. The Great Northern Railway had completed its transcontinental route near the southern boundary of the future park in 1893 and began to develop tourism, using references to "America's Alps" and the "Little Switzerland of America." The Great Northern Railway eventually built backcountry chalets and hotels that were one day's horse ride apart throughout the mountains. Among the early tourist attractions were guided hikes to the glaciers, continuing George Bird Grinnell's fascination with glaciers.

The Retreat of Glaciers

The glaciers that so fascinated early tourists were largely a legacy of the Little Ice Age (LIA), a period of colder temperatures and greater precipitation that lasted for several hundred years but ended around 1850 in the northern Rocky Mountains of the United States. Climate reconstructions of the GNP region show a 70-year period (about 1770–1840) of cool, wet summers and above-average winter snowfall that led to a rapid growth of glaciers just prior to the end of the LIA (Pederson et al. 2004). Before the LIA, the glaciers in GNP were smaller, alpine glaciers that fluctuated in size during various climatic changes over the past 7,000 years (Carrara 1989). It is unlikely that glaciers persisted during the rapid warming that occurred at the end of the Pleistocene (ca. 11,500 years ago) but instead reformed during a cooler period. Most of the landscape of GNP was created by the valley glaciers and icefields that were up to 1,500 meters thick during the last glacial maximum of 18,000 years ago.

At the end of the LIA, there were approximately 150 mountain glaciers in GNP, based on the terminal and lateral moraines left behind (Carrara 1989). Tree-ring-based climate reconstruction, sediment records, and historic photographs indicate the glacier termini began retreating and the ice mass began thinning between 1860 and 1880 (Pederson et al. 2004). Studies by early USGS scientists, including W. C. Alden, summarized the status of glaciers in GNP (Alden 1914) and recorded rates of movement (Alden 1923). Morton Elrod, an early naturalist and chief of interpretation for GNP, recorded positions of ice fronts for several glaciers. Dyson (1948) updated Alden's early synopsis of the state of glaciers, and Arthur Johnson, another USGS scientist, worked on Grinnell and Sperry glaciers for several decades and clearly showed the retreat of both glaciers (Johnson 1980).

Multidecadal droughts from 1850 to 2000 superimposed on the long-term trend of a 1.6-degree Celsius temperature increase led to recession rates as high as 100 meters per year for several glaciers such as Agassiz (Pederson et al. 2004). Based on 1966 aerial photographs, the first comprehensive map of the regional glaciers was published in 1968. Only 37 glaciers were named out of a total of 84 perennial snow-and-ice bodies (Key et al. 2002), and it is likely that some of the remaining 47 snow-and-ice bodies may have qualified as glaciers. GNP public documents list "about 50" glaciers during this period. Key et al. (2002) estimated that 99 square kilometers of ice covered GNP in 1850 but that only 26 square kilometers remained by 1968.

Aerial photographs were acquired in late September 1998 of all GNP glaciers. The measurements of glacier area from these photographs were the first for all glaciers since 1966. The total area of glaciers in GNP was reduced to 17 square kilometers. Using the criteria of 0.1 square kilometer minimum area or visual evidence of crevasses in the ice surface indicative of downslope movement, only 27 glaciers of the original 150 were present in 1998. Other former glaciers appeared to have shrunk to the point of being minuscule and stagnant ice masses. Between 1993 and 1998, glaciers ranging in size from 0.15 to 1.72 square kilometers became 8 to 50 percent smaller, and the relative rate of shrinkage was greatest for the smaller glaciers; Red Eagle Glacier, for example, was reduced to half its size between 1993 and 1998 and no longer meets the criterion of 0.1 square kilometer for being considered a glacier.

A glacier margin survey was completed for Grinnell Glacier in 2001 and showed a loss of 0.17 square kilometers, or 19 percent, from 1993 to 2001. The margin survey of Grinnell was repeated in 2004, and a further loss of 0.4 square kilometer, or 5.6 percent, occurred in only three years. Between 2004 and 2006, Grinnell Glacier lost an additional 9 percent of its area. Many watersheds no longer contain any glaciers (Key et al. 2002), and the area covered by glaciers in any of the remaining watersheds does not exceed 3 percent. Furthermore, glaciers have thinned by hundreds of meters, and, like Grinnell Glacier, may have less than 10 percent of the ice volume that existed at the end of the LIA (Fagre 2007). The area within park boundaries covered by ice and permanent snow was reduced 82 percent, from 99 square kilometers to 17 square kilometers, between 1850 and 1998. Between 1998 and 2005, the remaining ice and permanent snow area was reduced to less than 16 square kilometers.

The Future of Glaciers in Glacier National Park

The Blackfoot–Jackson Glacier Basin of GNP had the largest glaciers, and these decreased from 21.6 square kilometers in 1850 to 7.4 square kilometers in 1979. Over this same period, global temperatures increased 0.45 ± 0.15 degrees Celsius. Analysis of glacial area extent per decade from 1850 to 1979 versus a variety of climatic drivers reveals that annual precipitation and summer mean temperature together explain 92 percent of the loss over time (Hall and Fagre 2003). Using this information to parameterize the simulation model GLACPRED, potential future glacier behavior under both a "climate as usual" and a "global warming" scenario was predicted. The per decade (1850–2100) results of GLACPRED indicated that all glaciers in the basin will disappear by the year 2030 if current trends continue under the global warming scenario. Even if no further climatic forcing occurs, the glaciers were predicted to be all but gone by 2100. These results were corroborated by several other computer models (e.g., Bahr et al. 1997), which estimate that all glaciers will be gone between 2030 and 2050 at current warming rates in the northern US Rocky Mountains. If some of the largest glaciers in GNP will be gone by 2030, it is likely the smaller glaciers will largely be gone as well.

More recent measurements of the Blackfoot–Jackson glaciers indicate that the area in 1998 (2.94 square kilometers) was substantially less than the model predicted for 2000 (3.89 square kilometers) and was only a little greater than the area predicted for 2010 (2.44 square kilometers). This indicated that the glaciers were being reduced to specific areas nearly 10 years earlier than predicted. Hall (pers. comm.) compared the predicted temperature in-

crease used in the model for 1990–2007 against the actual temperature increase in GNP for the same period and found the actual increase was twice the model-predicted rate. The trend toward accelerated retreat is also reflected in Sperry Glacier. Reardon et al. (2008) recently showed that the negative mass balances for 2005 and 2006 were 4.3 and 3.1 times greater than the mean negative mass balance for the period 1950–1960. Without a significant reversal in the upward trend in temperatures, the glaciers will continue to disappear, perhaps as early as 2020.

Early Photography

The earliest photographs taken of the area destined to become GNP date from 1861 when a joint US–British survey expedition was marking the boundary between the United States and Canada along the 49th parallel. These photographs were taken by the British surveyors and are of poor resolution, but they document a cold and stark landscape following the end of the LIA (ca. 1850). The American surveyors brought a landscape painter but none of the photographs or paintings show glaciers in whole or with sufficient resolution to be useful. Cameras were large, heavy, and unwieldy at the beginning of the twentieth century (Boyer et al., chapter 2). Glass plates for the negatives were problematic when transported by horses into rugged terrain or on the backs of climbers when they were photographing from a summit or high ridge. However, photographs do exist of Dr. Lyman Sperry and his team on Sperry Glacier in 1887, and a photograph by a member of G. B. Grinnell's team shows a panoramic view of Grinnell Glacier useful for repeating. Photographs were used to illustrate one of the earliest reports of the glacier status in GNP (Alden 1914). Repeat photography in GNP has been used to effectively show other types of environmental change in mountains. Alpine treeline changes, geomorphological events, the aftermath of and long-term recovery from wildland fire, and changes in grasslands have all been documented with repeat photography (Butler et al.

1994, Butler 1994, Butler and DeChano 2001, Roush et al. 2007, Shea and Key 2006).

A Photographic Focus on Glaciers

Few photographs of glaciers were taken prior to 1900. With the major push by the Great Northern Railway to promote tourism in the area, the number of images increased dramatically. Great Northern hired photographers, painters, and writers to entice a wealthy clientele, primarily from the East Coast, to come and explore the scenic wonders of the area. Although the area was under limited protection as a forest reserve, it wasn't until 1910 that GNP was formally designated as a park, six years before the National Park Service was established in 1916 to manage the growing number of parks that Congress had authorized. GNP also hosted a National Park Service photographer, George Grant, to begin documenting the park's scenery and natural resources. The nascent national park photographic efforts and the continued photography by the railroad and other commercial interests generated a considerable body of photographs that were eventually archived in various official collections.

Because the Great Northern Railway's advertising campaigns focused on alpine scenery and adventures, as well as the strong fascination with glaciers described earlier, many of the early photographs featured the glaciers specifically or incorporated them in panoramic photographs. T. J. Hileman, in particular, ascended many difficult mountains toting a heavy box camera and supplies to photograph mountain scenery, and he secured excellent photographs of the glaciers. Being charismatic landscape features, glaciers were also frequently photographed by the increasing number of tourists. Guided walks to glaciers began in the 1920s and continue today. Many of the photographs taken have been incorporated into the archives of GNP over the years as personal collections have been donated.

The result of this photographic activity beginning in the early 1900s is a collection of about 18,000 photographs currently archived at GNP. This collection spans the period from 1865 to 1950 but has a significant emphasis on the early years of the park. This is particularly valuable for documenting glacier retreat not only because it encompasses most of the post-LIA warming trend but also because it covers several phases of the Pacific Decadal Oscillation (PDO), a 20- to 30-year period of anomalous sea surface temperatures that drive above- or below-average snowfall in Montana (Selkowitz et al. 2002). Glaciers have been shown to respond to these PDO phases in addition to the long-term climatic trends of warming temperatures (Bitz and Battisti 1999).

As stated earlier, there was recognition that glaciers were retreating, and it was natural to use comparisons with glacier photographs in a systematic manner. Dyson (1941) used this approach to demonstrate changes in glaciers, followed by Johnson in describing Sperry and Grinnell glaciers, and then several uses of repeat photographs were made by Carrara (1989) and Carrara and McGimsey (1981).

The USGS Repeat Photography Project at Glacier National Park

A relative hiatus in glacier studies ended in the 1990s with the establishment of a Global Change Research Program (1991) and the pursuit of several approaches to investigating the rates of glacier retreat and their causes. New aerial photography was acquired in 1993, geographic information system (GIS) analyses gave new estimates of the perimeters and areas of the remaining glaciers (Key et al. 1997), and a summary of glacier studies and information was compiled and published (Key et al. 2002). Repeat photographs were included where available but only as an illustrative adjunct to other work. The present effort was formalized following the visit by then vice president Al Gore, who visited GNP on 2 September 1997 to use the retreating Grinnell Glacier as a backdrop for a speech, and associated media scrutiny, on the topic of global warming. In preparation for the visit, several photographs of Grinnell Glacier were found in the archives of GNP and quickly retaken in the days before. The resulting use of these in national media outlets (e.g., national network TV, several news magazines and newspapers) clearly underscored the utility of repeat photographs in drawing attention to, and making absolutely clear, that substantial glacier reductions had occurred. The systematic and methodical effort to conduct repeat photography was launched immediately after this event and its media aftermath in 1997.

Approaches and Methods

The initial step was to search the historical archives at GNP in detail. In 1997, no cataloging system existed that consistently cross-referenced glaciers so thousands of photographs and negatives had to be examined to find suitable photographs for repeating. Because most photographs were taken without glaciers or eventual scientific use in mind, there was a certain degree of serendipity in locating photographs that clearly showed glaciers in their entirety. The photographs needed to be of sufficient resolution, focused, and undamaged by fading, spilled chemicals, or (in the case of prints) tears to provide the detail needed for comparisons with a modern image.

It also was important to find photographs taken late in the melt, or ablation, season when the annual snow had melted from the landscape. This allowed the ice margins of the glacier to be clearly visible. Ideally, both the historic and modern photographs would be taken in August or September when the maximum melt for the summer had occurred and shortly before the first snow of autumn accumulated on the glacier.

In many cases, exact dates were not recorded with the photographs, and inferences had to be made based on vegetation phenology in the photograph or other visual clues as to probable month or time of season (e.g., early versus late season). Lack of geographic information was another problem, but the subject being photographed could often be determined from distinctive mountain shapes or other topographic features. Similarly, the location from which the photograph was taken had to be deduced from limited clues in the original photograph.

Systematic searches were made not only of GNP's archives but also of the K. Ross Toole Archive at the University of Montana, several personal collections, and the collection at the University of Minnesota where the Great Northern Railway was headquartered during the early history of GNP. Burlington Northern Santa Fe, the railroad company that now incorporates the former Great Northern, also has a collection. A museum log was kept to help distinguish the photograph collections searched and the search criteria already used. Finding historical photographs worthy of repeating is a dynamic process as needs and questions are refined and new collections become available (Boyer et al., chapter 2; Bierman, chapter 9). Only in 2007 was a collection of Arthur Johnson's photographs donated to GNP, opening up possibilities for new photographs to be evaluated for repeating.

Selected negatives and photographic prints were scanned to 300 dpi on a flatbed scanner and saved as TIFF files. As computer and scanning technologies improved between 1997 and 2010, higher-resolution scans and different image management software were used. Images were cropped and expanded where glaciers were a small part of the original image. Some images were manipulated (e.g., with Adobe Photoshop) to bring out contrast in the original image that would facilitate finding the camera station in the field. Photographic prints were produced at the highest possible resolution for locating the site of the original photograph, the camera station.

Camera stations were located by lining up features in the original photograph (also see Boyer et al., chapter 2; Klett, chapter 4). Whenever there was a combination of objects in the foreground, such as boulders, with distant features, such as mountain silhouettes, an exact match could usually be made. In many cases, the exact rock on which the historical photographer had placed his tripod was easily found. Difficulties occurred when key foreground features were trees or other vegetation that had either grown up to cover landscape views of interest or had disappeared, making the matching process difficult. In the field, it soon became evident that suitable foreground features or intersecting ridgelines in the view were valuable photographic components for relocating the camera station. Photographs with only distant features required the most time, up to a day, to finally match up because of rugged terrain. Few roads exist in GNP, and helicopter use is generally prohibited. Thus many locations of photographs proved to require backpacking for several days into remote parts of GNP to relocate a single camera station. This often involved extensive climbing, and care to avoid grizzly bears. Figure 7.1 shows many of the camera stations and the glaciers associated with them.

All photographs were matched between 1997 and 2004 with 35 millimeter single-lens reflex film cameras using two film types: black-and-white (Kodak TMAX 100 Professional film, ISO 100) and color transparency (Fujichrome Velvia for Professionals, ISO 50) film. Exposure bracketing of half to full f-stops was used for all photographs for both film types, resulting in no less than six exposures per repeat view but usually more. For some photographs where the target glacier was a small part of the original image, an additional series was taken with the glacier zoomed in on to create a series for use in future repeat photographs. A parallax-corrected f2.8 zoom lens with haze filter was used with the camera mounted on a heavy-duty tripod outfitted with bubble levels.

A Global Positioning System (GPS) was used at the camera station for a precise location. This was confirmed by comparing GPS-derived coordinates with locations determined in the field by compass triangulation using 7.5-minute quadrangle maps (USGS). A photograph was taken of the tripod location to ensure precise duplication of the camera station in the future. The height of the camera above the ground surface, the compass direction along the long axis of the zoom lens, and the angle from horizontal were all recorded at each camera station along with all data regarding the camera, film, exposure settings, weather conditions, photographers, archival information on the historical image, and other details. In 2005, a switch was made to digital single-lens reflex cameras with 10 megapixel capability, but most other aspects of the protocol remained the same. Photo editing software eliminated the need for shooting separate color and black-and-white photograph series, while digital camera technology eliminated the need for recording exposure settings on the data sheet. All digital files of both original and modern (repeated) photographs are maintained on redundant network servers and managed with an image cataloging software within the Repeat Photography Project as a separate archive.

Several difficulties in obtaining photographs are unique to northern temperate mountains. First is the very limited field season. It is necessary to wait until seasonal snow is gone from the landscape and reveals the boundary of the glacier but, ideally, one should wait until the end of the melt (or ablation) season and photograph right before winter snows start. Because many of the cirques that contain glaciers face north and northeast, there is little sun that reaches into the cirque because the sun angle has shifted too far south. Many deep cirques have deep shadows, making for photographs with great contrast and constraining the detail that can be captured.

The second problem is that air quality is usually best for photographs early in the season when glacier photography opportunities are negligible. By August and September, general dustiness, burning of crop fields to the west of GNP, and the likelihood of regional forest fires diminish the air quality enough to compromise the clarity of the view. Because we attempt to mimic lighting conditions of the historical image both seasonally and by time of day, a good repeat photograph may require waiting several years before both lighting and similar snowmelt conditions occur. As weather and air quality conditions allow, repeat photographs have been taken every summer since 1997 to add to the collection.

Selected Results

Since 1997, more than 60 camera stations have been established of 22 different glaciers. Figure 7.2 and plates 7.1–7.4 depict the dramatic changes that have

taken place over various time periods. Fagre (2003) reported that 13 of 17 glaciers showed obvious reductions in size when comparing historic to repeated images. However, from 2003 to 2007, even glaciers that seemed to resist retreat, such as Gem and Sexton glaciers, have begun retreating. In fact, Reardon et al. (2008) estimated the rate of annual ice loss for Sperry Glacier in both 2005 and 2006 was up

Figure 7.2. Boulder Glacier Ice Cave, Glacier National Park, Montana, USA.
A. (1932). Early park visitors made the long trek to see Boulder Glacier in the northwest corner of Glacier National Park. (G. A. Grant).
B. (1988). Just over 50 years after the original image was recorded, rapid glacial retreat has eliminated the ice, and vegetation has become established at the glacier's forefront. (J. DeSanto, 642 Boulder Gl 88, Archives & Special Collections, The University of Montana–Missoula).

to four times greater than earlier last century. It appears the rates of glacier recession are increasing. Field results of measured ice loss suggest that Hall and Fagre (2003) were conservative in estimating that glaciers would disappear from GNP by 2030 if the present rates of glacier retreat continue.

The paired photographs of the Boulder Glacier Ice Cave (fig. 7.2) are significant from both cultural and natural resource perspectives. The 1932 image illustrates the attraction that early tourists to GNP had for glaciers as a charismatic geological phenomenon. The tourists are part of a guided horse-packing trip to the GNP backcountry, and the furry chaps of the guide are visible on the figure closest to the ice cave. A mere 56 years later, all of the ice is gone, and vegetation has become established in the forefield of the glacier. This repeat photograph garnered the most media attention in 1997 and provided the impetus for establishing our current project.

Shepard Glacier (plate 7.1) clearly illustrates that glaciers have complex boundaries and features in these mountain environments. In 1913, the upper glacier portion in the wide cirque has crevassing indicative of fairly thick ice. The glacier flows down to the bench where ice previously broke off the glacier front and fell to the valley below. By 2005, however, only a remnant of debris-covered ice (darkly streaked) remains in the upper left part of the cirque, and bedrock is showing elsewhere. Shepard Glacier is one that is less than 0.1 square kilometer in area in 2005 and considered by some scientists to be too small to be defined as a glacier. The modern photograph underscores this point compellingly.

The paired photographs of Boulder Glacier on Boulder Pass in the northwest corner of GNP (plate 7.2) show the virtual disappearance of all ice where a substantial glacier existed around 1910. In the 1910 photograph, the glacier actually extends to the right over Boulder Pass, and a lobe flowed down toward the next drainage. Matching this photograph required several hundred meters of climbing to a ridge, a feat we were glad to do with a relatively lightweight camera instead of the cumbersome and heavy gear that Elrod carried to his camera station.

The series of photographs from the summit of Mt. Gould looking down on Grinnell Glacier (plate 7.3) are somewhat unique for northern Rocky Mountain glaciers because they show the reduction in ice height from 1938, when ice filled the cirque, to the melting of the glacier into the proglacial lake. The height of the cliff behind the lake is nearly 200 meters. Prior to 1938, the ice-surface elevation was high enough to connect with the upper band of ice, now known as Salamander Glacier. This series also illustrates the forward movement of the glacier into the lake. A dark, triangular rock pile can be seen in the middle of the ice in 1981, and by 1998 it has moved to the front margin of the glacier (the dark diamond shape protruding into the lake). It, and much of the glacier front, disappeared by 2006.

Disseminating Repeat Photographs to the Public

Web Site

A Web site for viewing the collection of repeat photographs and for downloading the images was launched in March 2006 in response to the overwhelming number of requests received (www.nrmsc .usgs.gov/repeatphoto). This Web site has a number of features designed to make the use of the imagery as convenient as possible (also see Klett, chapter 4; Bierman, chapter 9). First, a visual interface guides the user to photographs of specific glaciers and uses thumbnail images to facilitate quick viewing and choosing. Second, all the photographs are free of copyright restrictions, having either been taken by federal employees and therefore in the public domain, or the photographs are part of public collections for which copyright restrictions have been waived. Third, photographs are provided in their original digital formats, either as digital files from the camera or from scans of negatives or color slides (positive transparencies). This allows end users to do their own cropping and matching of photographs or to use them in other ways. The paired photographs are also provided in a PowerPoint slide show where we did the cropping, alignment, and contrast enhancement. This provides an immediately useful visual product for people to download and use in slide presentations, or for general viewing, and demonstrates how the repeat photographs can be used. To enhance user accessibility, the Web site added PDF versions of the paired images in 2008. Instructions for downloading, legal uses of the images, proper attribution for the images, links to other information resources, and other ancillary information are presented in a way designed to make navigation of the Web site intuitive. Selected additions to the collection will include a direct link to the University of Montana archives where high-resolution versions of the historic images can be obtained. Direct contact with the authors is available for special circumstances or needs. Other Web site refinements will soon include an interactive map interface that links directly to the downloadable image files when the user selects the glacier. The original images are offered in two resolutions depending on intended use. One is at low resolution (72 dpi), in JPEG format (*.jpg), for use on Web sites. The other is a high-resolution file (300 dpi), in TIFF format (Tagged Image File Format, *.tif). Black-and-white versions of the repeated photographs are available for all of the images. Due to poor color rendering inherent in scanning Kodachrome slides, only selected color images are served for the images repeated using Kodachrome film (repeats done in 1998–2004), whereas all images repeated since 2005 are available in their original digital format.

There has been extensive use of the Web site. Users have spanned the globe and included places such as North Korea and Qatar. Users from Japan and Germany have been the most frequent visitors after US users. By user type, the US government has recorded the most use, but commercial use (hits from *.com domains) are a close second. There has been active use by *.edu domains as well. A user form is included in the instructions and guidelines at the repeat photography Web site, but filling out the form is voluntary. However, from the forms returned, we know that the repeat photographs have been used in numerous media outlets (e.g., TV or various Web sites), print media (e.g., ranging from *Time* magazine to the RV quarterly *Airstream*), books (from those on debate to the insurance industry), educational curricula and lesson plans (from Russia to Brazil), scientific Web sites (including the Hadley Centre for Climate Research in the United Kingdom), and numerous other publications. In addition, links to our Web site have been created at hundreds of other Web sites, and the repeat photographs have been reposted on numerous other Web sites. All of these examples underscore the power of repeat photography to provide a compelling message to the lay public. A new development is the creation of several posters on the Web site that people can download and print on their own color printers.

Park Outreach

The interpretive and educational personnel of GNP have made climate change and global warming one of their main messages to park visitors. To that end, the repeat photography of retreating glaciers has been used in a variety of modes to educate visitors about the impacts of climate change on park resources. These include roadside (i.e., wayside) exhibits placed along the Going-to-the-Sun Road at pullouts with views of remnant glaciers, interpretive posters at visitor centers, and printed visual aids for ranger-led walks. Visitors receive a newspaper guide to GNP that contains information on climate change with repeat photographs of the subject glaciers. An introductory orientation movie is shown at regular intervals in the St. Mary visitor center and highlights the repeat glacier photographs. National Park Service Resource Bulletins, a talking points document, and a climate-change bulletin use the repeat photographs to visually complement the text-based messages of these documents. In addition to its scientific values, repeat photography has special power to make people immediately and intuitively understand the visual consequences of rapid environmental change.

Discussion and Conclusions

The Repeat Photography Project described in this chapter began as an adjunct to investigations of glacier retreat. It has evolved into an important tool for documenting the disappearance of glaciers and is a compelling means for communicating the magnitude and rates of glacier retreat. In the course of documenting retreating glaciers, the repeat photographs have also documented other ecosystem responses. Plate 7.4 clearly shows the response of high-elevation forests to the same changes in annual snowfall and increasing temperatures that have caused glaciers to retreat. Because of the perceived irony of having GNP lose its namesake glaciers, and because glaciers continue to fascinate people in a spectacular mountain landscape, the repeat photographs of GNP glaciers have become iconic symbols of global warming. With evidence of worldwide glacier recession and modeled predictions that all of GNP's glaciers will melt by the year 2030, the task of documenting glacial retreat through photography not only is scientifically important but becomes a valuable historic record for the day when glaciers will have virtually disappeared from GNP.

Acknowledgments

Many people have contributed their efforts to this project. We particularly thank GNP Archivist Dierdre Shaw, Ann Fagre, Somer Treat, and others who have worked in the archives over the past decade for helping to find original photographs and negatives. Mark Fritch and Donna McCrea of the K. Ross Toole Archives, University of Montana, are gratefully acknowledged. Lindsey Bengtson, Karen Holzer, Fritz Klasner, Michelle Manly, Chris Miller, Greg Pederson, and Blase Reardon have repeated photographs in the field and contributed greatly to this project. We thank Carl Key for his photograph of Grinnell Glacier in 1981, and Jerry DeSanto for his photo of Boulder Glacier in 1988. Vintage photographs are reproduced courtesy of Glacier National Park Archives unless otherwise noted.

Literature Cited

Alden, W. C. 1914. *Glaciers of Glacier National Park.* Washington, DC: US Department of the Interior, US Government Printing Office.

Alden, W. C. 1923. Rate of movement in glaciers of Glacier National Park. *Science* 57:268.

Bahr, D. B., M. F. Meier, and S. D. Peckham. 1997. The physical basis of glacier volume-area scaling. *Journal of Geophysical Research* 102:355–362.

Bitz, C. M., and D. S. Battisti. 1999. Interannual to decadal variability in climate and the glacier mass balance in Washington, western Canada, and Alaska. *Journal of Climate* 12:3181–3196.

Butler, D. R. 1994. Repeat photography as a tool for emphasizing movement in physical geography. *Journal of Geography* 93:141–151.

Butler, D. R., and L. M. DeChano. 2001. Environmental change in Glacier National Park, Montana: An assessment through repeat photography from fire lookouts. *Physical Geography* 22:291–304.

Butler, D. R., G. P. Malanson, and D. M. Cairns. 1994. Stability of alpine treeline in Glacier National Park, Montana, USA. *Phytocoenologia* 22:485–500.

Carrara, P. E. 1989. *Late quaternary glacial and vegetative history of the Glacier National Park region, Montana.* US Geological Survey Bulletin 1902. Washington, DC: US Government Printing Office.

Carrara, P. E., and R. G. McGimsey. 1981. The late-neoglacial histories of the Agassiz and Jackson glaciers, Glacier National Park, Montana. *Arctic and Alpine Research* 13:183–196.

Diettert, G. A. 1992. *Grinnell's Glacier: George Bird Grinnell and Glacier National Park.* Missoula, MT: Mountain Press.

Dyson, J. L. 1941. Recent glacier recession in Glacier National Park, Montana. *Journal of Geology* 49:815–824.

Dyson, J. L. 1948. Shrinkage of Sperry and Grinnell glaciers, Glacier National Park, Montana. *Geographical Review* 38:95–103.

Fagre, D. B. 2003. Global environmental effects on the mountain ecosystem at Glacier National Park. In *National Park Service Natural Resource Year in Review—2002*, ed. J. Selleck, 24–25. Publication D-2283. Washington, DC: US Department of the Interior, National Park Service.

Fagre, D. B. 2007. Adapting to the reality of climate change at Glacier National Park, Montana, USA. In *Proceedings of the First International Conference on the Impact of Climate Change: On High-Mountain Systems, Bogotá, Colombia, November 21–23, 2005*, 221–235. Bogotá, Colombia: Instituto de Hidrología, Meteorología y Estudios Ambientales IDEAM.

Hall, M. H. P., and D. B. Fagre. 2003. Modeled climate-induced glacier change in Glacier National Park, 1850–2100. *BioScience* 53:131–140.

Johnson, A. 1980. *Grinnell and Sperry glaciers, Glacier National Park, Montana: A record of vanishing ice.* US Geological Survey Professional Paper 1180. Washington, DC: US Government Printing Office.

Key, C. H., D. B. Fagre, and R. K. Menicke. 2002. Glacier retreat in Glacier National Park, Montana. In *Satellite image atlas of glaciers of the world, glaciers of North America—glaciers of the Western United States*, ed. R. S. Williams Jr. and J. G. Ferrigno, J365–J381. US Geological Survey Professional Paper 1386-J. Washington, DC: US Government Printing Office.

Key, C. H., S. Johnson, D. B. Fagre, and R. K. Menicke. 1997. *Glacier recession and ecological implications at Glacier National Park, Montana.* Final report of the workshop on long-term monitoring of glaciers of North America and northwestern Europe. USGS Open File Report 98-031.

National Park Service (NPS). 2002. *Park Profile, Glacier National Park.* Washington, DC: US Department of the Interior, National Park Service.

Pederson, G. T., D. B. Fagre, S. T. Gray, and L. J. Graumlich. 2004. Decadal-scale climate drivers for glacial dynamics in Glacier National Park, Montana, USA. *Geophysical Research Letters* 31 (L12203, doi:10.1029/2004GL0197770).

Reardon, B. A., J. T. Harper, and D. B. Fagre. 2008. Mass balance of a cirque glacier in the U.S. Rocky Mountains. Proceedings of the mass balance measurement and modeling workshop, Skeikampen, Norway, 26–28 March 2008. *Annals of Glaciology* 50:A074 1–5.

Rockwell, D. 2002. *Exploring Glacier National Park.* Guilford, CT: Globe Pequot Press.

Roush, W., J. S. Munroe, and D. B. Fagre. 2007. Development of a spatial analysis method using ground based repeat photography to detect changes in the alpine treeline ecotone, Glacier National Park, Montana, U.S.A. *Arctic, Alpine, and Antarctic Research* 39:297–308.

Selkowitz, D. J., D. B. Fagre, and B. A. Reardon. 2002. Interannual variations in snowpack in the Crown of the Continent Ecosystem. *Hydrological Processes* 16:3651–3665.

Shea, D., and C. Key. 2006. *Glacier's Eastside Grasslands: Photographic comparisons over time.* West Glacier, MT: Glacier National Park.

Chapter 8

Historical Arroyo Formation: Documentation of Magnitude and Timing of Historical Changes Using Repeat Photography

Robert H. Webb and Richard Hereford

Arroyos are rectangular-shaped ephemeral or perennial stream channels that are incised from a few to several tens of meters into valley alluvium; they are common landscape features in the southwestern United States (Cooke and Reeves 1976). Here we restrict our discussion of arroyo formation along watercourses to southern Utah and Arizona, which are representative of the region as a whole (Hereford 1984, Webb and Baker 1987, Hereford 2002). Arroyos occur in geomorphic settings ranging from broad alluvial valleys to narrow, usually deep bedrock canyons partly filled with alluvium. Typically they are alluvial valley axial channels that drain small- to intermediate-sized watersheds. The circumstances associated with arroyo downcutting and filling are among the most researched topics in the geomorphology of this region, with hundreds of references dating back to the end of the nineteenth century (Cooke and Reeves 1976, Graf 1983). Arroyos are broadly related to geomorphic processes affecting incision of river channels in other parts of the United States as well as worldwide (Schumm et al. 1984).

Despite intensive research, the primary cause of arroyo downcutting remains elusive. Following Cooke and Reeves (1976), Graf (1983) presented a table of the various combinations of climatic effects and land-use practices that have been implicated in this process. Because arroyo downcutting generally occurred within a few decades of the introduction of livestock onto rangelands in the region, livestock grazing in particular and poor land-use practices in general are often blamed for historic channel erosion (Elliott et al. 1999). However, arroyos have repeatedly cut and then filled during the Holocene (Waters and Haynes 2001, Hereford 2002) when livestock were not present and human activity was minimal to nonexistent. This suggests that nonanthropogenic factors, such as the influence of climate variation and its control on rainfall and runoff, are major causative factors of arroyo downcutting over the long term. Climatic explanations include periodic drought that reduces vegetation cover, wet conditions with regional storms and floods, fluctuations in groundwater tables related to drought, and various combinations of these factors (Cooke and Reeves 1976, Graf 1983). Sediment-transport processes within ephemeral channel systems, commonly referred to as "complex response" (Schumm et al. 1984), have also been invoked to place arroyo downcutting and filling into a time-independent geomorphic domain unrelated to climatic or land-use changes (Schumm and Hadley 1957).

The historic episode of arroyo incision in this region allows us to evaluate the sequence of changes that occur as channels downcut and begin to refill, without considering the exact causes for initial downcutting. In this chapter, we use examples of arroyo cutting documented with repeat photography to illustrate the sequence of arroyo downcutting, widening, and filling from the 1870s through 2006. The conceptual model is a distillation of all available information on channel change in the region, although it relies heavily on our experiences with Kanab Creek and the Escalante, Paria, Virgin, San

Pedro, and Santa Cruz rivers (fig. 8.1). Finally, we cite examples of specific historical changes in river systems that were contemporaneous with variations in regional climate.

A Conceptual Model of Arroyo Downcutting and Filling

Several conceptual models have been proposed for the stages of channel downcutting and widening, whether the channels are typical arroyos or have been subjected to other severe disturbances, ranging from deforestation, fire, mining activities, or channel modification. Judson (1952), summarizing work

Figure 8.1. Map of the southwestern United States showing the locations of rivers and camera stations discussed in this chapter. Numbers indicate figure numbers.

begun by early researchers, including Bryan (1925) and Hack (1942), proposed a generalized conceptual model of arroyo cutting and filling. More recently, Hupp and Simon (1991) and Hupp (1999) discuss the response of channels to engineering modification as initially observed in Tennessee, and their conceptual model generally follows what we have observed after the initial phase of arroyo downcutting in the Southwest. The general sequence of changes developed here synthesizes our observations of ar-

royo processes from the start of historic arroyo downcutting around 1862 until the start of the twenty-first century, including backfilling beginning in the early 1940s. We note that post-downcutting changes are coincident with documented episodes of climate and runoff variation.

A five-stage sequence of changes (fig. 8.2) shows the relation between the timing of arroyo development and climatic variability, which also affects the establishment of riparian vegetation (Webb et al.

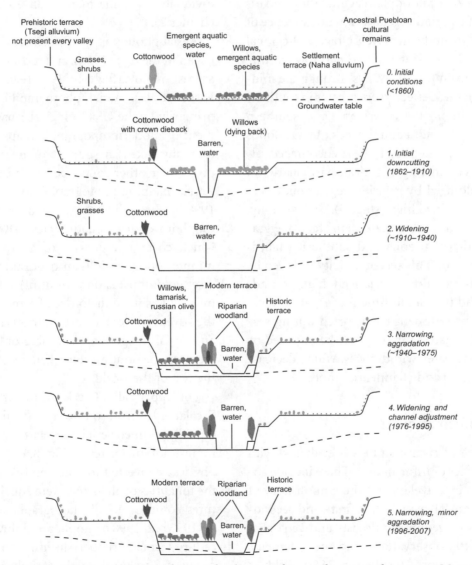

Figure 8.2. Schematic diagram showing the stages of arroyo development from settlement to the present. Many notable exceptions have occurred to this generalized model, particularly in some areas where stages 4 and 5 did not occur. The dashed line shows the water level in the alluvial aquifer for perennial streams.

2007). The initial stage (stage 1) of rapid downcutting and coalescence of discontinuously entrenched mainstem channel segments that resulted in deep, narrow channels occurred during either a single event or a series of floods associated with regional storms or extreme events. During the early-twentieth-century pluvial (Woodhouse et al. 2005), which spanned the period of about 1905 through about 1940, channels generally widened, damaging infrastructure and agricultural lands (stage 2). The midcentury drought, from the early 1940s through 1976, was a period of floodplain aggradation and establishment of riparian vegetation in the presence of relatively high groundwater levels (stage 3). Regional flooding caused reversion to channel downcutting and widening from the late 1970s through the mid-1990s (stage 4); these effects were greater in southern Arizona than in northern Arizona and southern Utah. Prolonged drought and the lack of large floods from about 1996 to the present have encouraged relatively minor channel aggradation, particularly on floodplains colonized by riparian vegetation, where groundwater remained high (stage 5). Arroyo filling, although of much smaller scale than stage 3 aggradation in northern Arizona and southern Utah, is presently occurring. This recent activity is driven by fewer regional-scale floods, relatively high groundwater levels, and the establishment of riparian vegetation that traps sediment transported during seasonal floods. The latest infilling may be impeded by excessive groundwater withdrawals, which decrease riparian vegetation and channel roughness.

Sources of Information

Historic change of river channels is evaluated with numerous sources of information. These include anecdotal reports by travelers and the ethnohistory of Native Americans (Bryan 1925), first- and second-hand accounts of geomorphologists and geologists (Davis 1903), the observations and measurements of early land surveyors (Betancourt 1990, Webb et al. 1991), tree-ring dating of plants established on floodplains (Hereford 1984, 2002; Friedman et al.

2005), analysis of historical aerial photography (beginning about 1935; Parker 1995), and periodic surveys of channel dimensions (Hack 1942, Leopold 1976). Repeat photography has been used extensively to document channel changes from initial downcutting through the current state of arroyo filling (Graf 1983; Webb 1985; Webb and Baker 1987; Webb et al. 1991, 2007; Hereford 1987, 1993; Hereford et al. 1996; Hereford and Webb 2003). Historical photographs of river channels in the Southwest are available from numerous archives and private individuals (e.g., Bierman et al. 2005; Boyer et al., chapter 2).

Aerial photography of watercourses in the region began in the early 1930s, and most of the region had at least one overflight by 1940 (e.g., Hereford 1993, Parker 1995). Aerial photographic surveys, done primarily by the US Geological Survey for development of modern topographic maps as well as photography taken in support of numerous land-use projects or other concerns, began in the late 1940s. Aerial mapping photography flown circa 1993 and 1998 has been digitally rectified, creating digital orthophoto quarter quadrangles (DOQQs), each of which cover one-quarter of a typical USGS 7.5-minute topographic map quadrangle (rockyweb.cr.usgs.gov/nmpstds/doqstds.html). The resolution on DOQQs is on the order of 1 meter. Finally, with the advent of Google Earth, various types of satellite and aerial imagery are available online for specific areas of the southwestern United States as well as the rest of the world.

For Upper Valley Creek, the major tributary of the Escalante River in southern Utah (fig. 8.1), we used digitally scanned aerial imagery to delineate changes in an 8-kilometer reach (fig. 8.6). Color photography was converted to black-and-white images, and the tonalities of all images were equalized. Using features common to all photographs and identifiable on 1993 and 1998 orthophotography, we used a procedure termed "rubber sheeting" to approximately correct for image distortion (rubber sheeting is less accurate than more intensive image rectification used in digital orthophotograph preparation in that it does not account for distortion owing to variations in elevation or nonlinear distortion associated with the camera lens). Finally, we interpreted and quantified the area of several alluvial features: the arroyo, which is the area between the highest walls of the channel; floodplain deposits, typically bearing visible riparian vegetation; and the channel, which may or may not have signs of running water or recent flood impacts.

Stages of Arroyo Development

Stage 1: Initial Downcutting (1862–1910)

In most places, the initial downcutting was through late Holocene alluvium deposited from circa AD 1400–1880, which is present in the alluvial valleys of most arroyo systems in Arizona and Utah (fig. 8.1, Hereford 1987, 2002). Deposition of this valley fill is a recent geological phenomenon that took place for the most part during the Little Ice Age (LIA) epoch of global climate variability (Grove 1988, Hereford 2002) with little if any human influence in the Southwest. Furthermore, it is the youngest of several valley-fill alluviations during the Holocene on the southern Colorado Plateau and much of the Southwest (Hack 1942, Hereford 2002).

A compilation of the year of initial downcutting on selected watercourses in the region indicates that arroyos formed between 1862 and 1909, with most downcutting commencing in the late 1870s through the mid-1890s (Webb 1985). The earliest date of initial downcutting is 1862, which is from the Santa Clara River just upstream from St. George, Utah (fig. 8.1). Incision relates to an unusually intense and widespread storm during the winter of 1861–1862 (Engstrom 1996). The latest date is 1910, which is from Polacca Wash on the Navajo Indian Reservation in northern Arizona (Hack 1942). Everitt (1979) reports that Bull Creek, a small tributary of the Fremont River, began downcutting in 1935, but this unusually late date is probably related to upstream headcut propagation from the mainstem river, which began downcutting and widening in 1896.

Few photographs are available to document geomorphic change during the period of initial downcutting; nevertheless, enough early photography exists to conduct repeat photography to document the extent of local channel changes. Photographs from the early 1870s show the pre-downcutting condition of Kanab Creek (fig. 8.3) and the Escalante River (Webb 1985). Other sites where conditions prior to stage 1 incision are captured on film include the Paria River in the Lonely Dell area (Hereford and Webb 2003: figs. 1–4), the Virgin River in Zion National Park (Hereford et al. 1996), and the Santa Cruz River at Tucson, Arizona (Betancourt 1990). The amount of incision into the older valley fill is highly variable; the bed of Kanab Creek, for example, deepened 15–20 meters in the 1880s (Webb et al. 1991), whereas the incision of the Escalante and Santa Cruz rivers was between 5 and 10 meters (Webb 1985, Betancourt 1990).

Historic downcutting (stage 1 incision) throughout the region began during the unusual precipitation during the winter of 1861–1862 (Webb 1985, Engstrom 1996), which was the forerunner of a period of frequent large floods that lasted from 1910 until the early 1940s. In southwest Utah, perhaps over most of the southern Colorado Plateau, and much of the western United States, rain fell almost continuously for 40–45 days from late December 1861 into February 1862, something not repeated since. Climate records, moreover, show that warm-season rainfall was unusually high on the southern Colorado Plateau (Hereford and Webb 1992) at the inception of global warming following the LIA until the early 1940s.

The unusually wet weather during historic arroyo cutting was preceded by two to three decades of the driest climate since 1700 in the upper Virgin River basin (Hereford et al. 1996). To what extent this drought affected vegetation is unknown; however, numerous landscape photographs taken as early as 1863 compared with photographs taken around 1900 provide little evidence of change in hillslope vegetation cover, although riparian vegetation changed substantially during and after arroyo cutting (Webb

Figure 8.3. Kanab Creek at Kanab, Utah, USA.

A. (1871). This view to the east across Kanab Creek at Kanab, Utah, was taken during the winter of 1871–1872. Its channel was shallow, perhaps 2 meters deep and 17–30 meters wide. The vegetation present was *Salix exigua* (coyote willow), *Artemisia tridentata* (big sagebrush), and *Atriplex canescens* (four-wing saltbush) (Webb et al. 1991). (J. K. Hillers, from Dutton 1882).

B. (4 November 1990). The bed of Kanab Creek lowered about 25 meters and its width increased to 80–115 meters beginning in 1882. Because of the extensive channel change, this camera station may be as much as 7 meters west of the original camera station. A low floodplain was deposited and riparian vegetation became established, including nonnative *Tamarix ramosissima* (tamarisk), *Populus fremontii* (cottonwood), and *Salix exigua*. (R. H. Webb).

C. (27 April 2000). The floodplain has aggraded and riparian vegetation has grown and thickened in the intervening decade, including nonnative *Tamarix* and *Elaeagnus angustifolia* (Russian olive) intermixed with *Populus*, *Salix gooddingii* (black willow), *S. exigua*, and *Artemisia tridentata*. (R. H. Webb, Stake 2055).

et al. 1991, Hereford et al. 1996, Webb et al. 2007). A second severe regional drought occurred from 1891 through 1904, preceding the early-twentieth-century pluvial (Woodhouse et al. 2005). The effect of this drought on stream channels in the region is not well known, although cottonwood trees on the highest terraces in southern Utah date from this period (Hereford et al. 1996).

Paleoflood studies of the Southwest generally do not document an increase in the frequency of large floods at the beginning of historic arroyo cutting (Ely et al. 1993). However, studies of individual basins (Kanab Creek and the Escalante and Paria rivers), based primarily on high-resolution tree-ring dating of flood scars, show that extreme floods did not occur during most of the LIA (Webb and Baker 1987, Webb et al. 1991). Beginning in the late 1800s, the magnitude and frequency of floods increased, largely coincident with the initiation of arroyo cutting. This increased flood frequency was evidently caused by a corresponding increase in the frequency and intensity of El Niño Southern Oscillation events that peaked in the early 1900s (Hereford 2002) and dominated the early-twentieth-century pluvial (Woodhouse et al. 2005).

Stage 2: Widening (about 1910–1940)

Repeated flooding during stage 2 caused considerable channel erosion, particularly widening of arroyo channels with modest or little increase in depth. The net effect of channel widening is the creation of a trapezoidal or rectangular channel cross section within the confines of the expanding arroyo walls (fig. 8.2). Throughout the region, photographs from the early 1900s to around 1940 generally show a wide, mostly flat-floored channel almost completely devoid of vegetation. Numerous photographic examples of stream-channel geometry exist (Webb et al. 2007) in large part because the erosion attracted photographers who wanted to document the effects on water-supply structures, agricultural fields, and other infrastructure and because regional aerial photography surveys commenced in the 1930s. These examples include Kanab Creek (fig. 8.4A, Webb et al. 1991), the Virgin River (Hereford et al. 1996), Escalante River (fig. 8.5A), Upper Valley Creek (the principal tributary of the Escalante River, fig. 8.6A), Little Colorado River (Hereford 1984), Gila River (Burkham 1972), San Pedro River (Hereford 1993), and the Santa Cruz River (figs. 8.7A and 8.8B; Betancourt 1990, Parker 1995).

For Upper Valley Creek in the Escalante River basin, southern Utah (fig. 8.1), aerial photography, beginning in 1940, documents the extent of channel widening (fig. 8.6A) as originally reported by Webb (1985). In 1940, in the reach from just upstream

from Pet Hollow to just downstream from Main Canyon, the channel appears to be wide and generally of rectangular cross section (fig. 8.6A). From georeferenced imagery, the area between the arroyo walls was 280,000 square meters (table 8.1), of which only 1,000 square meters were clearly visible vegetated floodplain. At the mouth of Main Canyon, as well as just downstream and across from the mouth of Pet Hollow, flood flows have carved meander bends into the arroyo banks (fig. 8.6A).

On the Colorado Plateau, the early-twentieth-century pluvial occurred from 1905 through 1940 (Hereford and Webb 1992, Hereford et al. 2002); in southern Arizona, this period extended from 1905 through about 1930 with some high rainfall years through 1942 (Webb and Betancourt 1992). Cloudburst type summer storms were common throughout Arizona and southern Utah (Woolley 1946, Butler and Marsell 1972). Large regional storms caused extensive flooding in 1906, 1909, 1911, and 1915–1916; regional flooding was less extensive between 1920 and 1940 (Webb et al. 2007).

Stage 3: Narrowing and Aggradation of Low Floodplains (about 1940–1975)

Hereford (1984, 1986, 1987, 1993, 2002) reported that low floodplain deposition within arroyos commenced throughout the region shortly after 1940. On the Paria River just upstream from its junction with the Colorado River, repeat photography places the beginning of aggradation after October 1939 and before 1951; tree-ring dating of the floodplain deposits shows that aggradation began in the early 1940s (Hereford 1986). For Kanab Creek, a 1939 image shows incipient floodplain development (fig. 8.4). Overall, stage 3 aggradation and the spread of riparian vegetation were far and away the most profound events affecting these channels since initial downcutting and widening.

On Kanab Creek, the channel changed from one that was relatively wide with low relief to one with a low floodplain and dense stands of riparian vegetation (fig. 8.4B), primarily *Tamarix ramosissima* (tamarisk). Similar low floodplains developed along

Figure 8.4. Kanab Creek at Tiny Canyon, Utah, USA.

A. (1939). This downstream view shows that Kanab Creek at its confluence with Tiny Canyon (right midground) is wide with a barren, trapezoidal channel. Arroyo downcutting had lowered the bed elevation from the high bench shown prominently at far left. Channel widening ended just after this photograph was taken. (H. E. Gregory 950, USGS Photographic Library).

B. (9 October 1984). A floodplain formed in the channel supports a mixture of *Tamarix* and native riparian vegetation that obscures the floodplain, which at this time was 2 to 4 meters above the main channel. Small *Populus* appear above the dense *Tamarix* in the foreground. The channel of Kanab Creek, which is perennial here, appears on the extreme left side of the view. Large floods ended about 1941 in southern Utah, and the channels in the region narrowed in response to decreased floods as well as the influx of riparian vegetation. (R. H. Webb).

C. (27 April 2000). The new grove of *Populus* in figure 8.4B has grown to the extent that part of the far arroyo wall is obscured. The *Tamarix* in the foreground are becoming senescent; some are falling over as a result of the large amount of growth in recent decades. The formerly sharp lines on the alluvial terraces in the midground and background are now rounded. (R. H. Webb, Stake 1227).

the Escalante River (fig. 8.5B), although Webb (1985) reported germination of *Pinus ponderosa* (Ponderosa pine) on low floodplains as early as 1932. Low floodplain development is clearly visible in mid-twentieth-century aerial photography of the Santa Cruz River (Parker 1995), although when these floodplains began developing is not well docu-

mented. Local channel avulsions still occurred during the relatively small floods, notably documented for the Santa Cruz River between 1953 and 1960 (Parker 1995).

On Upper Valley Creek, aerial photography spanning the period of channel narrowing documents the small change in area between the arroyo walls

Figure 8.5. Escalante River at Escalante, Utah, USA.

A. (July 1932). This view, taken shortly after an extremely large flood damaged the town of Escalante, shows a wide channel devoid of riparian vegetation. The bed of this arroyo is several meters below the top of its former floodplain following the downcutting associated with another large flood in 1909. (R. W. Bailey 270658, used by permission, Utah State Historical Society, all rights reserved).

B. (2 October 1985). Riparian vegetation, mostly nonnative *Tamarix* and *Elaeagnus*, has encroached onto the once barren channel, and a low floodplain is present. Native *Salix exigua* may be present along the channel in the center distance. (R. H. Webb).

C. (5 October 1999). Dense riparian vegetation and a combination of *Populus* and nonnative *Tamarix* and *Elaeagnus* block the view of the background cliffs. The channel has narrowed considerably because the dense vegetation trapped sediment that accumulated on floodplains. (D. P. Oldershaw, Stake 1241).

compared with floodplain development and channel narrowing (table 8.1). From 1940 through 1974, the area of the arroyo increased from 280,000 to 300,000 square meters, while the floodplain area increased from 1,000 to 96,000 square meters and channel area decreased from 276,000 to 205,000 square meters. Also based on repeat aerial photography, the channel of the Little Colorado River changed from one essentially free of vegetation to one with extensive tamarisk groves on low floodplains from the mid-1930s to 1954. Repeat photog-

raphy documents this change near Cameron, Arizona (Webb et al. 2007). Along with the spread of vegetation into the channel, floodplain alluvium up to 5 meters thick was eventually deposited (Hereford 1984). Comb Wash in southeast Utah had a broad, flat channel in the early 1900s that was partially filled by floodplain alluvium beginning in the early 1940s and lasting until about 1980 (Hereford 1987).

The midcentury drought occurred throughout the Southwest with some regional variation from the

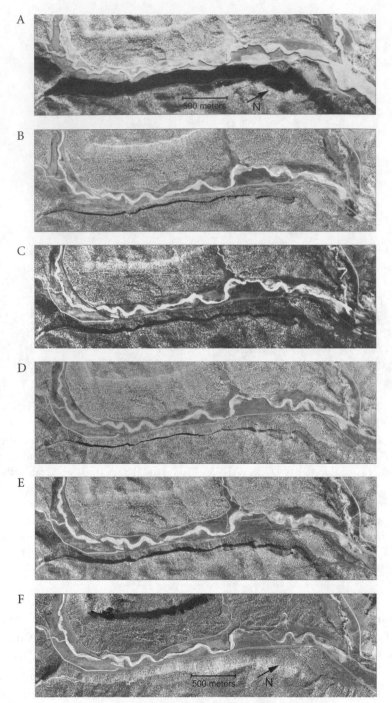

Figure 8.6. Repeat vertical imagery of Upper Valley Creek, Escalante River, Utah, USA.

A. (6 October 1940; image CSJ-1-111, USGS). See text for a discussion of this series of aerial images.

B. (29 June 1958; image 6-11, USGS).

C. (2 September 1974; image 2-17-45, Bureau of Land Management).

D. (7 September 1993; digital orthophotograph from the Canaan Creek NW and Wide Hollow Reservoir SW quadrangles, USGS).

E. (19 September 1998; digital orthophotograph from the Canaan Creek NW and Wide Hollow Reservoir SW quadrangles, USGS).

F. (25 June 2007; Quickbird imagery image 5553274070625181200 1J from Spot 5 satellite, Google Earth).

Figure 8.7. Santa Cruz River south of Martinez Hill, Arizona, USA.

A. (June 1942). Springs in the Santa Cruz River south of and at Martinez Hill (the camera station) prompted Spanish missionaries to establish the Mission San Xavier del Bac nearby in the early 1700s. By 1940, an arroyo had downcut and widened upstream, but the riparian ecosystem developed around the perennial flow continued to thrive. The channel is lined with *Populus*, visible at lower right, with an extensive *Prosopis* (mesquite) bosque in the background. (Photographer unknown, Arizona Game and Fish Department).

B. (14 November 1978). By 1978, *Populus* had disappeared and the bosque was dying because of excessive groundwater withdrawal (Webb et al. 2007), which had eliminated perennial flow. The 1977 flood widened the channel considerably here (Parker 1995), and a low terrace with sparse vegetation remains on river left (across channel at right). (R. M. Turner).

C. (5 April 1989). The 1983 flood, the largest recorded for the Santa Cruz River, widened this reach considerably and caused collapse of one of the two Interstate 19 bridges just downstream. The low terrace, which was on the inside of a sweeping bend, is much smaller. (R. M. Turner).

D. (25 November 2002). The 1993 flood again widened the channel, removing the low floodplain and leaving a barren, ephemeral channel. Groundwater lowering limits reestablishment of riparian species, and bank protection just downstream at the Interstate 19 bridges minimizes channel shifting. (R. M. Turner, Stake 937).

mid-1940s through 1975 to 1978 (Hereford and Webb 1992, Hereford et al. 2002, Turner et al. 2003). The drought was most severe from around 1951 through 1955 in most of the region. Relatively wet El Niño conditions, particularly in 1952, 1957–1958, 1963, 1969, and 1973, interrupted its multidecadal duration. For the most part, large floods were infrequent during this drought, which favored the establishment of vegetation and sediment accumulation on floodplains in most channel systems.

Table 8.1. Change in the arroyo walls and channel area of Upper Valley Creek, Escalante River basin, Utah

Year	Area between Arroyo Walls (square meters)	Floodplain Area (square meters)	Area of Active Channel (square meters)
1940	280,000	1,000	276,000
1958	301,000	88,000	213,000
1974	300,000	96,000	205,000
1993	292,000	150,000	142,000
1998	296,000	133,000	163,000
2007	298,000	169,000	129,000

Change between	Arroyo Wall Area (percent)	Floodplain Area (percent)	Channel Area (percent)
1940–1958	9	8,700	−23
1958–1974	0	8	−4
1974–1993	−3	56	−31
1993–1998	1	−11	15
1998–2007	1	28	−21
1940–2007	8	1,700	−53

Stage 4: Renewed Widening and Channel Adjustment (1976–1995)

During stage 4, channels were widened and shifted course in most arroyo systems. Regionally, these adjustments began in the mid- to late 1970s, extending into the early 1980s, and continued intermittently until the late 1990s. The net effect was incision of the stage 3 floodplains, forming cutbanks and widening the earlier channels. Although the widening and channel adjustment were far less extensive than those of stages 1 and 2, the earlier stage 3 floodplains were transformed into terraces that in most cases have not been completely overtopped by subsequent floods.

Of all the rivers in the region, the Santa Cruz River in southern Arizona was most affected by the changes in flood frequency (Webb and Betancourt 1992) that produced the stage 4 channel adjustments (Parker 1995). For one reach of the Santa Cruz River upstream from Tucson, 44 percent of the width increase between 1936 and 1986 was directly attributable to two floods, in 1977 and 1983 (fig. 8.7, Parker 1995). Downcutting of the Santa Cruz River in Tucson was as much as 5 meters between 1960 and 1986, again mostly during the 1977 and 1983 floods. As a

result of flooding and channel change, particularly widening, bank protection in the form of soil cement was installed within the city of Tucson beginning in 1982 (fig. 8.8). Although little channel change occurred during the 1983 flood in the reaches where soil cement was installed, large changes occurred elsewhere (Parker 1995).

At Upper Valley Creek, stage 4 adjustments produced only modest channel change that was mostly associated with the flood of record at the gaging station (early 1950s through the present). In August 1998, a flood of 127 cubic meters per second was recorded at the gaging station, and considerable channel widening occurred upstream (fig. 8.6E). In the study reach, between 1993 and 1998, the change in area between arroyo walls was insignificant while the area of the channel floodplain decreased from 150,000 to 133,000 square meters (−11 percent) and the area of the channel increased from 142,000 to 163,000 square meters (+15 percent, table 8.1).

A well-documented episode of increased precipitation from circa 1976 through 1996 (Hereford et al. 2002, Turner et al. 2003) was largely coincident with stage 4 channel adjustments. In some precipitation records, this wet episode rivals the early-twentieth-

Figure 8.8. Santa Cruz River at the Congress Street Bridge, Tucson, Arizona, USA.
A. (23 December 1914). Repeated floods and channel change in the early twentieth century (Betancourt 1990) caused fords across the Santa Cruz River near downtown Tucson to be replaced with steel-and-wood bridges. This bridge on Congress Street was built following the floods of 1904 and 1905. The short approach structure shown in the center of this view failed shortly after this photograph was taken. (G. G. Sykes, Arizona Historical Society/Tucson, Sykes Collection PC240).
B. (November 1926). The replacement bridge, a reinforced arched concrete span, lasted through much of the first half of the twentieth century. Taken during a relatively small flood in 1926, this image shows flow in a relatively wide channel with some laid over riparian vegetation in the foreground. (Photographer unknown, Arizona Historical Society/Tucson 28765).
C. (12 September 1983). The arch bridge was replaced in the 1970s by a simple reinforced concrete span that accommodated four traffic lanes and a center lane. The flood of 1977 (Aldridge and Eychaner 1984) caused significant channel change in Tucson, prompting installation of bank protection (right midground). The channel was engineered according to the natural channel sinuosity, but the cross section was changed to trapezoidal, eliminating low floodplain deposits. (R. M. Turner).
D. (17 June 2008). Channel aggradation followed the 1993 flood, resulting in reduction in channel area and large increases in low floodplains that support dense riparian vegetation, both native and nonnative species. The Congress Street Bridge withstood both the 1983 and 1993 floods, the largest in the 91-year record. (R. H. Webb, Stake 1084).

century pluvial in precipitation amounts. The hydrologic effect of this wet episode across the region was numerous, relatively large, channel-changing floods, although a couple of notable events occurred well before 1976. For the Virgin River, where channel change commenced earlier than in other rivers in the region, a flood in December 1966 caused significant channel change (Butler and Mundorff 1970). In northern Arizona in particular, floods associated with Tropical Storm Norma in 1970 (Roeske et al. 1978) may mark the beginning of this period of channel adjustment. In southern Arizona, these include floods of 100-year magnitude that occurred in 1977 (Aldridge and Eychaner 1984), 1978–1979 (Aldridge and Eychaner 1984, Aldridge and Hales 1984), 1983 (Roeske et al. 1989), and 1993.

Stage 5: Narrowing, Renewed Aggradation (1996–2006)

The two decades spanning the turn of the twenty-first century are mostly characterized by channel narrowing and renewed aggradation of low floodplains, although the magnitude of change varies regionally. Interrupted by high winter rainfall during the El Niño conditions in 1997–1998 and 2004–2005, drought conditions largely prevailed from 1996 through 2007 in the southwestern United States (Webb et al. 2007). The net effect of these conditions was few large floods across the region and infrequent runoff, most of which occurred during local summer thunderstorms.

Changes along Upper Valley Creek are representative of channel changes in much of southern Utah. The floodplain area increased by 28 percent and the channel area decreased by 21 percent from 1998 through 2007 (table 8.1, fig. 8.5C). Although channel narrowing generally occurred during this period in this region (e.g., figs. 8.3C and 8.4C; Webb et al. 2007), several notable exceptions are known. Several large floods occurred in the Fremont River in south-central Utah, north of the Escalante River basin, most recently in 2006. Certain ephemeral and perennial streams on the Navajo Nation in northeastern Arizona also had large floods that removed most of the low floodplain deposits during this period.

Installation of soil cement along the Santa Cruz River through Tucson in 1982 (fig. 8.8C) followed the original sinuous channel. The soil cement prevents channel widening, and concrete sills (grade-control structures) installed across the channel bed prevent downcutting. Following the 1993 flood, the second-largest event in a 91-year record, aggradation of low floodplains occurred, reducing the width of the active channel by as much as 50 percent (R. H. Webb, unpublished data). Establishment of vegetation appears to be trapping sediment during floods that overtop these floodplains, creating a positive feedback mechanism that encourages aggradation under the regime of reduced flood frequency.

Discussion and Conclusions

In the arid and semiarid regions of southern Utah and Arizona, intermediate-sized watercourses with ephemeral or perennial flow have changed historically, downcutting into incised channels termed arroyos. Using a variety of photographic evidence combined with field studies, the history of channel change can be described by a five-stage sequence of initial downcutting, widening, channel narrowing and floodplain aggradation, renewed widening and/or downcutting, and renewed channel narrowing and floodplain aggradation. These stages fall within well-defined periods with specific climatic characteristics and do not appear to be related to large-scale land-use practices in the region.

Although significant temporal deviations are known, all of the documented watercourses appear to pass through these five stages of channel change, although some stages (notably stage 4) may be rather minor in some drainage basins. The Virgin River in southwestern Utah, in particular, has changes that significantly precede the others in the region by as much as 20 years. In other drainage basins, particularly ones in the northeastern parts of the region, channel changes may lag in time behind changes in most of the watercourses.

As this chapter demonstrates, repeat photography is an invaluable tool in working out site-specific channel change histories in alluvial channels, particularly in the era preceding aerial photography and satellite-based remote sensing. One of the few limitations of this technique is that few landscape photographers were in this once remote and inaccessible area during or before the time of initial downcutting (stage 1). Thus we have only a few photographs of the alluvial channels before the onset of regionwide downcutting in about 1880. Notable exceptions occur in the case of Kanab Creek (fig. 8.3, Webb et al. 2007) and the Santa Cruz River (Betancourt 1990). Combined with repeat aerial photography, replication of oblique ground photographs provides the documentation needed for environmental reconstruction of channel processes in this region.

Acknowledgments

All photographs are courtesy of the Desert Laboratory Repeat Photography Collection unless otherwise noted. Peter G. Griffiths and Raymond M. Turner reviewed the manuscript.

Literature Cited

Aldridge, B. N., and J. H. Eychaner. 1984. *Floods of October 1977 in southern Arizona and March 1978 in central Arizona.* US Geological Survey Water-Supply Paper 2223. Alexandria, VA: US Government Printing Office.

Aldridge, B. N., and T. A. Hales. 1984. *Floods of November 1978 to March 1979 in Arizona and west-central New Mexico.* US Geological Survey Water-Supply Paper 2241. Alexandria, VA: US Government Printing Office.

Betancourt, J. L. 1990. *Tucson's Santa Cruz River and the arroyo legacy.* Unpublished Ph.D. dissertation, University of Arizona, Tucson.

Bierman, P. R., J. Howe, E. Stanley-Mann, M. Peabody, J. Hilke, and C. A. Massey. 2005. Old images record landscape change through time. *GSA Today* 15:4–10.

Bryan, K. 1925. Date of channel trenching (arroyo cutting) in the arid Southwest. *Science* 62:338–344.

Burkham, D. E. 1972. *Channel changes of the Gila River in Safford Valley, Arizona, 1846–1970.* US Geological Survey Professional Paper 655-G. Washington, DC: US Government Printing Office.

Butler, E., and R. E. Marsell. 1972. *Cloudburst floods in Utah, 1939–69.* US Geological Survey–Utah Division of Natural Resources Cooperative-Investigations Report No. 11. Salt Lake City.

Butler, E., and J. C. Mundorff. 1970. *Floods of December 1966 in southwestern Utah.* US Geological Survey Water-Supply Paper 1870-A. Washington, DC: US Government Printing Office.

Cooke, R. U., and R. W. Reeves. 1976. *Arroyos and environmental change in the American South-West.* Oxford, UK: Clarendon Press.

Davis, W. M. 1903. An excursion to the Plateau Province of Utah and Arizona. *Bulletin of the Museum of Comparative Zoology at Harvard College* 42:1–49.

Dutton, C. E. 1882. *Tertiary history of the Grand Canyon District.* US Geological Survey Monograph No. 2. Washington, DC: US Government Printing Office.

Elliott, J. G., A. C. Gellis, and S. B. Aby. 1999. Evolution of arroyos: Incised channels of the southwestern United States. In *Incised river channels: Processes, forms, engineering and management,* ed. S. E. Darby and A. Simon, 153–185. New York: John Wiley and Sons.

Ely, L. E., Y. Enzel, V. R. Baker, and D. R. Cayan. 1993. A 5000-year record of extreme floods and climate change in the southwestern United States. *Science* 262:410–412.

Engstrom, W. N. 1996. The California storm of 1862. *Quaternary Research* 46:141–148.

Everitt, B. L. 1979. The cutting of Bull Creek arroyo. *Utah Geology* 6:39–44.

Friedman, J. M., K. R. Vincent, and P. B. Shafroth. 2005. Dating floodplain sediments using tree-ring response to burial. *Earth Surface Processes and Landforms* 30:1077–1091.

Graf, W. L. 1983. The arroyo problem—paleohydrology and paleohydraulics in the short term. In *Background to paleohydrology,* ed. K. J. Gregory, 279–302. New York: John Wiley and Sons.

Grove, J. M. 1988. *The Little Ice Age.* New York: Methuen.

Hack, J. T. 1942. *The changing physical environment of the Hopi Indians of Arizona.* Peabody Museum Papers 25, no. 1. Cambridge: Harvard University Press.

Hereford, R. 1984. Climate and ephemeral-stream processes: Twentieth-century geomorphology and alluvial stratigraphy of the Little Colorado River, Arizona. *Geological Society of America Bulletin* 95:654–668.

Hereford, R. 1986. Modern alluvial history of the Paria River drainage basin. *Quaternary Research* 25:293–311.

Hereford, R. 1987. The short term: Fluvial processes since 1940. In *Geomorphic systems of North America,* ed. W. L. Graf, 276–288. Geological Society of America Centennial Special Paper 2. Boulder, CO.

Hereford, R. 1993. *Entrenchment and widening of the upper San Pedro River, Arizona.* Geological Society of America Special Paper 282. Boulder, CO.

Hereford, R. 2002. Valley-fill alluviation (ca. A.D. 1400–1800) during the Little Ice Age, Paria River basin and southern Colorado Plateau, United States. *Geological Society of America Bulletin* 114:1550–1563.

Hereford, R., G. C. Jacoby, and V. A. S. McCord. 1996. *Late Holocene alluvial geomorphology of the Virgin River in the Zion National Park area, southwest Utah.* Geological Society of America Special Paper 310. Boulder, CO.

Hereford, R., and R. H. Webb. 1992. Historic variation of

warm-season rainfall, southern Colorado Plateau, southwestern U.S.A. *Climatic Change* 22:235–256.

Hereford, R., and R. H. Webb. 2003. *Map showing Quaternary geology and geomorphology of the Lonely Dell Reach of the Paria River, Lees Ferry, Arizona with accompanying pamphlet, Comparative landscape photographs of the Lonely Dell area and the mouth of the Paria River.* US Geological Survey, Geologic Investigations Series Map I-2771, scale 1:5,000. Flagstaff, AZ.

Hereford, R., R. H. Webb, and S. Graham. 2002. *Precipitation history of the Colorado Plateau region, 1900–2000.* US Geological Survey Fact Sheet 119-02. Flagstaff, AZ.

Hupp, C. R. 1999. Relations among riparian vegetation, channel incision processes and forms, and large woody debris. In *Incised river channels: Processes, forms, engineering and management,* ed. S. E. Darby and A. Simon, 219–245. New York: John Wiley and Sons.

Hupp, C. R., and A. Simon. 1991. Bank accretion and the development of vegetated surfaces along modified alluvial channels. *Geomorphology* 4:111–124.

Judson, S. 1952. Arroyos. *Scientific American* 187:71–76.

Leopold, L. B. 1976. Reversal of erosion cycle and climatic change. *Quaternary Research* 6:557–562.

Parker, J. T. C. 1995. *Channel change on the Santa Cruz River, Pima County, Arizona, 1936–86.* US Geological Survey Water-Supply Paper 2429. Denver, CO: US Government Printing Office.

Roeske, R. H., M. E. Cooley, and B. N. Aldridge. 1978. *Floods of September 1970 in Arizona, Utah, Colorado, and New Mexico.* US Geological Survey Water-Supply Paper 2052. Washington, DC: US Government Printing Office.

Roeske, R. H., J. M. Garrett, and J. H. Eychaner. 1989. *Floods of October 1983 in southeastern Arizona.* US Geological Survey Water-Resources Investigations Report 85-4225-C. Tucson, AZ.

Schumm, S. A., and R. F. Hadley. 1957. Arroyos and the semiarid cycle of erosion. *American Journal of Science* 255:164–174.

Schumm, S. A., M. D. Harvey, and C. C. Watson. 1984. *Incised channels: Morphology, dynamics and control.* Littleton, CO: Water Resources Publications.

Turner, R. M., R. H. Webb, J. E. Bowers, and J. R. Hastings. 2003. *The changing mile revisited: An ecological study of vegetation change with time in the lower mile of an arid and semiarid region.* Tucson: University of Arizona Press.

Waters, M. R., and C. V. Haynes. 2001. Late Quaternary arroyo formation and climate change in the American Southwest. *Geology* 29:399–402.

Webb, R. H. 1985. *Late Holocene flooding on the Escalante River, south-central Utah.* Unpublished Ph.D. dissertation, University of Arizona, Tucson.

Webb, R. H., and V. R. Baker. 1987. Changes in hydrologic conditions related to large floods on the Escalante River, south-central Utah. In *Regional flood-frequency analysis,* ed. V. Singh, 306–320. Dordrecht: D. Reidel.

Webb, R. H., and J. L. Betancourt. 1992. *Climatic variability and flood frequency of the Santa Cruz River, Pima County, Arizona.* US Geological Survey Water-Supply Paper 2379. Washington, DC: US Government Printing Office.

Webb, R. H., S. A. Leake, and R. M. Turner. 2007. *The ribbon of green: Change in riparian vegetation in the southwestern United States.* Tucson: University of Arizona Press.

Webb, R. H., S. S. Smith, and V. A. S. McCord. 1991. *Historic channel change of Kanab Creek, southern Utah and northern Arizona.* Grand Canyon Natural History Association Monograph Number 9. Grand Canyon, AZ.

Woodhouse, C. A., K. E. Kunkel, D. R. Easterling, and E. R. Cook. 2005. *The 20th century pluvial in the western United States.* Geophysical Research Letters 32. dio:1029/2005GL022413.

Woolley, R. R. 1946. *Cloudburst floods in Utah, 1850–1938.* US Geological Survey Water-Supply Paper 994. Washington, DC: US Government Printing Office.

Chapter 9

Clear-Cutting, Reforestation, and the Coming of the Interstate: Vermont's Photographic Record of Landscape Use and Response

Paul Bierman

In the early to mid-1800s, the men and women of New England cleared what was once continuous forest cover off the rocky, glacial landscapes of northeastern North America, and thus ran one of the largest, most significant landscape-scale experiments in human history. The experiment had no plan nor any coordinated leadership, but, in a matter of decades, settlers nearly stripped the forests from the hills and valleys of the northeastern United States. Then, half a century or more later, with similarly little planning and coordination, many of the cleared fields were abandoned and an equally extensive and unplanned experiment in reforestation and landscape response ran its course.

This regional-scale vegetation disturbance took place on a part of North America that had been continuously forested for more than 10,000 years since the climate warmed and the last of the continental glaciers had melted back into Canada (Webb et al. 1987). In the uplands, forest clearance exposed thin soils developed on glacial till to the direct impact of heavy rainfall. The effect was exacerbated by the hooves of grazing animals, a pounding that compacted the soils and, as long as it continued, pre-

vented regrowth of the forests. In the lowlands, sandy glacial outwash or river terraces left behind as streams readjusted after deglaciation were stripped of trees. As stumps were pulled or roots rotted away, the strength imparted to the soil by those roots, termed effective cohesion, was lost (Bierman et al. 2005) and otherwise stable hillslopes failed in landslides and gullies (fig. 9.1).

The impact of land-use practices was not lost on those living in the mid-1800s. Artwork from the time clearly records landscape-scale responses to human actions including clear-cutting and road building (plate 9.1). With the popularization, during the 1850s, of photography as a medium for recording both human and landscape conditions, the deforested, pre–Civil War, New England landscape was clearly and extensively documented (fig. 9.2). Such documentation was both intentional in images where landscapes were the subject and unintentional in images where the landscape was an incidental background to portraits, images of homes, and depictions of rural life (fig. 9.3). By 1882, George Perkins Marsh had documented the effects of unsustainable land-use management practices and decried

Figure 9.1. Near Tunbridge, Vermont, USA.
(1905). Deforestation catalyzed erosion by removing not only the trees but their roots, which for millennia had bound weak soils on steep slopes (Bierman et al. 2005). In Tunbridge, Vermont, valley clearance exposed the steep slopes between sandy river terraces and led to gully erosion and small-scale landsliding. The sediment eroded from the steep slopes accumulated below on cone-shaped alluvial fans; the horse stands on one of these fans. Trenching revealed layers of sand and soil buried in such fans. These layers clearly show a peak in sedimentation (and thus hillslope erosion) during settlement and postsettlement times (Bierman et al. 1997, Jennings et al. 2003). (Photographer unknown, LS07574, courtesy Vermont Historical Society).

in print the lack of concern over and the impact of human-induced erosion (Marsh 1882).

In this chapter, we use both original and repeat images of Vermont to document how the landscape has responded to human actions that include deforestation, reforestation, and road building. We use three examples to demonstrate how repeat photography of the same site at different times can be used to document—both qualitatively and quantitatively—landscape response to human actions with the thought that, by examining past landscape responses, we can better inform future land-use decisions. The images and research reviewed in this chapter are part of the Landscape Change Program,

a digital image archive (www.uvm.edu/landscape) described in the next section.

The Landscape Change Program

Since 1999, the University of Vermont has hosted a Web-based archive of landscape imagery known as the Landscape Change Program. Funded primarily by the National Science Foundation, the Landscape Change Program contains (as of summer 2010) almost 34,000 images of Vermont landscapes documenting over 200 years of landscape change and human–landscape interaction. More than 10 percent of

Figure 9.2. Near Duxbury, Vermont, USA.
(ca. 1875). Early stereo photograph of the cleared Vermont landscape near the town of Duxbury, Vermont. Such images show the degree to which New England slopes were cleared of trees during much of the 1800s. (Photographer unknown, LS03480).

Figure 9.3. Clarendon Springs, Vermont, USA.
(ca. 1880). Image of hotel at Clarendon Springs, Vermont. Looking closely at the main image, one can see the barren slopes and clear evidence of gullying and deposition of sediment on an alluvial fan. In this image, erosion and sediment transport are not the subject but rather incidentally captured as part of the distant background. (Photographer unknown, LS03684).

the archive is rephotography, useful for depicting change over time.

The archive is unusual in that it contains imagery useful for documenting changes in the natural and built environments of a rural area in contrast to image repositories or rephotography efforts documenting urban or well-known park landscapes such as Yellowstone (Meagher and Houston 1998) or Grand Canyon (Webb 1996). The archive is student and community centered, having been built over time by a series of student interns and projects. There is strong community outreach and involvement in both image collection and image description; nearly 3,000 images now have public comments. All images in the archive are described with at least a narrative paragraph and a set of Library of Congress Table of Authorities keywords. The archive is publicly accessible and fully searchable (www.uvm.edu/landscape). Users can download 800-pixel-wide images and use a zoom tool to investigate details in the images. High-resolution scans are available by permission of the source archives.

Examples from Vermont

The images of the Landscape Change Program provide a rich visual archive of changing landscape conditions over time. The following three examples show the power of different types of imagery taken over time for different purposes.

Deforestation and Reforestation

Pollen records, preserved in ponds and bogs, clearly indicate that many different tree species cloaked Vermont's hillslopes within several thousand years after the last glacial ice melted away from Vermont. Although the species distribution was well known, their prevalence on the landscape was not. Recent historical research (Cogbill et al. 2002) has determined the presettlement forest composition for New England. Prior to the settlement of many New England towns, the land was divided by surveyors who used distinctive trees as boundary points. These

"witness" trees represent a sample of the presettlement forest. Cogbill et al.'s (2002) regional-scale compilation suggests that the presettlement forest in Vermont was dominated by beech with significant amounts of maple, hemlock, and spruce. There were gradients that reflected latitude and elevation, with spruce and fir more prevalent in the mountains and oak and pine more prevalent to the south.

Initially, the settlers cleared these northern hardwood and softwood forests for agriculture, including both crops and grazing; later, higher-elevation sites and those with poor soil were cleared for timber resources and for fuel (Cogbill et al. 2002). In some cases, stumps were pulled to clear fields (fig. 9.4); in other cases, where timber was the object, not land clearance, the stumps were left in place (fig. 9.5). Because photography came to Vermont after the old-growth forests had been cleared, many of the images of active clearance from the late 1800s (such as fig. 9.5) show the cutting of second-growth trees.

River Channel Response: The 1927 Flood

Rivers are dynamic landscape elements that respond over varying timescales to hydrologic and sedimentologic changes in their source basins. The widespread and rapid deforestation of New England slopes changed both the timing and magnitude of runoff events and made copious quantities of sediment available for transport. Sequences of photographs, taken over time, can be used to examine the physical riverine response to landscape change. We start by analyzing a single massive disturbance, the 1927 flood (National Weather Service 2002).

Characterizing Flood Hydraulics and Effects

In 1927, a November flood, with peak flows more than two times higher than any other recorded event, struck Vermont. The flood destroyed more than a thousand bridges and caused significant channel change and channel bank erosion. The preceding October had been very wet, leaving soils saturated. The storm dropped up to 220 millimeters of rain in central Vermont with at least 125 millimeters falling over most of the state (National Weather

Figure 9.4. Highgate Falls, Missisquoi River, Vermont, USA.
(ca. 1880). Stump fences were a means by which to reuse what otherwise would have been wasted resources (the stumps). Highgate Falls is pictured along the Missisquoi River in northwestern Vermont. The stump fence defines an enclosure in the lower right of the image. Unstable, deforested slopes show in the distance. (Truax, LS04684).

Figure 9.5. (ca. 1870). This historic stereoview of a valley bottom clear-cut at Champlain Spring, Highgate, Vermont, shows shallow planar landslides in the background, hosted in fine-grain glacial deposits, on a cleared slope. These slides were likely catalyzed by loss of effective root strength after the slope was cleared (Bierman et al. 2005). In the middle ground are many stumps and much slash, the remains of cutting second-growth timber. A spring house is at the center of the image. Note the size of the stump that the man is sitting on; it remains from the old growth, presettlement forest that once covered Vermont lowlands like this. (Photographer unknown, LS03668).

Service 2002). Within days of the devastating flood, the US Army flew over Vermont photographing the damage. Of the 90 images taken, 67 are extant. During the summer of 2004, these 67 were rephotographed to show the changes in riparian corridors, development, and channel characteristics. We also examined hundreds of ground-level images taken both during and after the flood (Stanley-Mann et al. 2004).

Modern rephotography of flood and postflood images in the archive allows us to quantify changes that have occurred since 1927 (plate 9.2). Examination of the 67 pairs of aerial images shows that between 1927 and 2004, forest cover increased in 70 percent of the images, new roads were built in almost 60 percent, development altered the landscape in almost 50 percent, and vegetation cover in riparian zones increased in over 60 percent of the images. These changes have differing effects on surface-water hydrology, with reforestation tending to reduce peak flows and storm-flow volumes, whereas development and road building both tend to increase runoff and storm peaks (Dunne and Leopold 1978).

More generalizable is the identification of flood heights from historic photographs, because river stages are critical data for flood hazard evaluations. In bedrock channels, flood heights recorded by photographs can allow calculation of the channel roughness coefficients (n), which are required for modeling flood flows with the Manning equation (Dunne and Leopold 1978). Roughness coefficients of bedrock channels during exceptionally high flows are not well known (Wohl 1998, Reusser et al. 2004). In order to calculate n, we use USGS data to constrain peak-flood discharge (Q), survey data to determine channel geometry and slope (S), and historical photographs to define a water level at the presumed flood peak (plate 9.3). Combining the latter two data sources, we estimate the wetted perimeter at the time of flooding to calculate the hydraulic radius, R. We solve for n using values for all other variables in Manning's equation using

$$n = R^{1.67} \cdot S^{0.5}/Q. \tag{1}$$

Assuming that the flood image (plate 9.3) was taken at peak flow (3,340 cubic meters per second), Manning's n for this bedrock channel during the peak of the 1927 flood was 0.020. This value is similar to our contemporary estimate (0.017) made at lower flow (510 cubic meters per second). Uncertainties in this approach include random errors in channel cross-section surveys and discharge estimates as well as potential underestimation of the wetted perimeter and consequent overestimation of n if the image used to determine flow depth (and thus wetted perimeter) were not taken at peak flow. It is important to note that the example we give is of a bedrock reach, where we can reasonably assume that the channel geometry has not changed since the 1927 flood. Such an approach would fail in an alluvial system where channel scour during the flood and subsequent changes from later events would alter channel geometry as shown in the section that follows.

Characterizing Channel Change over Time

Rephotography of the oblique aerial imagery from the 1927 flood indicated that some parts of the river system were more susceptible to change than others. For example, the Winooski River forms the boundary between the cities of Winooski and Burlington. Here the river is alluvial and is downstream from the last set of bedrock falls. For the next 20 or so river kilometers, this major river flows across a broad alluvial lowland, termed the Intervale, before emptying into Lake Champlain.

Although several image pairs showed significant changes in river-channel morphology in the 77 years between the original 1927 images and the 2004 rephotography, one image pair (plate 9.4) was located in an area where mapping and other historic imagery provided more than a 100-year record of channel change over time (fig. 9.6). Analysis of all the records (fig. 9.6), catalyzed by the observations made on the image pair (plate 9.4), demonstrated that the flood of 1927 removed a midchannel island, which subsequently reformed. The channel-narrowing, documented in the paired imagery (plate 9.4), began before 1910 (fig. 9.7).

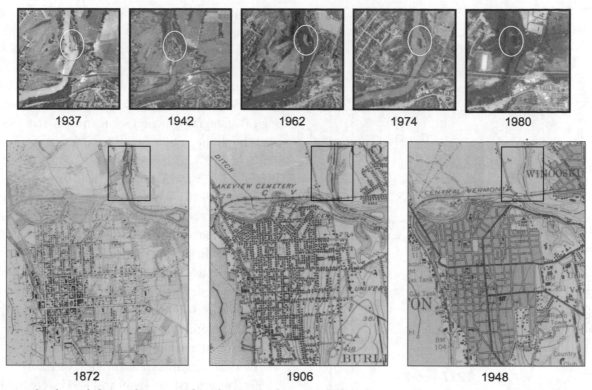

Figure 9.6. The channel change documented in plate 9.4 can be quantified over time using maps and aerial photographs. There is a rich cartographic and aerial photographic record of landscape change in parts of New England. In parts of Burlington, Vermont, surveyed maps date back over 135 years. Upper images are aerial photographs with the island circled (see plate 9.4). Lower maps have the approximate area of the aerial photographs boxed. (Courtesy Documents and Maps, Bailey/Howe Library, University of Vermont).

One can speculate on why the Winooski River (as pictured in plate 9.4) responded the way it did to changes in watershed conditions as Vermont was settled and cleared by westerners. Field reconnaissance suggests that the channel narrowing observed near the bridge (fig. 9.7) is the result of incision over time that has left terraces composed of historic sediment several meters above the channel and beyond what today is bankful stage. Trenching studies elsewhere in the Winooski Intervale indicate that 1 meter to as much as several meters of historic alluvium were deposited during the 1800s after the Winooski watershed was deforested (Thomas 1985). Presumably, this widespread alluvial deposition reflects increased sediment supply from the cleared and eroding uplands (Bierman et al. 1997). Map analysis indicated that the Winooski River delta prograded into Lake Champlain in the mid- to late 1800s before

sediment yields dropped (due to reforestation) and longshore drift moved much of the sediment away from the river mouth (Severson 1991). This dynamic landscape history and fluvial response can be represented graphically (fig. 9.8).

Coming of the Interstate Highways

Vermont remains a place where residents take pride in the statistic that more kilometers of the state's roads are dirt than paved. With few exceptions, the paved roads are narrow, two-lane affairs that twist their way up and down steep hills and through narrow valleys. Until the 1930s, many of Vermont's roads were unimproved, and in spring, when the snow melted and the ground thawed, the roads turned to swamps of mud, stranding cars and bringing commerce and transportation to a near standstill.

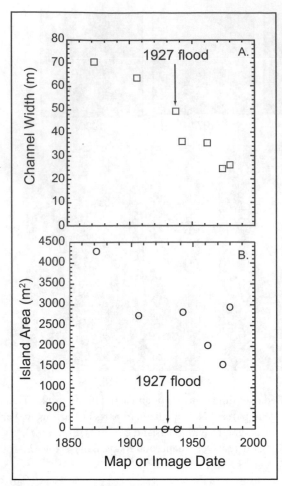

Figure 9.7. Changes to the Winooski River, Vermont, USA. A. Both aerial photographic imagery and a variety of map products demonstrate that the channel of the Winooski River at the location shown in plate 9.4 has narrowed over time .
B. The adjacent island, which existed prior to its removal by the flood of 1927, had reappeared by 1948.

The late 1920s and early 1930s saw the first of two major changes in road networks and thus the landscape of Vermont. Photographers from the Vermont Highway Department traveled the state, documenting the changes as roads were straightened and improved, guardrails were added, bridges rebuilt, and drainages rerouted (fig. 9.9). Presumably these images were used to show the public the effectiveness of their tax dollars. This was rephotography on a short timescale, but it documented significant and rapid changes that affected both the built and the natural environment. The resulting pairs of images show a landscape in transition in response to the growing importance of the automobile.

The revolution in transportation brought about by the internal combustion engine and the creation of a modern road network, as well as other technological changes, spawned some imaginative views of what might happen in small rural towns. During the 1920s or 1930s, a series of postcards was created that depict small, rural Vermont towns sometime in the future (fig. 9.10). These images do not feature the automobile; rather, all of the extant cards are dominated by imagery of mass transit: trains, subways, buses, and dirigibles. With three quarters of a century perspective, one can look back and see how this vision evolved. One finds a transportation landscape not at all like what the artist imagined but rather a landscape dominated by single-occupancy vehicles and the interstate highways that carry them.

Interstate highways came to Vermont in the 1960s. Two interstates (I-89 and I-91) were built and extended a little more than 480 kilometers within the state. Construction began in the 1950s, and the last section was completed in 1978. The process of planning, constructing, completing, and opening the highways was extensively documented, primarily by one photographer, Donald Wiedenmayer. His work, comprising nearly 40,000 negatives taken from 1961 to 1972, documents in detail the landscape change brought about by massive earthmoving, paving, and blasting activities. Channels were redirected, slopes were reshaped, and bedrock was blasted to carve exits through outcrops and carry four lanes of traffic over raging rivers.

Within Wiedenmayer's work are sets of images that were reshot over several years from exactly the same locations. Some of these image sets (e.g., fig. 9.11) contain up to four images. These image sets clearly show the scale and impact of interstate highway construction on the landscape. Entire slopes are moved, rivers are dammed, banks are rip-rapped, and the hydrology is radically changed. Our experience is that sequential presentation of these image sets is most effective at awakening audiences to the

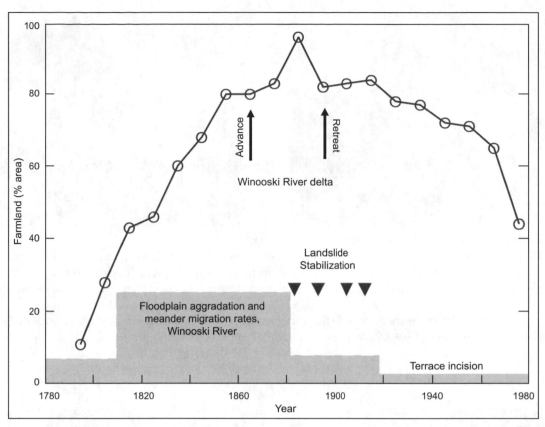

Figure 9.8. Graph summarizing Vermont landscape history and fluvial response. The temporal change in percent of farmland in Chittenden County, Vermont, is shown by open circles connected by a solid line. Arrows indicate the advance and retreat of the Winooski River delta into Lake Champlain (Severson 1991). The timing of landslide stabilization along a tributary of the Winooski River was determined by coring trees occupying landslide scars (Baldwin et al. 1995). Timing of floodplain aggradation was suggested by Thomas (1985). (Figure modified from Bierman et al. 1997).

immensity of earth movement and landscape change occasioned by highway construction. People who lived through the decades it took to build the highways tend to view the change as gradual; those whose knowledge of the landscape postdates highway construction are typically shocked at the scale of change. It appears that such sequential rephotography of massive earthmoving projects has the potential to inform and perhaps sway public opinion.

Discussion and Conclusions

The photographic record is a rich archive that reveals how the landscape responds, over time, to both human and natural influences. Repeat photography of the same location allows both qualitative and quantitative evaluation of landscape change. The image record and rephotography over time provide a unique means by which to study landscape response to widespread but uncoordinated, human-induced disturbances such as land clearance. One example presented in this chapter demonstrates that river channel width responded gradually to changing sediment and water loads; the same channel also changed shape abruptly in response to a discrete event, in this case, the rare, high-magnitude, 1927 flood.

Repeat photography is also a powerful educational and political tool (see also Moore, chapter 19;

A

B

Figure 9.9. Road project at Alburg, Vermont, USA.

A. (1937). Photographers hired by the State of Vermont documented road reconstruction projects that changed the face of the state and enabled rapid movement by automobile. These views are from Alburg, Vermont. The first image shows a muddy rutted road before reconstruction and paving. (Photographer unknown, LS08600, Vermont State Archives and Records Administration).

B. (1937). The same road after reconstruction. Note that earthmoving activities here are quite limited; paving is the greatest change. (Photographer unknown, LS08600, Vermont State Archives and Records Administration).

Figure 9.10. Ascutneyville, Vermont, USA.

(circa 1925). What might the future of transportation have looked like to someone living in Vermont in the 1920s or 1930s? Here is one of a series of photographic postcards with hand-drawn imagery overlain. These futuristic views feature various forms of mass transit, none of which ever came to the hamlet of Ascutneyville, Vermont, pictured here. (Photographer unknown, LS01320).

Figure 9.11. Bridge construction on the Winooski River, Vermont, USA.

A. (11 May 1961). The coming of the interstate highway system fundamentally changed the Vermont landscape. A Vermont Agency of Transportation photographer, Donald Wiedenmayer, documented the construction of a bridge over the Winooski River about 10 kilometers upstream from Lake Champlain over the course of three years. The first image is a preconstruction view, showing the Winooski River from its upper bank, which is heavily wooded. (D. Wiedenmayer).

B. (11 October 1961). An image of the bridge construction, in which much of the local vegetation was removed and the riverbed altered. (D. Wiedenmayer).

C. (11 July 1962). The grading of the roadbed. (D. Wiedenmayer).

D. (1 October 1964). The completed roadway. Both the channel-bordering slope and the riparian zone have been completely remolded by construction. The channel is now obstructed by bridge supports and the grass and wooded riparian zone has been replaced by stone rip-rap. (D. Wiedenmayer, LS00376, all images from Vermont State Archives and Records Administration).

Hoffman and Todd, chapter 5; and Klett, chapter 4). In environmental decision making, the emotional or affective power of imagery complements other means of information delivery. When the scale of landscape change is significant, such as the road-building example presented in this chapter, the visual data provided by repeat photography become a particularly germane part of the decision-making process allowing stakeholders to see by analogy the landscape-scale impacts of their land use choices.

Acknowledgments

The Landscape Change Program has been and is supported by grants from the National Science and Lintilhac foundations. Winooski River rephotography was completed by Elizabeth Stanley-Mann as part of an undergraduate research project at the University of Vermont. Vintage photographs and maps are reproduced courtesy of the University of Vermont, Bailey-Howe Library, Special Collections, unless otherwise noted. We thank Jamie Russel for scanning maps.

Literature Cited

Baldwin, L., P. Bierman, A. Schwartz, A. Church, and P. Larsen. 1995. The effects of colonial disturbance and subsequent reforestation on the Vermont landscape. *Geological Society of America Abstracts with Programs* 27:28.

Bierman, P. R., J. Howe, E. Stanley-Mann, M. Peabody, J. Hilke, and C. A. Massey. 2005. Old images record landscape change through time. *GSA Today* 15:4–10.

Bierman, P. R., A. Lini, P. T. Davis, J. Southon, L. Baldwin, A. Church, and P. Zehfuss. 1997. Post-glacial ponds and alluvial fans: Recorders of Holocene landscape history. *GSA Today* 7:1–8.

Cogbill, C., J. Burk, and G. Motzkin. 2002. The forests of presettlement New England, USA: Spatial and compositional patterns based on town proprietor surveys. *Journal of Biogeography* 29:1279–1304.

Dunne, T., and L. B. Leopold. 1978. *Water in environmental planning*. New York: W. H. Freeman and Company.

Jennings, K., P. Bierman, and J. Southon. 2003. Timing and style of deposition on humid-temperate fans, Vermont, United States. *Geological Society of America Bulletin* 115:182–199.

Marsh, G. P. 1882. *The Earth as modified by human action*. New York: Scribner and Sons.

Meagher, M., and D. B. Houston. 1998. *Yellowstone and the biology of time: Photographs across a century*. Norman: University of Oklahoma Press.

National Weather Service. 2002. *The flood of 1927*. www.erh.noaa.gov/btv/events/27flood.shtml (accessed 10 June 2009).

Reusser, L. J., P. R. Bierman, M. J. Pavich, E.-a. Zen, J. Larsen, and R. Finkel. 2004. Rapid late Pleistocene incision of Atlantic passive-margin river gorges. *Science* 305:499–502.

Severson, J. P. 1991. *Patterns and causes of 19th and 20th century shoreline changes of the Winooski Delta*. MS thesis, Field Naturalist Program, University of Vermont, Burlington.

Stanley-Mann, E., J. Hilke, P. Bierman, and I. A. Worley. 2004. *Repeat photography documents landscape change 75 years after an horrendous flood*. Geological Society of America Denver Annual Meeting, Denver, CO.

Thomas, P. A. 1985. *Archaeological and geomorphological evaluation: Burlington M5000(3) Northern Connector material supply/disposal area, Howe Farm floodplain*. Report 54. Burlington: University of Vermont, Consulting Archaeology Program.

Webb, R. H. 1996. *Grand Canyon, a century of change*. Tucson: University of Arizona Press.

Webb, T., P. Bartlein, and J. E. Kutzbach. 1987. Climatic change in eastern North America during the past 18,000 years: Comparisons of pollen data with model results. In *North America and adjacent oceans during the last deglaciation*, ed. W. F. Ruddiman and H. E. Wright, 447–462. Boulder, CO: Geological Society of America.

Wohl, E. E. 1998. Uncertainties in flood estimates associated with roughness coefficient. *Journal of Hydraulic Engineering* 124:219–223.

PART III

Applications in Population Ecology

Both Gruell (foreword) and Webb et al. (chapter 1) remind us that repeat photography was recommended as one technique to document plant populations around the turn of the twentieth century, and its applications to population ecology persist. Two key questions are addressed with repeat photography: plant demography, in terms of longevity and recruitment, and directional trends in specific species.

Both Bullock and Turner (chapter 10) and Hoffman et al. (chapter 11) use repeat photography to ask these questions with respect to very different types of plants, with a commonality restricted to long-term changes in arid environments. Bullock and Turner (chapter 10) summarize several long-term studies of population change in the Sonoran Desert of the United States and Mexico with an emphasis on columnar cacti and one unique hemisuc-

culent species. Their work advances the technique of estimating demographics using repeat photography as well as positing that long-term changes in certain key species, notably *Prosopis* (mesquite), can be resolved from repeat photographs.

Hoffman et al. (chapter 11) concentrate on two key tree *Aloe* species from southwestern Africa, one of which is considered to be a critically endangered species. Their work could provide a monitoring protocol for long-term assessment of the status of endangered species, particularly in areas where long-term data have not been collected and the species is clearly recognizable in photographs. They also show that interpretations of change, where dire predictions of loss are made, can be simplistic, and that repeat photography can reveal patterns that may be indicative of future changes as rare species respond to climate change or variability.

Chapter 10

Plant Population Fluxes in the Sonoran Desert Shown by Repeat Photography

Stephen H. Bullock and Raymond M. Turner

Long-term studies of desert perennial plants are still uncommon compared to the area they occupy and compared to plants of more humid temperate and tropical regions. The discouraging difficulty of waiting years for visible change can be addressed in part by retrospective long-term study, based on documentation by other people. A detailed record of de facto permanent plots is provided by repeatable and dated photographs, whatever the purpose of the original photographer. Here we review advances, problems, and potentials for plant demography based on repeat photography in the Sonoran Desert of northwestern Mexico and the southwestern United States. We treat individual species in much less detail than elsewhere in order to bring more attention to comparisons, to the interaction of methodological and conceptual issues, to heterogeneity and variability in different dimensions and scales, and to opportunities for research and application.

There is an appreciable number of population and individual growth studies for long-lived perennial plants at sites radiating from the Desert Laboratory in Tucson, but remarkably few take advantage of repeat photography. The monumental and outstanding exceptions (Hastings and Turner 1965b, Turner 1990, Parker 1993, Turner et al. 2003, Webb et al. 2007) have made this a prime area for demog-

raphy and desert vegetation studies with repeat photography. In the Arizona and Sonora areas of the Sonoran Desert, the present database for repeat photography extends back to 1880, deriving from the work of 39 photographers (pre-1978), notably D. T. MacDougal, H. L. Shantz, J. R. Hastings, and R. M. Turner. Matched sites now number about 667 exclusive of largely riparian views, grasslands, and nondesert woodlands (fig. 10.1). Many views have been matched twice and some have been matched many times.

In Baja California, the original photograph database starts with the transpeninsular trek of botanist T. S. Brandegee (1889). Notable contributors were E. A. Goldman (in 1905, Nelson 1922), L. M. Huey (from 1923 to 1940), and J. R. Hastings and R. M. Turner (from 1963 to 1972). The work of a total of 42 photographers has been used to make matches at 557 sites from 25.98° to 32.34° N (fig. 10.1), representing observation intervals of 27 to 112 years.

At a range of spatial scales, photographic coverage of the Sonoran Desert is, unsurprisingly, quite uneven (fig. 10.1). How well repeat photography represents the geomorphological landscape is a worthy question that has not been addressed (but see Webb et al., chapter 1). However, despite the bias to views along traveled paths, a great variety of landscapes are

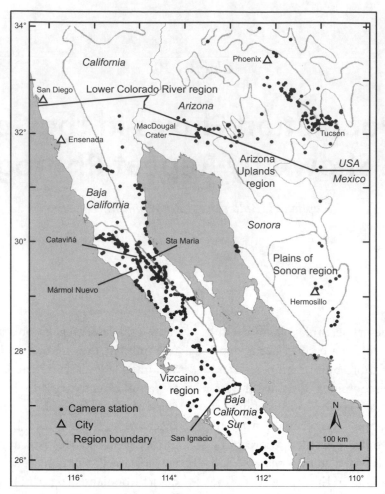

Figure 10.1. Map of the study region, with repeat photography sites, reference localities, and some subdivisions of the Sonoran Desert.

included, due to the photographers' varied interests and changes in routes following change in modes of travel and in patterns of land use. One aspect of heterogeneity we consider here consists of the contrasts in geomorphology, climate, vegetation, and land use between the Vizcaíno, Arizona Uplands, and Lower Colorado River regions of the Sonoran Desert (Shreve 1964), which reflect on the subjects and methods of study and results.

Physical Environment

The broad expanse of the Sonoran Desert covers a range of geomorphic and climatic settings, and these vary at different scales between and within regions of the desert. The Arizona Uplands and Lower Colorado River regions cover broad bajadas and plains with a sparse constellation of hills and mountains, mostly drained by substantial river systems except for largely endoreic basins along the peninsular ranges. Variations in substrate, drainage, and microclimate across the bajadas and hills affect the distribution and dynamics of plant populations (Shreve 1942, Yang and Lowe 1956, Niering et al. 1963, Parker 1987, McAuliffe 1994, Parker 1995). The Baja California peninsula has a tectonic history very distinct from that of the continent. In the landscape of the Vizcaíno region, watersheds are small and its modest valleys are few, with just one large plain

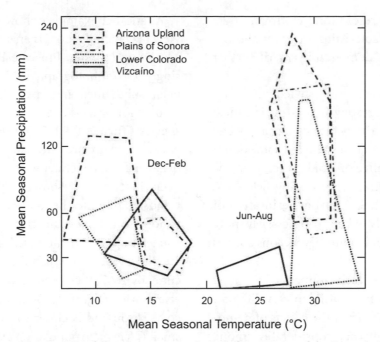

Figure 10.2. Mean winter and summer rainfall–temperature regimes in four regions of the Sonoran Desert (polygons encompass many stations; after Turner and Brown 1982).

where wind dominates reworking of the surface, and substantial areas with rocky or petrocalcic substrates (INEGI 1995, 2001).

Much of the climatic variation has been aptly summarized by Hastings and Turner (1965a), Turner and Brown (1982), and Turner et al. (2003), with the summer–winter contrast of temperature–precipitation normals (fig. 10.2). Temperatures are more moderate on the Pacific slope of the peninsula where cool marine breezes are common. The seasonality of rainfall tends to grade from winter to summer from north to south and west to east. Hurricanes incur occasionally from the south, and severe freezes nip at the north or northeast fringe and at higher elevations. None of these features show a single pattern of historical variation across the Sonoran Desert (Comrie and Glenn 1998, Bullock 2003, Turner et al. 2003).

In tandem with variations in the physical environment, there are marked variations in floristic and physiognomic characteristics (Shreve 1964). Toward the desert's margins in the northeast, desert scrub grades into grassland and some mountains support

conifer forests. On the peninsula, the Vizcaínoan associations grade into California Mediterranean scrub to the north (Brandegee 1893, Shreve 1936, Peinado et al. 1995). To the east and south the blend is with tropical thorn scrub and deciduous forest (Shreve 1934, 1964). Notably, while phytogeography has long been sketched, contrasts among the regions in population and community dynamics have yet to be determined. A major factor in this regard, besides substrate, climate, and logistics, is the contrast in the history of land use, as we discuss in this chapter.

Repeat Photography and Plant Demography

Demographic analyses based on repeat photography face several and varied problems of uncertainty (Regan et al. 2002). Of course, the limited perspective of the camera requires cautions in interpreting the photographs, as individuals become apparent or hidden due to other changes in the view, such as growth or death of other plants, lighting changes, or differences in focus or resolution (Hoffman et al., chapter 11). Other characteristics of repeat photography and desert plant demography merit further comment.

In arid regions, a problem exacerbated in, but not unique to, repeat photography is that during most months of most years, a majority of shrub species appear as barren twig structures, leaving some doubt whether the plants are alive or dead. Species may be identified in the photographs, or during fieldwork, but even an exhaustive field effort at verification of live or dead status is likely to have progressively more errors among smaller and more herbaceous species and individuals.

"Plot" size and shape vary dramatically among photographic views, often resulting in very different local population samples. Thus the effects of demographic stochasticity will vary considerably among sites. Bullock et al. (2005) attempted to evaluate such effects by comparing analysis of a data set limited to sites with minimum population sizes of 5, 15, or 50 individuals per species. These restrictions gave rise to data set sizes of 72, 62, and 26, respectively, for *Fouquieria columnaris* (cirio), and 48, 33, and 13 for *Pachycereus pringlei* (cardón). The difference between the species largely reflects the restriction of initial censoring to sites with at least five individuals. Remarkably, regression parameters and confidence limits were not strikingly affected by such apparently large changes in local population and in number of sites analyzed. This robustness may have been due, in part, to the sites representing a wide variety of time intervals.

Of course, it is not necessary that one photographic view be analyzed as one site. One view may be subdivided if different landscape units (e.g., geomorphologic, hydrologic, land use) are present, or data can be joined if views are close together and represent the same landscape unit. As an example, Bullock et al. (2005) divided land units and combined views in using 107 old photographs to study 77 sites. This can greatly reduce the problem of stochasticity due to small numbers of individuals per view.

As another example of how a single photographic view may include more than one habitat, the floor of MacDougal Crater, Pinacate Preserve, as seen in a single 1907 photograph by D. T. MacDougal, has been subdivided into two geomorphic units for studying *Carnegiea gigantea* (saguaro) and other species (Turner 1990, 2007). Because of the differences in soil texture across the crater floor and in the distance from the crater wall, the *C. gigantea* populations in two sites showed different demographic trends. At one site, the soils are finer and receive little runoff from the distant crater wall. Here, the *C. gigantea* population increased almost twofold from 1907 to 1959, after which there was a prolonged decline. By 2007, there were fewer individuals than 100 years earlier (fig. 10.3). By contrast, the other site has coarser soil and is closer to the crater wall. Here, a small *C. gigantea* population expanded by a factor of about four from 1907 to 1959 and continued to increase slightly until 1972. Subsequent decline left the population smaller than in 1959 but still far above its 1907 value (fig. 10.3). Despite the decline over the last decades, this plot has maintained a more robust population than near the crater center.

As usual for plant demography, individual age is rarely known. Seedlings and very small plants are usually not included because they are hidden or too small to be distinguished. Thus the large mortality expected for very young plants is simply excluded, and recruitment refers to newly visible plants, generally well established and several years old. For very old plants as well, the expectation of increased mortality will often be difficult to demonstrate, because sample size becomes very limited. Notable exceptions are provided by invasive, dense, short-lived populations, as with *Cylindropuntia fulgida* (jumping cholla), which established rapidly in one study area, then disappeared in a few decades (Tschirley and Wagle 1964, McClaran et al., chapter 12). For very long-lived species, exceptional conditions may be easier to study than "normal" attrition: one case is the abrupt loss of old *Larrea tridentata* (creosotebush) growing under conditions favoring the rapid establishment of its root parasite, *Krameria parvifo-*

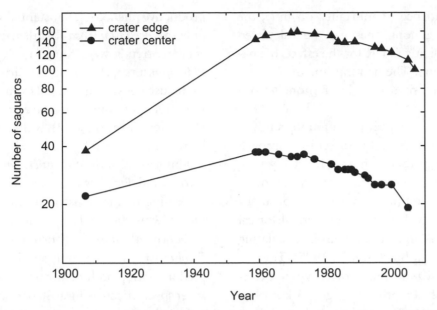

Figure 10.3. Graphs showing variation in number of *C. gigantea* plants through one century as seen in photographs of two contrasting areas on the floor of MacDougal Crater, Pinacate Preserve, Sonora, Mexico. (Crater center data from Turner 2007; crater edge data from Turner 1990 and new data after 1989).

lia (range ratany) (Turner et al. 2003). Demographic studies by repeat photography can aspire to encompass both the recruitment and the death of many individuals (with intermediate records), although the only known component of individual age is the lapse between photographs.

Time

Characterizing survivorship is an essential and difficult task. Age, alone, is generally a strong predictor of survivorship because of the inexorable (but not constant) accumulation of external and internal effects, and their myriad nature. But age data are rarely available for individual plants in nature. However, elapsed time of several years to decades is widely available through repeat photography. For a broad view of survivorship in two species, Bullock et al. (2005) lumped individuals of all sizes together at each site (following common practice in forest ecology) and obtained a wide variety of time intervals by using many, widely scattered populations as repli-

cates. The simplistic assumption of constant survivorship, however, was relaxed by using the Weibull function, the parameters of which can flex the curve to any of Deevy's types or intermediates (Pinder et al. 1978). Of course, there is no single value of survivorship or mortality in this formulation. However, the Weibull function does allow longevity to be calculated once it is defined (e.g., the time for survivorship to drop below 1 percent [10/1,000] or 0.1 percent), and the function's parameter values can be compared among populations or species. In the cases of *F. columnaris* and *P. pringlei*, time interval explained 45 percent and 35 percent of the variation in survivorship among sites with more than five individuals. The longevity (0.1 percent) estimates were 388 and 723 years, respectively.

These analyses across sites assumed homogeneity of mortality risk within sites (Sheil and May 1996), although growth studies suggest that cryptic subpopulations may exist in consequence of microsite variations in soil (e.g., petrocalcic or calcic horizons) and competitors (Escoto-Rodríguez and Bullock 2002). On the other hand, by estimating the broad-scale dependence of mortality on time, the residual variation among sites can be calculated and then used explicitly to test for the effects of spatially variable factors. The assumption of uniform mortality across all sizes may be a more serious problem.

To the degree that a repeat photography study is done in a short span of years, the foregoing approach might confuse life history trends in mortality with historical variations. On the other hand, in a modeling exercise, the assumption of historical constancy allows more flexibility than assuming both historical and life history constancy, as happens in calculating one average rate (e.g., Bowers et al. 1995). There is evidence from *C. gigantea* that mortality is not constant with age (size as a proxy for age: Parker 1993, Pierson and Turner 1998). Another obvious class of exceptions to historical constancy is the effects of catastrophic weather (see Effects of the Physical Environment: Climate and Substrate subsection).

Recruitment is typically considered to be limited to sporadic favorable periods, but time was a remarkably strong factor in the cases of *F. columnaris* and *P. pringlei*, accounting for 71 percent and 66 percent of the variation among sites (Bullock et al. 2005). Of course, much of the multiannual variation in both rainfall and recruitment will be masked as the intervals between repeat photographs are extended to several decades. Another lesser problem in calculating recruitment rates is the arbitrary decision on how to represent the number of reproductive individuals, considering the effects of mortality, recruitment, and maturation across several decades.

Population Trends

Repeat photographs covering several species, many sites, and two substantial time periods have been analyzed for apparent changes in species biomass (table 10.1). This regional or metapopulation perspective found that despite many cases of apparent stasis, unbalanced fluxes were typical. The patterns were not clearly organized by life form, size, or vegetative or re-

productive phenology. Substantial stability appeared only in *Encelia farinosa* (brittlebrush) while heterogeneous trends were found in *C. fulgida* (cholla), *Platyopuntia* spp. (nopal), and *Larrea tridentata* (creosotebush). A strong reversal from downward to upward trend was found for *Agave palmeri* (mescal) and *Dasylirion wheeleri* (sotol), while moderate or rapid increases continued for *Ambrosia deltoidea* (triangle-leaf bursage), *Parkinsonia microphylla* (foothill palo-verde), and *Prosopis* sp. (mesquite), and a notable slackening of increases occurred in *Acacia neovernicosa* (whitethorn), *Fouquieria splendens* (ocotillo), and *Isocoma tenuisectus* (burroweed). *C. gigantea* and *F. columnaris* were the only species to show decreases predominating both before and after 1963, with the latter species decreasing at 90 percent of sites over the twentieth century. In contrast, *Prosopis* increased at a great majority of sites, and *P. pringlei* increased widely in both Sonora and Baja California. The regional stability of *E. farinosa* is notable, especially given its relatively short lifespan (less than or equal to 32 years in plot-based studies, Goldberg and Turner 1986) and the large changes documented at particular sites (Turner 1990, Turner et al. 2003).

Intensive studies of *C. gigantea* were provoked by its widespread decline and showed the plausibility of several different explanations (Niering et al. 1963). However, long-term, locally intensive studies of its demography have allowed estimates to be made of the variation in recruitment rates and population structure over the past two centuries, and these reconstructions favor the interpretation that the largest regional fluctuations have been consequences of variation in climate at the scale of decades (Turner 1990, Pierson and Turner 1998). A related hypothesis was presented for *F. columnaris* and *P. pringlei*, linking their contrasting balances of recruitment and mortality to dependencies on different seasonal precipitation regimes (Bullock et al. 2005).

The case of *F. columnaris* merits more comment, and some interpretations regarding management. Despite the remarkable universality of its decline, and longevity (0.1 percent) estimated to be only

Table 10.1. Percentage of Sonoran Desert study sites with decreasing, constant, or increasing populations of some medium- to large-size plants as shown in repeat photography studies[1]

Species	Earlier Change Percent			Number of Years[2] (No. Sites)	Later Change Percent			Number of Years[2] (No. Sites)
	Neg.	None	Pos.		Neg.	None	Pos.	
Acacia neovernicosa	0	5	95	72 (19)	6	72	22	32 (18)
Agave palmeri	57	29	14	72 (7)	0	30	70	32 (10)
Ambrosia deltoidea	0	64	36	52 (11)	0	62	38	14 (13)
Carnegiea gigantea	46	28	26	47 (82)	48	37	15	32 (91)
Cylindropuntia fulgida	27	26	47	35 (16)	6	53	41	32 (17)
Dasylirion wheeleri	62	13	25	71 (8)	33	14	56	32 (9)
Encelia farinosa	10	66	34	30 (21)	5	85	10	28 (21)
Fouquieria columnaris	83	8	8	33 (12)	10	0	0	33 (12)
Fouquieria splendens	4	28	68	53 (58)	17	50	33	32 (63)
Isocoma tenuisectus	11	11	78	52 (9)	38	31	31	33 (13)
Larrea tridentata	40	35	25	72 (20)	12	67	21	32 (24)
Pachycereus pringlei BC	14	14	71	28 (7)	25	0	75	33 (12)
Sonora	0	44	56	30 (18)	0	39	61	32 (18)
Parkinsonia microphylla	7	25	68	48 (68)	2	42	56	32 (77)
Platyopuntia spp.	13	55	32	47 (38)	4	22	74	32 (46)
Prosopis (2 spp.)	0	5	95	53 (149)	4	22	74	32 (159)

[1]Data are from Turner et al. (2003) and Bullock et al. (2005), based on qualitatively assessed population biomass except for *Fouquieria columnaris* and *Pachycereus pringlei* in Baja California (BC), which were based on numbers of individuals.
[2]Interval is from median year of older photograph to median year of the newer. For the first interval, the range of years for the older photograph was 1883–1959 and 1933–1985 for the newer photograph, with a median of 1962–1963 except for *Ambrosia deltoidea* (1980). For the second interval, the range of years for the later photograph was 1984–2000 with a median of 1994–1995 except for *Encelia farinosa* (1990).

about half that of *C. gigantea*, there were variations among sites in rates of mortality of *F. columnaris*, and some recruitment did take place. We have already suggested that the decline is a result of low recruitment (Bullock et al. 2005), and that expected recruits from the 1980s and 1990s may be present but not yet visible. In some localities there was overexploitation for wood, but a loose net of protection has existed since the mid-1990s: the government has often rejected permit requests, the species was listed as a national priority for recovery (and exploitation), and it was listed in a Convention on International Trade in Endangered Species (CITES) treaty appendix II (www.cites.org). Two management issues—whether intentionally felled plants are useless or not and whether natural falls can be distinguished from assisted ones—have not been addressed with disinterested experiments. Also, average growth rates (Humphrey and Humphrey 1990, Escoto-Rodríguez and Bullock 2002) suggest that extremely long cutting cycles would be required to maintain

population stability. However, growth studies (via repeat photography) have also shown both geographic and microsite heterogeneity that suggests some small plants are very old and some large plants are very young (Escoto-Rodríguez and Bullock 2002). If the latter were removed and *F. columnaris* were again established on those microsites, the cutting cycle might be greatly shortened. Additionally, it could be argued that slow-growing plants might be allowed a century or so of reproduction, such that plants over some threshold height (depending on, among other factors, soil and climate variation) could be cut. The technical elements regarding these issues have been much affected by repeat photography studies and would benefit from their intensification, extension, and further improvements of methods.

Remarkably, there has been scant effort to exploit repeat photography to analyze some conspicuous matters in community and population ecology in our region. For example, the local coordination of changes among many species (Turner 1990) has not been studied at a regional scale. Nor has there been a focus on patch dynamics or microsite replacement patterns, which are supposed to be a conspicuous aspect of community dynamics resulting from shrub–succulent interactions (Vandermeer 1980, Aguiar and Sala 1999), nor has the technique been used for moderate- to long-term study of disease spread or effects of pests (such as rat burrows in large cacti).

Another overdue but more complex development would be linkage of repeat photography with individual-based or cellular models of multipopulation dynamics (Wiegand et al. 1995) or hierarchical landscape models (Bestelmeyer et al. 2006). Of course, the temporal and spatial dimensions of repeat photography databases would provide a great resource for such models, but technical innovations would be very useful to allow three-dimensional reconstruction from rapid and inexpensive field methods.

Growth in Size of Individual Plants

Study of individual growth rate is a particularly useful but underused application of repeat photography. Growth of C. fulgida was studied using measurement of individual current height and distance to the camera for the cacti (Tschirley and Wagle 1964). In the Baja California landscape, with more boulders and complex horizons, current height of the plants and apparent height of landscape features, as seen from the camera and measured at each plant, were used to study growth of F. columnaris (Escoto-Rodríguez and Bullock 2002, fig. 10.4). A simpler technique was used for the fan palms Washingtonia robusta (palma verde) and Brahea armata (palma ceniza) (Bullock and Heath 2006, fig. 10.5): assuming crown diameter remains nearly constant after aboveground trunk growth is initiated, height was scaled by individual crown diameter, thus correcting for large differences among individuals in distance to the camera. Field measurement of the crowns al-

lowed consideration of systematic biases in this method: crown diameter generally continued to increase in juveniles and decreased slowly in larger palms (with occasional traumatic, major reductions), while the rate of change (in relation to estimated height) showed a slow, long-term, linear decline.

Height growth rate of C. fulgida was reported to decline linearly with plant size, and a similar pattern appeared to hold in F. columnaris. However, direct measurement of many more small plants than were available in the repeat photographs showed that, in F. columnaris, main axis growth increased rapidly to a maximum at moderate height, then decreased slowly (Ramírez Apud López 2002). This latter pattern was previously shown for Carnegiea from field measurements at MacDougal Crater (Turner 1990) and also was found for Carnegiea at Tucson, with only differences in detail (Pierson and Turner 1998). The two fan palms showed a great dispersion among individuals of height growth rates, although the data seemed to suggest that extremely tall plants did not grow at either near-zero or very high rates (Bullock and Heath 2006).

Reconstruction of population-age structure is an extremely useful tool for interpreting secular changes in mortality and recruitment. Repeat photography, with appropriate field calibration, provides a key data source that should be more widely exploited to examine the geographic, secular, and individual variation of growth at the scale of decades. Such techniques extended the sample size and time period for measurement of F. columnaris growth, and radically changed the growth-based estimates of its longevity.

People in the Landscape

Domestic animals are usually considered the medium of the most widespread and long-term effects of people in the Sonoran Desert landscape other than climate change, although agricultural and urban transformation are more important in the

Figure 10.4. Near Cataviñá, Baja California, Mexico.
A. (1949). This view shows a stand of *Fouquieria columnaris* (cirio) and *Pachycereus pringlei* (cardón). (H. Aschmann, CICESE I0140).
B. (12 October 1998). Some of the individuals have died in the intervening 49 years between photographs, notably the large *F. columnaris* at the far left and far right. The apparent height of rocks and the horizon, as seen from the camera position and measured from the base of each plant, can be used to establish the height of the surviving plants visible in the old photograph. (S. H. Bullock, CICESE I1021).

Lower Colorado Valley, and more recently in Sonora. On the east side of the Sonoran Desert, cattle arrived in the mid-seventeenth century and were abundant before 1900 (Turner et al. 2003), whereas, in the Vizcaíno region, colonization was scant and mostly delayed to the mid- to late-nineteenth century. Only the flow of federal subsidies in the 1960s and 1970s encouraged widespread efforts in an inhospitable terrain. Generally higher rainfall, flatter land, and better water availability along longer courses probably account for much of the greater debate over the impacts of cattle in the Arizona/Sonora than the Vizcaíno region. These physical boundary conditions have certainly also affected the distribution and abundance of native and exotic grasses and herbs, and hence the spread of both cattle and fires, all of which affect the community dynamics of native woody plants. Balancing and intertwining these factors with climate and other variables for the Arizona/Sonora region (and further east) is a sustained academic industry (Humphrey 1987, Bahre 1991, Turner et al. 2003, among many others). Across the gulf, forage production has only sporadically been good (Gabb 1867), exceptionally productive sites are few (Paulín-Ramírez et al. 1981), and settlement is scant (Deasy and Gerhard 1944, Aschmann 1959). Some large woody plants have seen locally significant use in ranch structures, but a regional impact is

Figure 10.5. Misión Santa María, Baja California, Mexico.
A. (1949). This photograph, near Misión Santa María, which shows a wet hillside, has two palm species, *Washingtonia robusta* and *Brahea armata*. (H. Aschmann, CICESE I0139).
B. (30 October 1996). Palm height can be approximated, without field measurement of reference points or distance to the camera, as the ratio of apparent height to crown width. (R. H. Webb, CICESE I0483).

not apparent (Moran 1968; fig. 4 in Bullock et al. 2005).

In the Vizcaíno region, the late-nineteenth-century mining boom required large amounts of structural timber and fuel wood for the mines, steam-powered stamp mills, and smelters, particularly in the Calmallí district and at three other widely separated operations (Southworth 1899, Böse and Wittich 1912–1913, Chaput et al. 1992). The succulent giants *F. columnaris* and *P. pringlei* usually may have escaped the axe, but *Prosopis* woodlands lining some arroyos may have been prime targets in a landscape of scant wood (Engerrand and Paredes 1912–1913, Aschmann 1972). The impacts of these boom-and-bust mines were probably more lasting than those of a long-lasting, millless onyx mine, despite a populous settlement fueling its kitchens with locally abundant *Prosopis* (Pedro Maclish, Rancho San Pablo, San Agustín, Baja California, pers. comm., 2004). Many small settlements have waxed and waned or been extinguished in this region. Written history is largely lacking, and in some cases the extent or degree of development of settlements is only brought to light by repeat photography (fig. 10.6).

Another case of substantial impact is the exploitation of *Yucca schidigera* (palmilla) stems, which contribute to a variety of products ranging from digestive aids to soap (Castellón-Olivares et al. 2002).

Across numerous stands of hundreds of hectares, a large proportion of stems may be cut to ground level, cut leaves are left in situ, and small trucks wander extensively to load the heavy trunks. Unfortunately, these stands are not well represented in old photographs, and there has been little interest in documentation to improve management.

Effects of the Physical Environment: Climate and Substrate

Climate variations have been major elements in the explanation of population change in Sonoran Desert perennials. Prolonged or pronounced drought (Pierson and Turner 1998), exceptionally wet seasons or series of them, severe freezes (e.g., Bowers 1980–1981, Glinski and Brown 1982, Parker 1993), excessive soil moisture, and hurricanes (Clark and Ward 2000) have all had a role. Further methodological work is needed on translating the explanations into numerical and testable hypotheses, even if fuzzy, of increased mortality or recruitment. In some cases the conceptualization of stressors has been affected by repeat photography studies, as in the comparison of the effects of length and depth of drought (Turner 1990). Hurricane Nora's impact was assessed in two areas within or close to the probable cross-peninsula

Figure 10.6. Mármol Nuevo, Baja California, Mexico.
A. (1911). Mármol Nuevo was a sizeable settlement that essentially disappeared from local oral history, probably undermined by bad water and difficulties in transporting the onyx blocks. (From Engerrand and Paredes 1912–1913).
B. (14 October 2002). Over the course of the 91 years between photographs, the shrub cover has increased considerably. (S. H. Bullock, CICESE I1196).

track, showing about 4 percent and 0.8 percent mortality in *F. columnaris* and *P. pringlei* among plants greater than 1.5 meters tall (Bullock et al. 2005).

The historical lack of many climate stations is a substantial obstacle to more quantitative models. This may be remedied in some degree by proxy records and regional interpretations. For the northern Vizcaíno region, an estimation of rainfall over 147 years was derived from the recent relation to precipitation in San Diego, California (about 440 kilometers NNW; Bullock 2003). The risk of overinterpreting such a relation is reduced by focusing attention on particularly wet years or a series of these, rather than on the full range of precipitation, or on trying to estimate average conditions. Larger-scale, more-distant but regionally robust, and much longer term records might also be used (Webb and Betancourt 1992, Meko et al. 1995). Long-term records of summer rainfall that can be applied, interpolated, or interpreted locally remain as problematic as they are desirable, although much less so in Sonora and Arizona than on the peninsula.

The distribution and population dynamics of plants are clearly affected by geophysical characteristics, including soils. Interest in this subject regarding Sonoran Desert plants has focused on soil age and texture (McAuliffe 1994, Parker 1995; see previous

discussion regarding *C. gigantea* in MacDougal Crater). We have recently used repeat photography to apply multiple regression in analysis of both direct and proxy variables taking advantage of the long and varied observation intervals, and the wide dispersion of sites (Bullock et al. 2005). Our approach was to use survivorship and recruitment data, after first removing the effects of time as indicated earlier, in regression analysis or analysis of variance with geophysical variables. A similar approach was used by Parker (1993) for *C. gigantea* and *Stenocereus thurberi* in southern Arizona, but she incorporated estimated age-structure, which would not be reliable for *F. columnaris* at least (Bullock et al. 2004). Our analysis used several geophysical variables: elevation, distance to the Pacific coast and latitude (all proxies of climate), and soil-chemistry features, in particular the concentrations of N and P and ratios of Ca:Mg and Na:K (Graham and Franco-Vizcaíno 1992, Franco-Vizcaíno et al. 1993). We also used categorical variables, each with two to four classes: landform, topographic protection from high winds, soil texture, slope gradient, exposure, soil stability, alluvium age, and geology. In the cases of *F. columnaris* and *P. pringlei*, several features showed significant effects and $r^2 > 0.1$, thus making notable additions to the time factor. The variables of significance differed between species and between survivorship and recruitment, except that southerly exposures favored recruitment in both species. Recruitment of *F. columnaris* was affected by the soil's parent rock and *P. pringlei* survivorship by soil texture. The old versus new alluvium contrast failed, perhaps because the categories were too heterogeneous at a regional level, as did the nutrient factors, perhaps due to the small number of sites with such data.

Discussion and Conclusions

We conclude that repeat photography is a significant tool for weighing many of the hypotheses of change because it can readily cover substantial lapses of time and many sites with a range of conditions. Constancy in population size or structure is probably uncommon if not absent among desert perennials, although repeat photography has documented some remarkable cases of population stasis and individual longevity. The patterns and causes of flux are probably heterogeneous among species at any locality or among localities for any species (Watson et al. 1997, Petraitis and Latham 1999). The demonstrated potential of repeat photography for broad geographic coverage adds a perspective essential for use of the results in management. Evolving techniques of measurements and their combination with other methods of study hold promises of substantially more power. An increasing importance for repeat photography is assured by the demands for analyses of recent and current trends of change that are inexpensive and rapid as well as locally precise, historically deep, and geographically broad.

Literature Cited

Aguiar, M. R., and O. E. Sala. 1999. Patch structure, dynamics and implications for the functioning of arid ecosystems. *Trends in Ecology and Evolution* 14:273–277.

Aschmann, H. 1959. The Central Desert of Baja California: Demography and ecology. *Ibero-Americana* 42:1–315.

Aschmann, H. 1972. Recovery of desert vegetation. In *International geography, 1972*, Volume 1, ed. W. P. Adams and F. M. Helleiner, 631–633. Toronto: University of Toronto Press.

Bahre, C. J. 1991. *A legacy of change: Historic human impact on vegetation of the Arizona borderlands*. Tucson: University of Arizona Press.

Bestelmeyer, B. T., D. A. Trujillo, A. J. Tugel, and K. M. Havstad. 2006. A multi-scale classification of vegetation dynamics in arid lands: What is the right scale for models, monitoring, and restoration? *Journal of Arid Environments* 65:296–318.

Böse, E., and E. Wittich. 1912–1913. Informe relativo a la exploración de la región Norte de la costa occidental de la Baja California. *Parergones del Instituto Geológico de México* 4:307–533 (plates LXXXVIII–CXII).

Bowers, J. E. 1980–1981. Catastrophic freezes in the Sonoran Desert. *Desert Plants* 2:232–236.

Bowers, J. E., R. H. Webb, and R. J. Rondeau. 1995. Longevity, recruitment and mortality of desert plants in Grand Canyon, Arizona, U.S.A. *Journal of Vegetation Science* 6:551–564.

Brandegee, T. S. 1889. A collection of plants from Baja California, 1889. *Proceedings of the California Academy of Sciences, Series 2* 2:117–215.

Brandegee, T. S. 1893. Southern extension of California flora. *Zoe* 4:199–210.

Bullock, S. H. 2003. Seasonality, spatial coherence and history of precipitation in a desert region of the Baja California peninsula. *Journal of Arid Environments* 53:169–182.

Bullock, S. H., and D. Heath. 2006. Growth rates and age of native palms in the Baja California desert. *Journal of Arid Environments* 67:391–402.

Bullock, S. H., N. E. Martijena, R. H. Webb, and R. M. Turner. 2005. Twentieth century demographic changes in cirio and cardón in Baja California, México. *Journal of Biogeography* 32:127–143.

Bullock, S. H., R. M. Turner, J. R. Hastings, M. Escoto-Rodríguez, Z. Ramírez Apud López, and J. L. Rodríguez-Navarro. 2004. Variance of size-age curves: Bootstrapping with autocorrelation. *Ecology* 85:2114–2117.

Castellón-Olivares, J. J., A. Rublúo-Islas, J. Sepúlveda-Betancourt, and G. Ruiz-Campos. 2002. Environmental effects on biomass productivity of wild populations of *Yucca schidigera* in Baja California, Mexico. *Southwestern Naturalist* 47:576–584.

Chaput, D., W. H. Mason, and D. Z. Loperena. 1992. *Modest fortunes: Mining in northern Baja California.* Los Angeles, CA: Natural History Museum of Los Angeles County.

Clark, W. H., and D. M. Ward Jr. 2000. Hurricane impacts in the central desert of Baja California norte, Mexico. *Haseltonia* 7:81–85.

Comrie, A. C., and E. C. Glenn. 1998. Principal components-based regionalization of precipitation regimes across the southwest United States and northern Mexico, with application to monsoon precipitation variability. *Climate Research* 10:201–215.

Deasy, G. F., and P. Gerhard. 1944. Settlements in Baja California: 1768–1930. *Geographical Review* 34:574–586.

Engerrand, J., and T. Paredes. 1912–1913. Informe relativo a la parte occidental de la región Norte de la Baja California. *Parergones del Instituto Geológico de México* 4:278–306 (plates LXVI–LXXXVII).

Escoto-Rodríguez, M., and S. H. Bullock. 2002. Long-term growth rates of cirio (*Fouquieria columnaris*), a giant succulent of the Sonoran Desert in Baja California. *Journal of Arid Environments* 50:593–611.

Franco-Vizcaíno, E., R. C. Graham, and E. B. Alexander. 1993. Plant species diversity and chemical properties of soils in the central desert of Baja California, Mexico. *Soil Science* 155:406–416.

Gabb, W. M. 1867. Exploration of Lower California. In *Resources of the Pacific slope, appendix A: Sketch of the settlement and exploration of Lower California*, ed. J. R. Browne, 82–112. New York: D. Appleton and Co., 1869.

Glinski, R. L., and D. E. Brown. 1982. Mesquite (*Prosopis juliflora*) response to severe freezing in southeastern Arizona. *Journal of the Arizona–Nevada Academy of Science* 17:15–18.

Goldberg, D. E., and R. M. Turner. 1986. Vegetation change and plant demography in permanent plots in the Sonoran Desert. *Ecology* 67:695–712.

Graham, R. C., and E. Franco-Vizcaíno. 1992. Soils on igneous and metavolcanic rocks in the Sonoran Desert of Baja California. *Geoderma* 54:1–21.

Hastings, J. R., and R. M. Turner. 1965a. Seasonal precipitation regimes in Baja California, Mexico. *Geografiska Annaler* 47 (ser. A):204–223.

Hastings, J. R., and R. M. Turner. 1965b. *The changing mile.* Tucson: University of Arizona Press.

Humphrey, R. R. 1987. *90 years and 535 miles: Vegetation changes along the Mexican border.* Albuquerque: University of New Mexico Press.

Humphrey, R. R., and A. B. Humphrey. 1990. *Idria columnaris*: Age as determined by growth rate. *Desert Plants* 10:51–54.

INEGI. 1995. *Síntesis geográfica del estado de Baja California Sur.* Aguascalientes, México: Instituto Nacional de Estadística, Geografía e Informática, Aguascalientes.

INEGI. 2001. *Síntesis de información geográfica del estado de Baja California.* Aguascalientes, México: Instituto Nacional de Estadística, Geografía e Informática, Aguascalientes.

McAuliffe, J. R. 1994. Landscape evolution, soil formation, and ecological patterns and processes in Sonoran Desert bajadas. *Ecological Monographs* 64:111–148.

Meko, D., C. W. Stockton, and W. R. Boggess. 1995. The tree-ring record of severe sustained drought. *Water Resources Bulletin* 31:789–801.

Moran, R. 1968. Cardón. *Pacific Discovery* 21:2–9.

Nelson, E. W. 1922. Lower California and its natural resources. *Memoirs of the National Academy of Sciences* 16:1–194.

Niering, W., R. H. Whittaker, and C. H. Lowe. 1963. The saguaro: A population in relation to its environment. *Science* 142:15–23.

Parker, K. C. 1987. Site-related demographic patterns of organ pipe cactus populations in southern Arizona. *Bulletin of the Torrey Botanical Club* 114:149–155.

Parker, K. C. 1993. Climatic effects on regeneration trends for two columnar cacti in the northern Sonoran Desert. *Annals of the Association of American Geographers* 83:452–474.

Parker, K. C. 1995. Effects of complex geomorphic history on soil and vegetation patterns on arid alluvial fans. *Journal of Arid Environments* 30:19–39.

Paulín-Ramírez, O., A. Navarro-Córdova, V. F. Morales-Guiza, J. R. Rosiñol-Monges, A. Preciado-Torres, C. A. Aguirre-Wallace, G. Peralta-Castro, and J. A. Camacho-Camacho. 1981. *La determinación regional de los coeficientes de agostadero para el estado de Baja California.* Ensenada, Baja California: Secretaria de Agricultura y Recursos Hidráulicos.

Peinado, M., F. Alcaraz, J. L. Aguirre, J. Delgadillo, and I. Aguado. 1995. Shrubland formations and associations in Mediterranean–desert transitional zones of north-western Baja California. *Vegetatio* 117:165–179.

Petraitis, P. S., and R. E. Latham. 1999. The importance of scale in testing the origins of alternative community states. *Ecology* 80:429–442.

Pierson, E. A., and R. M. Turner. 1998. An 85-year study of saguaro (*Carnegiea gigantea*) demography. *Ecology* 79:2676–2693.

Pinder, J. E., III, J. G. Wiener, and M. H. Smith. 1978. The Weibull distribution: A new method of summarizing survivorship data. *Ecology* 59:175–179.

Ramírez Apud López, Z. 2002. *Modelos de crecimiento para la estimación de la edad con aplicación en el cirio* (Fouquieria columnaris). Master's thesis. Facultad de Ciencias, Universidad Autónoma de Baja California, Ensenada, Baja California.

Regan, H. M., M. Colyvan, and M. A. Burgman. 2002. A taxonomy and treatment of uncertainty for ecology and conservation biology. *Ecological Applications* 12:618–628.

Sheil, D., and R. M. May. 1996. Mortality and recruitment rate evaluations in heterogeneous tropical forests. *Journal of Ecology* 84:91–100.

Shreve, F. 1934. Vegetation of the northwestern coast of Mexico. *Bulletin of the Torrey Botanical Club* 61:373–380.

Shreve, F. 1936. The transition from desert to chaparral in Baja California. *Madroño* 3:257–264.

Shreve, F. 1942. The deserts of North America. *Botanical Review* 8:195–246.

Shreve, F. 1964. Vegetation of the Sonoran Desert. In *Vegetation and Flora of the Sonoran Desert*, Volume 1, ed. F. Shreve and I. Wiggins, 6–186. Stanford, CA: Stanford University Press.

Southworth, J. R. 1899. *El territorio de la Baja California México ilustrada: Su agricultura, comercio, minería é industrias en Inglés y Españo.* Repr. 1989, La Paz: Government of Baja California Sur.

Tschirley, F. H., and R. F. Wagle. 1964. Growth rate and population dynamics of jumping cholla (*Opuntia*

fulgida Engelm.). *Journal of the Arizona Academy of Science* 3:67–71.

Turner, R. M. 1990. Long-term vegetation change at a fully protected Sonoran Desert site. *Ecology* 71:464–477.

Turner, R. M. 2007. Desert on the march. *Journal of the Southwest* 49:141–163.

Turner, R. M., and D. E. Brown. 1982. Sonoran desert-scrub. *Desert Plants* 4:181–221.

Turner, R. M., R. H. Webb, J. E. Bowers, and J. R. Hastings. 2003. *The changing mile revisited: An ecological study of vegetation change with time in the lower mile of an arid and semiarid region.* Tucson: University of Arizona Press.

Vandermeer, J. H. 1980. Saguaros and nurse trees: A new hypothesis to account for population fluctuations. *Southwestern Naturalist* 25:357–360.

Watson, I. W., M. Westoby, and A. M. Holm. 1997. Continuous and episodic components of demographic change in arid zone shrubs: Models of two *Eremophila* species from Western Australia compared with published data on other species. *Journal of Ecology* 85:833–846.

Webb, R. H., and J. L. Betancourt. 1992. *Climatic variability and flood frequency of the Santa Cruz River, Pima County, Arizona.* US Geological Survey Water-Supply Paper 2379.

Webb, R. H., S. A. Leake, and R. M. Turner. 2007. *The ribbon of green: Change in riparian vegetation in the southwestern United States.* Tucson: University of Arizona Press.

Wiegand, T., S. J. Milton, and C. Wissel. 1995. A simulation model for a shrub ecosystem in the semi-arid karoo, southern Africa. *Ecology* 76:2205–2221.

Yang, T. W., and C. H. Lowe. 1956. Correlation of major vegetation climaxes with soil characteristics in the Sonoran Desert. *Science* 123:542.

Chapter 11

Repeat Photography, Climate Change, and the Long-Term Population Dynamics of Tree Aloes in Southern Africa

M. Timm Hoffman, Richard F. Rohde, John Duncan, and Prince Kaleme

Repeat photography is a useful tool for evaluating hypotheses for landscape changes caused by global warming. Climate change scenarios for southern Africa suggest that it is the western part that will experience the greatest change, with hotter and drier environments predicted over the next 100 years (Hewitson et al. 2005, MacKellar et al. 2007). Such changes are likely to have dire consequences for the region's biota (Midgley and Thuiller 2007). A detailed study of mortality levels within populations of *Aloe dichotoma* in Namibia and South Africa by Foden et al. (2007) suggests that the impact of climate change is already evident in the region. Similar interpretations for the decline in populations of *Aloe pillansii* in the Richtersveld have also been made (Rogers 2004). The developing range shifts in these two species are now considered "fingerprints" of anthropogenic climate change in both the scientific and the popular media (Wilson and Law 2007).

In this chapter, we show, like Bowers et al. (1995) and Bullock and Turner (chapter 10), how the use of repeat photography adds depth to our understanding of the population demographics of long-lived species such as *Aloe dichotoma* and *A. pillansii* and contributes to the debate on climate change. First, we derive annual mortality, recruitment, and population change rates for the two species from repeat

photographs. We synthesize the results from several sites and provide examples of both expanding and contracting populations. Next, we use repeat photography to determine growth rates and the potential longevity of the two species. Finally, we discuss our results in the light of previous interpretations of population changes in *A. dichotoma* and *A. pillansii* and in the context of the ongoing climate change debate for southern Africa.

Methods

The Tree Aloes: *Aloe dichotoma* and *A. pillansii*

While the two arborescent desert aloes of southern Africa, *Aloe dichotoma* and *A. pillansii*, are broadly similar in appearance, they have very different distributions (Reynolds 1950). *A. dichotoma* is widely distributed in the arid and semiarid winter and summer rainfall regions of the western part of South Africa and Namibia over an area of more than 500,000 square kilometers. Rainfall varies from about 100 to 350 millimeters per year across its range, and very large populations of many thousands of individuals can be seen at some localities. In

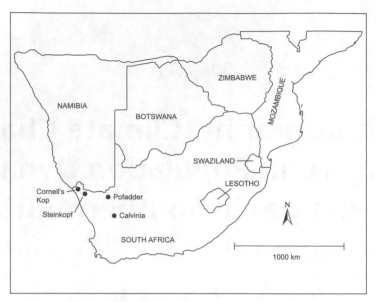

Figure 11.1. Location of main photograph sites shown in this analysis.

contrast, populations of *A. pillansii* are confined to a small, very arid (annual rainfall is generally below 100 millimeters) region of less than 200 square kilometers in the winter rainfall region of northwestern South Africa (the Richtersveld) and southwestern Namibia. There are probably fewer than 5,000 individuals of this species alive today (Bolus et al. 2004), and distinct populations may be composed of only a few individuals.

Mortality, Recruitment, and Population Change

Historical photographs of *Aloe dichotoma* and *A. pillansii* populations were obtained from a wide range of sources, including herbarium sheets, books, research publications, private slide collections, and the archives of the South African National Biodiversity Institute (SANBI). A total of 27 historical photographs of *A. dichotoma* populations from 11 different sites within four broad regions of their range in South Africa were rephotographed in this demographic analysis, and four photographs of *A. pillansii* were matched at a single site in the Richtersveld (Cornell's Kop) (fig. 11.1). The original and repeat photographs were scaled to the same size and

cropped to match each other as closely as possible. Then, by superimposing the images on top of each other and by reducing the opacity of one of the images, the number of aloes in the original image could be counted as well as the number that had died and the number that had recruited in the intervening years. Each pair of photographs was analyzed in this way, and the mortality, recruitment, and change in a population were calculated and expressed as an annual percentage as follows:

$$\text{Mortality} = (N_t - N_n/N_o)^{1/y} - 1, \qquad (1)$$

$$\text{Recruitment} = (N_n + N_o/N_o)^{1/y} - 1, \qquad (2)$$

$$\text{and Population change} = (N_t/N_o)^{1/y} - 1, \qquad (3)$$

where N_o is the number of individuals in the original photograph, N_t is the number of individuals in the repeat photograph, N_n is the number of new individuals in the repeat photograph, and y is the number of years between photographs.

Growth Rate and Longevity

To determine the growth rate and longevity of the two species, additional images of individual *Aloe di-*

chotoma and *A. pillansii* trees were rephotographed in the field. Because of the differences in architecture and growth habit between the two species, the analysis for *A. dichotoma* was more qualitative than that for *A. pillansii*. Information derived from interviews with local landowners about the growth rate and longevity of well-known, individual *A. dichotoma* trees on their farms was combined with observations from multiple matched photograph pairs to get an impression of the life history of this species from pre-reproductive to senescent individuals.

Since *Aloe pillansii* grows upright, it was possible, following Duncan et al. (2006), to develop a more quantitative method for calculating the annual growth rate and longevity of this species (see also Escoto-Rodríguez and Bullock 2002). To do this, 17 historical images of individual trees were rephotographed with a calibrated ranging rod in the repeat image so that the present-day height could be accurately determined. The two images were then rescaled so that the permanent features of the two photographs (usually a rock) were exactly the same size. By using a digital ruler, relative measurements of each tree in the original and repeat image could be made. Because the exact present-day height of the tree was known, the height of the tree in the original photograph could be determined using ratios and an annual growth rate calculated by dividing the difference in height by the number of years separating the two time periods. In addition to the assessment of growth rates in adult plants, the heights of 38 *A. pillansii* seedlings, less than 1 meter in height, were measured on Cornell's Kop from 2004 to 2007, and an annual growth rate for this size class was calculated from these measurements.

Results

Mortality, Recruitment, and Population Change

An analysis of 27 matched photograph pairs of the mortality, recruitment, and population change within *Aloe dichotoma* populations revealed distinct regional differences (table 11.1). Mortality was higher in the eastern and northern populations, while recruitment was higher in the south and to a lesser extent in populations from the west. The southern and western regions showed a positive change in their populations over time, while populations in the northern region shrunk by an average of 3.3 percent per year. Figure 11.2 shows a northern population near Pofadder, which has lost more than half of the individuals in view over just 41 years with no new individuals evident in the repeat photograph. However, a more detailed survey of the southern and northern slopes of the hill did record 30 individuals less than 1 meter in height that do not appear in the matched photograph. Southern populations, such as the one represented in figure 11.3 near Calvinia, have all expanded, and a number of young, pre-reproductive individuals are evident in the repeat photograph as well as in the field survey, which revealed 30 individuals less than 1 meter in height in a 25- by 100-meter transect directly away from the camera station.

The four views of different parts of the single *Aloe pillansii* population growing on Cornell's Kop all show a decline in the population over time (table 11.2). Fewer individuals growing on the cooler southern slopes have died, whereas the greatest proportion of individuals that have died was recorded in a photographic view of the hotter, northwest-facing slope. Figures 11.4 and 11.5 provide examples of what declining populations of *A. pillansii* look like in matched photograph pairs. Although no new individuals are evident in the repeat photographs, 38 seedlings, all less than 1 meter in height and not visible in the matched photographs, have been counted on Cornell's Kop.

Growth Rate and Longevity

A qualitative assessment of the growth rate and longevity of *Aloe dichotoma* using historical photographs and field observations is summarized in table 11.3. All evidence suggests that this species grows

Table 11.1. Mean (plus standard deviation) mortality, recruitment, and population change values derived from changes in the counts of *Aloe dichotoma* individuals in 27 repeat photographs at 11 sites clustered within four regions of South Africa

Region	No. Sites	Total No. Photos	No. of Years between Photos	No. of Aloes in Original Photos	No. of Aloes in Repeat Photos	No. of Aloes Not Surviving	No. of New Aloes in Repeat Photos	Mortality	Recruitment	Population Change
			$[y]$	$[N_o]$	$[N_t]$	$[N_d]$	$[N_n]$	%	%	%
Southern	4	6	81(16)	59(56)	142(131)	28(27)	111(106)	−0.8(0.7)	1.3(0.7)	1.1(0.6)
Western	2	5	40(7)	72(74)	85(96)	12(10)	25(33)	−0.6(0.1)	0.6(0.5)	0.1(0.7)
Eastern	3	9	66(17)	269(191)	139(96)	194(137)	64(44)	−2.1(0.5)	0.4(0.2)	−1.0(0.1)
Northern	2	7	48(9)	58(23)	19(23)	45(9)	6(8)	−3.8(1.4)	0.2(0.3)	−3.3(2.1)

A

B

Figure 11.2. Near Kabas, Poffadder, South Africa.
A. (5 February 1961). In this northwest-facing, quartzite hillslope near Kabas, Poffadder, 44 *Aloe dichotoma* trees can be counted in the original photograph. (J. P. H. Acocks, 6414, South African National Biodiversity Institute).
B. (29 October 2002). Twenty-three of the 44 individuals present in 1961 have survived on this hillslope. Since no new individuals are visible, the recruitment rate is 0.0 percent, and the annual mortality and population change rate are the same at −1.57 percent. (M. T. Hoffman, 149).

rapidly in the first part of its existence, reaching 2 meters or more in height with three or more main branches between 20 and 40 years of age (fig. 11.6). Although there is considerable variation in architecture, even within a single population, reproductive maturity seldom occurs before individuals have branched and extended beyond 2 meters in height. However, once individuals have branched dichotomously and trees have assumed a round-topped shape, historical photograph analysis suggests that continued extension of the end shoots is slowed (figs. 11.3 and 11.7). Trees retain more or less the same shape for the next 50 to 100 years, although

branches do extend in length, and leaf whorls become smaller and contain fewer leaves with age. By the time individuals reach an age of 150–200 years, only a few small leaves remain at the tips of branches. As the trees senesce, whole branches are shed (plate 11.1) and the trees eventually collapse leaving the light, porous stem standing or lying on the ground for a decade or longer.

Quantitative estimates from matched photographs of 17 separate adult *Aloe pillansii* trees revealed a mean annual growth rate of 0.0163 ± 0.0101 meter with a range from 0 to 0.0345 meter. No correlation of growth rate with original height, number

A

B

Figure 11.3. Near Calvinia, South Africa.

A. (1904). On this northwest-facing, relatively unstable, scree slope near Calvinia, 13 *Aloe dichotoma* trees can be counted in the original photograph. (R. Marloth, 1025, South African National Biodiversity Institute).

B. (15 October 2002). Six of the original trees have survived, giving an annual mortality rate of –0.63 percent, and 38 new individuals were recruited in the intervening years. The annual values for recruitment and population change are 1.4 percent and 1.28 percent, respectively. (R. F. Rohde, 137).

Table 11.2. Counts of *Aloe pillansii* individuals from four different repeat photograph views on Cornell's Kop with annual estimates of percent mortality, percent recruitment, and percent population change

Station No.	Aspect	No. of Years between Photos [y]	No. of Aloes in Original Photos [N_o]	No. of Aloes in Repeat Photos [N_r]	No. of Aloes Not Surviving [N_d]	No. of New Aloes in Repeat Photos [N_n]	Mortality %	Recruitment %	Population Change %
225	Northwest	57	12	4	8	0	–1.91	0	–1.91
265	Northeast	67	13	5	8	0	–1.42	0	–1.42
271	North	55	37	17	20	0	–1.40	0	–1.40
299d	Southeast	50	10	8	2	0	–0.45	0	–0.45

of branches, slope, or aspect was found. Two examples of individual trees used in the growth rate analysis are shown in plates 11.2 and 11.3. The growth rate of individuals less than 1 meter, however, is nearly 2.5 times faster. The mean annual growth rate for the 38 individuals less than 1 meter in height that were monitored on Cornell's Kop between 2004 and 2007

was 0.0415 ± 0.0253 meter (range of 0 to 0.09 meter). Because of the difference in growth rate between young plants and adult plants, it is difficult to determine an accurate age for individual plants. By combining the two measurements, however, we suggest that it takes between 50 and 75 years for an individual to reach between 2 and 3 meters in height, when

A

B

Figure 11.4. Cornell's Kop, South Africa.

A. (1937). In this northeast-facing population of *Aloe pillansii* on Cornell's Kop, 15 trees are visible. (G. W. Reynolds, South African National Biodiversity Institute).

B. (1 April 2004). Of the 15 individual trees in the original photograph, 6 survived over the 67 intervening years. No new individuals are visible in the repeat photograph. The annual mortality and population change rate are the same at −1.36 percent. (M. T. Hoffman, 266).

A

B

Figure 11.5. Cornell's Kop, South Africa.

A. (1946). In this northwest-facing population of *Aloe pillansii* on Cornell's Kop, 12 trees are visible. (G. W. Reynolds, South African National Biodiversity Institute).

B. (6 September 2003). Of the original 12 individual trees, 4 survived and no new individuals are visible in the repeat photograph. The annual mortality and population change rate are the same at −1.91 percent. (R. F. Rohde, 225).

branching usually starts. Thereafter, growth is slowed considerably and each additional meter takes about 60 years to grow. Using these calculations, the tallest plant on Cornell's Kop (8 meters) could, therefore, be close to 400 years of age.

Discussion and Conclusions

Repeat photography is a powerful tool for understanding the long-term population dynamics of individual species such as the two tree aloes investi-

Table 11.3. Three main age classes of *Aloe dichotoma* with associated subcategories and their age, height, branching, and reproduction characteristics

Phase and Description	Years	Description
Pre-reproductive		
Seedling	1–10	Less than 0.5 meter in height, no branches or flowers
Juvenile	10–20	0.5–1 meter in height, no branches or flowers
Preadult	20–40	1–2 meters in height with 1–3 branches but still no flowers
Adult		
Young adult	30–50	Usually greater than 2 meters in height and usually with more than 3 branches
		Reproductively mature individuals
Mature adult	50–150	Greater than 2 meters in height and usually with many branches and a large leaf whorl at the end of branches
		Usually produces an inflorescence per leaf whorl
Senescent		
Senescent adult	150–200	Smaller and fewer leaves per leaf whorl cluster
		No longer any branches and no green material
		Some flowers but not always

Figure 11.6. Aloe dichotoma near Westerberg, South Africa. (18 November 2002). Local landowner indicating the size (about 0.75 meter) of the now mature *Aloe dichotoma* individual growing behind him when he was a young boy in the early 1950s. (M. T. Hoffman, 3497).

gated in this study. Besides determining a minimum age for long-lived species directly from the repeat photo pairs themselves (e.g., Bowers et al. 1995, Escoto-Rodríguez and Bullock 2002, Turner et al. 2003, Webb et al. 2004), we have shown how the images can be analyzed to derive annual estimates of mortality, recruitment, and change. When enough images are combined across the geographic range of the species, a detailed assessment of the history, growth, and conservation status of the species and its populations can be made. This information is crucial for assigning a species to a particular rank within the International Union for the Conservation of Nature and Natural Resources (IUCN) Red List category (Golding 2002).

However, there are clear limitations to this approach, and detailed sampling of some populations added key insights to our understanding of population processes, particularly of recruitment, as many young plants are too small to be seen in the repeat photograph. Therefore, a combination of transect surveys together with the photographic analysis is essential. However, repeat photography is valuable for obtaining long-term population data without having to measure plots for many decades.

In addition to generating useful population-level statistics, repeat photography can also be used to

A

B

Figure 11.7. Near Coboop, Poffadder, South Africa.
A. (8 October 1948). This *Aloe dichotoma* individual, near Coboop, Poffadder, was a healthy adult plant with a rounded, dense canopy. (J. P. H. Acocks, 5399, South African National Biodiversity Institute).
B. (30 October 2002). After more than 50 years, the shape of the canopy has remained unchanged, although individual branches have lengthened and leaf clusters at the end of the branches have become much reduced. (M. T. Hoffman, 154).

derive growth rate and longevity estimates of long-lived desert plants (Escoto-Rodríguez and Bullock 2002). Previous ^{14}C dating of *Aloe dichotoma* (Vogel 1974) suggests that very large individuals could be as old as 145 years, and we suggest this could be adjusted upward. A more likely age for large, senescent individuals is in the region of 150 to 200 years and possibly even older.

For *A. pillansii*, the method for determining growth rate and longevity is significantly more robust. Escoto-Rodríguez and Bullock (2002) have shown for *Fouquieria columnaris* (cirio) that while height growth slows in older plants there is a shift to growth of secondary stems. Thus age estimates based on total stem growth might reflect more accurately the total growth of individuals over time.

Besides the development of novel techniques, what has this analysis revealed about the long-term population dynamics of the two tree aloes and particularly in the context of recent climate change de-

bates for the region? Similar to the study by Foden et al. (2007) our analysis shows that populations of *Aloe dichotoma* in the eastern and northern region of South Africa are declining, whereas those in the west, and particularly those in the south, are expanding. However, we suggest that populations in the east and north have been declining for at least the last 100 years and must have established themselves at such high numbers in these areas over hundreds if not thousands of years under climatic conditions different from those that exist today, particularly in terms of the amount and seasonality of precipitation.

Chase and Meadows (2007) have synthesized the paleoclimatic record and show that during the last glacial maximum, the winter rainfall region occurred far to the east of where it occurs today. Conditions at this time would have been very similar to those that prevail in the southern part of the species' current distribution range, where its populations are expanding. We suggest that the spread of this species across the eastern part of South Africa would have been favored under these conditions. The Little Ice Age, which occurred roughly from 1300 to 1800 in southern Africa (Tyson et al. 2000), would also have favored an expansion of populations in the northern and eastern regions. Furthermore, there is no evidence that rainfall has decreased in these regions in the last 100 years, and although mean annual temperatures have increased between 0.25 and 0.74° Celsius for the period from 1950 to 2000, there is little evidence of large-scale deaths of mature adult plants. Anecdotal evidence also suggests that the severe droughts of the 1930s and 1940s were far more devastating for eastern and northern populations than has been the case over the last 30 years.

Aloe pillansii has a far more restricted distribution than *A. dichotoma* and is confined to about 15 populations in South Africa (Bolus et al. 2004) and 18 in Namibia (Swart and Hoffman 2006). What is intriguing is that little or no recruitment has occurred in the drier southern Namibian populations for at least the last 50 years, while in South Africa, there is a "missing" 2- to 3-meter-tall cohort in many of the populations, including the one we have examined in some detail on Cornell's Kop (Bolus et al. 2004). Is this the result of recent anthropogenic climate change, or do these patterns reflect much longer population processes? We suggest that it is the latter. The absence of 2- to 3-meter-tall plants can be explained best by one of two factors. The first possibility is that this is the cohort most likely to have been targeted by unscrupulous plant collectors during the middle part of the twentieth century when there were no laws prohibiting collecting. This might be the case for Cornell's Kop, which is easily accessed and is also on a major tourist route to the Richtersveld. However, many of the populations, particularly in southern Namibia, are extremely remote, and this is unlikely to explain the absence of 2- to 3-meter-tall plants in this region.

The second and more plausible reason why this size class is missing is simply that conditions favorable for recruitment were not suitable during the middle part of the twentieth century. Recruitment

intervals of between 30 and 70 years appear to be the norm for this species, and since individuals can live for 400 years or more, it should be expected that gaps in the size class distribution would occur. Many young plants (less than 1 meter in height) occur in the Cornell's Kop population as well as in several other South African populations (Bolus et al. 2004). This suggests that recruitment over the last 20 to 30 years has indeed occurred. In our view, the death of old, senescent individuals is not a response to recent climate change impacts or increased drought stress, but simply the result of normal population-level processes that operate over hundreds if not thousands of years and are to be expected in such a long-lived desert species. Through the use of repeat photographs, we have been able to catch a glimpse of these long time frames and extend our understanding backward in time beyond that which is possible from field surveys alone.

Acknowledgments

This work was carried out under the project BIOTA Southern Africa, which is sponsored by the German Federal Ministry of Education and Research under promotion number 01 LC 0024A and the EU-funded project WADE (Contract No. 506680). The Mazda Wildlife Vehicle Fund kindly supplied a courtesy vehicle.

Literature Cited

Bolus, C., M. T. Hoffman, S. W. Todd, E. Powell, H. Hendricks, and B. Clark. 2004. The distribution and population structure of *Aloe pillansii* in South Africa in relation to climate and elevation. *Transactions of the Royal Society of South Africa* 59:133–140.

Bowers, J. E., R. H. Webb, and R. J. Rondeau. 1995. Longevity, recruitment and mortality of desert plants in Grand Canyon, Arizona, USA. *Journal of Vegetation Science* 6:551–564.

Chase, B. M., and M. E. Meadows. 2007. Late Quaternary

dynamics of southern Africa's winter-rainfall zone. *Earth Science Reviews* 84:103–138.

Duncan, J. A., M. T. Hoffman, R. F. Rohde, E. Powell, and H. H. Hendricks. 2006. Long-term population changes in the giant quiver tree, *Aloe pillansii* in the Richtersveld, South Africa. *Plant Ecology* 185:73–84.

Escoto-Rodríguez, M., and S. H. Bullock. 2002. Long-term growth rates of cirio (*Fouquieria columnaris*), a giant succulent of the Sonoran Desert in Baja California. *Journal of Arid Environments* 50:593–611.

Foden, W., G. F. Midgley, G. Hughes, W. J. Bond, W. Thuiller, M. T. Hoffman, P. Kaleme, L. Underhill, A. Rebelo, and L. Hannah. 2007. A changing climate is eroding the geographic range of the Namib Desert tree aloe through population declines and dispersal lags. *Diversity and Distributions* 13:645–653.

Golding, J., ed. 2002. Southern African plant red data lists. *Southern African Botanical Diversity Network Report* 14:1–237.

Hewitson, B., M. Tadross, and C. Jack. 2005. Scenarios from the University of Cape Town. In *Climate change and water resources in southern Africa: Studies on scenarios, impacts, vulnerabilities and adaptation*, ed. R. E. Schulze, 39–56. WRC Report 1430/1/05. Pretoria: Water Research Commission.

MacKellar, N. C., B. C. Hewitson, and M. A. Tadross. 2007. Namaqualand's climate: Recent historical changes and future scenarios. *Journal of Arid Environments* 70:604–614.

Midgley, G. F., and W. Thuiller. 2007. Potential vulnerability of Namaqualand plant diversity to anthropogenic climate change. *Journal of Arid Environments* 70:615–628.

Reynolds, G. W. 1950. *The aloes of South Africa*. Cape Town, South Africa: Balkema.

Rogers, D. 2004. *Pillansii* in peril. *Africa Geographic* 12(5):24.

Swart, E., and M. T. Hoffman. 2006. *The population status of Aloe pillansii L. Guthrie in southern Namibia*. Unpublished report submitted to the South African National Biodiversity Institute, Pretoria.

Turner, R. M., R. H. Webb, J. E. Bowers, and J. R. Hastings. 2003. *The changing mile revisited: An ecological study of vegetation change with time in the lower mile of an arid and semiarid region*. Tucson: University of Arizona Press.

Tyson, P. D., W. Karlen, K. Holmgren, and G. A. Heiss. 2000. The Little Ice Age and medieval warming in South Africa. *South African Journal of Science* 96:121–126.

Vogel, J. C. 1974. The life span of the Kokerboom. *Aloe* 12:66–68.

Webb, R. H., J. Belknap, and J. S. Weisheit. 2004. *Cataract Canyon: A human and environmental history of the rivers in the Canyonlands*. Salt Lake City: University of Utah Press.

Wilson, J., and S. Law. 2007. *A brief guide to global warming*. London: Robinson.

PART IV

Applications in Ecosystem Change

By far, the most sustained research effort using the technique of repeat photography has been in the documentation of ecosystem changes. Documentation of landscape changes using repeat photography began around the turn of the twentieth century and has continued unabated to the present (Gruell, foreword; Webb et al., chapter 1). Veblen (chapter 13), Nyssen et al. (chapter 14), and Lewis (chapter 15) show the breadth of this application on three continents in three types of ecosystems, ranging from forested ecosystems in South America to riparian ecosystems in Australia. Nyssen et al. (chapter 14), in particular, combine repeat photography with other metrics of land-use impacts to provide a semiquantitative look at landscape change in Ethiopia.

McClaran et al. (chapter 12) compare the results of permanent ecological study plots with what appears in repeat photography, aerial photography, and satellite imagery. They show that repeat photography documents certain ecosystem elements also seen in permanent plots, only over a wider area. Although repeat photography, aerial photography, and satellite imagery have different scales at different resolutions, the longer time period offered by repeat photography may be as important as the quantitative potential for vertical imagery in ecosystem-change applications.

The causes of ecosystem change can be elusive to determine, particularly where a pervasive land-use practice, such as grazing, occurs simultaneously with climatic change (Lewis, chapter 15). McClaran et al. (chapter 12) show that the two effects can be separable in controlled landscape experiments, yet unforeseen and unexplainable changes still occur. Where land-use practices overwhelm climate effects, such as in the case of elephants confined to a relatively small place in Kenya (Western, chapter 16), analyses using geographic information system (GIS) technology can shed light on the causes of change and its future trajectory. Wildland fire, particularly when repeated and especially where it was historically rare, induces ecosystem recovery that may be extremely slow or even result in directional ecosystem change (Turner et al., chapter 17). This type of ecosystem response is also known in forest ecosystems (Veblen, chapter 13), albeit with a faster recovery time, and Lewis (chapter 15) discusses the ecological implications of fire suppression on woody plant growth in the outback of northern Australia. These chapters show the importance of fire, in terms of ecosystem alteration and maintenance, in a way that is mostly conjectural without photographic evidence.

Long-term ecosystem changes documented with repeat photography reflect the combined effects of a number of causes, but with the potential for directional climate change in the future, this technique is perhaps the best available to determine baseline information on the stability of vegetation assemblages if species-specific changes are desirable. Bullock and Turner (chapter 10) show that semiquantitative analyses can document directional changes that cannot be determined from remote sensing and can only be quantitatively documented using permanent plots of limited spatial extent (McClaran et al., chapter 12). Particularly for continents with limited permanent plot data, such as Africa and South America, repeat photography likely will be the method of choice for documenting long-term ecosystem change in the future.

Temporal Dynamics and Spatial Variability in Desert Grassland Vegetation

Mitchel P. McClaran, Dawn M. Browning, and Cho-ying Huang

The conversion of open grasslands to shrub-savannas since the late nineteenth century is a worldwide phenomenon (van Auken 2000, Wessman et al. 2004, Sankaran et al. 2005), and some of the earliest uses of repeat photography illustrate this pattern in the southwestern United States (Wooten 1916, Parker and Martin 1952, Hastings and Turner 1965). Most interpretations of this change focus on reduced fire occurrence, declines in grass and increased dispersal of shrub seeds by domestic livestock, and changing weather patterns (Scholes and Archer 1997, McClaran 2003, Gibbens et al. 2005). The ability of repeat photography to evaluate these interpretations hinges on the frequency and spatial distribution and resolution of the repeat photography as well as the richness of ancillary information about weather, soils, and land use.

Frequency, or the time between repeat photographs, influences the ability to detect the timing of plant establishment and death. Multidecade and century-long frequencies represent the presence of and size changes in long-lived plants like trees and shrubs, but are unlikely to document the timing of plant establishment or death. Seasonal and annual-scale frequencies are most likely to document establishment and death, but maintaining this schedule is difficult. Intermediate-frequency repeat photogra-

phy, on the order of 5 to 10 years, is more likely to capture these establishment and mortality events.

The spatial scale and distribution of repeat photography can describe interspecific differences and spatial anomalies in the vegetation changes. While repeat photography provides undeniable evidence of a change from grassland to shrub-savanna, there are questions about how well a few ground-level (about 1.5 meters above and parallel to ground surface) photographs represent the changes across the larger landscape. Ground-level photography is limited by the narrow field of view, the two-dimensional representation compresses the distance to far objects, and foreground objects can obscure distant objects. One solution is a widely distributed network of ground-level locations. Another solution is images from elevated positions such as oblique-ground locations on hilltops, airborne vehicles, and satellites, but species identification becomes less certain with increasing elevation. The optimum solution may be the combination of images from ground-level and elevated positions.

We used evidence from the Santa Rita Experimental Range (SRER) in Arizona to describe and interpret the temporal dynamics and spatial variability of vegetation change in the desert grassland. We used a widely distributed network of frequently repeated ground-level photography from locations that span different soils and geomorphic settings, repeat oblique-ground photographs from a hill summit, and repeat photography from aerial and orbiting satellite locations. In addition, we employed a long precipitation record and repeated vegetation measurements from nearby permanent transects. We focused on the large long-lived shrub *Prosopis velutina* (mesquite); the small short-lived shrub *Isocoma tenuisecta* (burroweed); *Opuntia engelmanni* and *O. phaeacantha* (prickly pear), *Cylindropuntia fulgida* and *C. spinosior* (cholla) cacti; and perennial grasses. We expected that species-specific precision will deteriorate as the spatial resolution decreases from ground to aerial and satellite images, that episodic and high-frequency vegetation dynamics will be missed with long periods between repeat photographs, and that combining information from all these sources provides the best description and in-

terpretation of species-specific and landscape-scale vegetation change.

Santa Rita Experimental Range

The 21,000-hectare SRER stretches across the western alluvial skirt of the Santa Rita Mountains, about 50 kilometers south of Tucson (31°50′ N, 110°53′ W). Elevation increases from about 900 to 1,450 meters, with a corresponding increase in annual precipitation from 275 to 450 millimeters (fig. 12.1). Between 1,100- and 1,200-meter elevation, the mean annual (1922–2006), summer (June–September), and winter (October–May) precipitation is 371, 212, and 159 millimeters, respectively (fig. 12.2). Winter snow is rare. Interannual coefficient of variation (CV) is greater in winter (CV = 43.3 percent) than in summer (CV = 31.6 percent), and their asynchrony

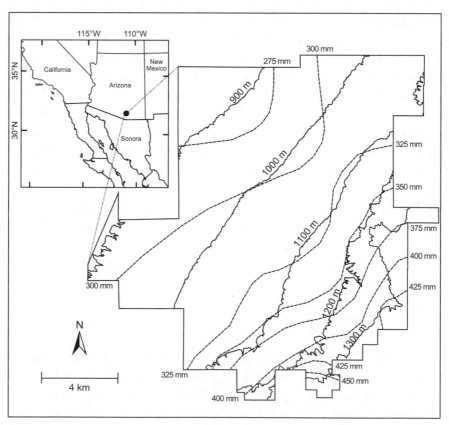

Figure 12.1. Elevation and annual precipitation gradients on the Santa Rita Experimental Range (from McClaran 2003).

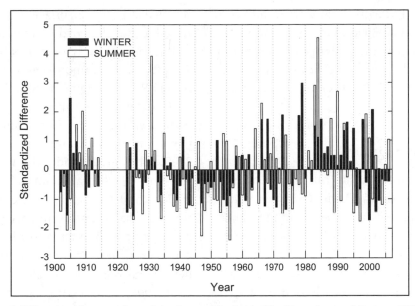

Figure 12.2. Standardized difference of seasonal precipitation on the Santa Rita Experimental Range, 1901–02 to 2005–06. Values from 1902–03 to 1913–14 are for McCleary Ranch, 1,200-meter elevation (Thornber 1910, Wooten 1916), values since 1922 are the average of four rain gauges: Box, Eriopoda, Road, and Rodent between 1,100- and 1,200-meter elevation (McClaran et al. 2002). Precipitation year is October to September, summer months are June through September. Standardized difference is the yearly value minus the long-term average, and divided by the standard deviation. Mean and standard deviation for McCleary Ranch were calculated separately from other rain gauges.

moderates the variation in annual precipitation (CV = 24.2 percent). Noteworthy features are very wet summers in 1931 and 1984; a prolonged dry period from 1932 to the late 1950s; wet conditions in the mid-1980s; and since 1994–1995, high interannual variability in winter precipitation (CV = 62.4 percent).

The approximately 500 plant species (Medina 2003) include short trees, shrubs, succulents, and herbaceous grass and dicot life forms. The physiognomy ranges from desert scrub at the lowest elevations to oak woodlands at the highest elevations. The most extensive vegetation is a mesquite–grass savanna, but desert grassland has become a popular name (McClaran 1995, 2003).

Established in 1902, the SRER is the oldest continuously operating rangeland research facility (McClaran et al. 2002) and among the five oldest biological field stations in the United States. Originally administered by the US Department of Agriculture, the SRER was transferred to the Arizona State Land

Department in 1988 and is administered by the University of Arizona, College of Agriculture and Life Sciences (Medina 1996). Hundreds of experiments and manipulations have been performed to evaluate cattle grazing practices, rodent influences, methods of vegetation control, seeding of plants, and general ecological studies (Medina 1996, McClaran et al. 2003).

Between 1880 and 1903, overgrazing of vegetation was common because unlimited open access prevented the control of cattle numbers, and recurring droughts limited forage production (Griffiths 1904, Bahre and Shelton 1996). From 1903 to 1916, cattle were excluded from all areas below 1,200 meters to allow recovery from overgrazing and to estimate a cattle carrying capacity (Wooten 1916). From 1916 to 1941, the stocking rate was about 0.12 animals per hectare per year, but since the 1950s it has been 0.02 to 0.06 animals per hectare per year (increasing with elevation), which translated to 40 to 60 percent utilization of grass production (Martin and

Cable 1974, Cable and Martin 1975, Martin and Severson 1988, Mashiri et al. 2008). Fires were probably common (10- to 20-year intervals) prior to the intensification of cattle grazing (Humphrey 1958, Bahre 1991), but since 1902, fires have been rare (Huang et al. 2007).

Repeat Ground Photography

The repeat ground photography collection contains images from 120 locations. David Griffiths made original images at 17 locations between 1902 and 1905, but the majority of original images were made between 1930 and 1948. Typically, there are 4 to 6 retakes of the original image, but more than 20 locations have at least 10 retakes. Nearly half the locations have images facing at least two separate directions. Several locations are on elevated hillslopes that provide an oblique perspective. Comprehensive repeat photography of the collection occurred in the 1940s and again in the 1980s (the latter by Robert Buttery), and Mitchel McClaran performed two comprehensive retakes in 1999–2000 and 2007.

The collection is available on the SRER Web site (ag.arizona.edu/srer), where locations are shown in map and table formats, Universal Transverse Mercator (UTM) coordinates are provided, and image captions and site descriptions can be queried.

Repeat Aerial Photography

We used repeat aerial photography representing a 150- by 150-meter area centered on repeat ground-level photography location 111 (31°48′38″ N, 110°53′07″ W) from four dates: 21 February 1936 (scale 1:31,000, black/white), 9 August 1966 (1:15,640, black/white), 12 October 1977 (1:24,000, natural color), and 2 June 1996 (1:40,000, color-infrared). The June 1996 photos were acquired as digital orthophoto quarter quadrangles (DOQQs) from the US Geological Survey. Film negatives for 1936, 1966, and 1977 images were acquired from the US National Archives and the USDA Forest Service Aerial Photography Field Office. Negatives were

scanned at 1,200 dpi, images were geometrically corrected to the 1996 orthorectified base image with a root mean-square error of less than 1 pixel, and resampled to a common spatial resolution (1 meter) using ERDAS Imagine, Version 8.7 (ERDAS 2002).

We delineated shrub patches in the 1936 and 1966 images with a supervised classification applied to the panchromatic band and a texture layer (Asner et al. 2003). Blue, green, and red image bands and green, red, and near-infrared image bands were used to classify the 1977 and 1996 images, respectively. Texture served as a contrast index within a focal window passed across the image to distinguish dark shrub canopies from the surrounding brighter soil or herbaceous vegetation (Nellis and Briggs 1989, Anys et al. 1994). The texture calculation used the variance within a 3- by 3-pixel (0.6-meter pixel size, 3.25 square meters) window prior to resampling 1936 and 1966 images. Texture bands were not used to classify the 1977 and 1996 multiband images because they overestimated shrub canopies.

Classification accuracies were assessed using the Cohen's Kappa ($K^\wedge$) statistic for a random sample of points stratified by image class (Cohen 1960, Congalton and Green 1999). Random points ($n = 150$) on photographic images were assigned manually to the reference classes "shrub" or "nonshrub" for 1936 and 1966 images, and "grass," "shrub," or "bare soil" for the 1977 and 1996 images, and classification accuracy was evaluated across the larger image. Only 90 random points were used for the 1977 image. The automated classifications produced good estimates of shrub canopy cover with $K^\wedge$ values of 0.75, 0.68, 1.00, and 1.00 for 1936, 1966, 1977, and 1996 classified images, respectively; and producer's accuracy values (probability of correctly classifying shrub canopies) were 87.2, 84.2, 74.1, and 94.0 for 1936, 1966, 1977, and 1996 classified images, respectively.

Repeat Satellite Multispectral Data

We used repeat satellite multispectral data representing a 150- by 150-meter area centered on repeat ground photography location 111 from three dates:

12 June 1990 (Landsat Thematic Mapper[TM]), 11 May 1996 (Landsat TM), and 7 May 2003 (Landsat Enhanced Thematic Mapper plus [ETM+]). TM and ETM+ are multispectral spaceborne sensors with six bands covering the visible, near-infrared, and short-wave infrared spectral regions with a nominal spatial resolution of 30 meters, plus one thermal 60-meter band.

We used a probabilistic mixture model (Asner and Lobell 2000) to estimate green vegetation cover based on the spectral signatures of the shrub and herbaceous species, litter, and bare soil (Huang et al. 2007). Shrub and cactus cover was represented in our TM and ETM+ images because they were acquired in the dry summer season before the arrival of the summer monsoon rains when the herbaceous vegetation is dormant, and therefore only shrubs and cacti were contributing to "green" vegetation (McClaran 1995). In the wet summer season (July through September), all the vegetation was green.

Permanent Transects

Measures of plant cover and density are available from 130 permanent transect locations that are separate from the repeat ground-level photography locations. Transects were established between 1956 and 1972 during four separate research campaigns (Martin and Ward 1970, Martin and Cable 1974, Cable and Martin 1975, Martin and Severson 1988). Transects are 30.4 meters long and 0.31 meter wide, and line intercept measured on the long axis represents grass basal cover and shrub canopy cover. Counts of plants rooted in the 9.28-square-meter area represent plant density. All previous measurements and UTM coordinates are available on the SRER Web site (ag.arizona.edu/srer).

We report changes in cover from 1960 to 2006 and density from 1972 to 2006 for *P. velutina*, *I. tenuisecta*, cacti, and perennial grasses on 60 transects between 900- and 1,200-meter elevation, on sandy loam upland and sandy loam deep ecological sites (Breckenfeld and Robinett 2003). None have burned since 1910 or earlier, no vegetation removal treatments

were performed, and average cattle utilization of grass forage was greater than 60 percent from 1916 to 1972 and 50–60 percent since 1972 (Martin and Ward 1970, Martin and Cable 1974, Cable and Martin 1975, Martin and Severson 1988, Ruyle 2003).

Temporal Dynamics and Spatial Anomalies

Prosopis velutina

Prosopis velutina is a long-lived (greater than 200 years), leguminous shrub that can grow more than 5 meters tall. Roots are both shallow and deep (0.25 to more than 3.0 meters), and some shallow roots may extend far (15 meters) from the trunk (Cable 1977). Growth begins in April after a winter deciduous period. Seeds can remain viable for 20 years (Martin 1970).

The increase of *P. velutina* since 1902 is the most conspicuous and long-lasting vegetation change above 1,000-meter elevation in desert grassland. According to early-twentieth-century reports and photographs, *P. velutina* occurred only as widely scattered plants in the grassland or as dense stands in the large washes and arroyos (figs. 12.3 and 12.4; Griffiths 1904, 1910, Thornber 1910, Wooten 1916). It was much more common in the lower-elevation desert scrub with other large shrubs such as *Acacia greggii* (catclaw acacia), *Parkinsonia florida* (blue palo verde), and *Larrea tridentata* (creosotebush).

The conversion from open grassland to *P. velutina*-grass savanna may have occurred in less than 50 years (figs. 12.3 and 12.4; Parker and Martin 1952) when there was a 33 percent increase in the area supporting dense (greater than 200 plants per hectare) stands of *P. velutina* (Humphrey and Mehrhoff 1958). The establishment of seedlings in the grassland by 1902 hastened the conversion (Griffiths 1910, Wooten 1916). Since 1972, the density has remained more than 200 plants per hectare, but the increase appears to have stabilized around 20 percent canopy cover (fig. 12.5; Browning et al. 2008).

Figure 12.3. Rodent Station livestock exclosure, Santa Rita Experimental Range, southern Arizona, USA.

A. (December 1922). Ground repeat photography location 111 on south boundary of Rodent Station livestock exclosure (1,086-meter elevation). Area right of the fence has not been grazed by livestock since 1903. In this view, there is one large *Celtis pallida* (desert hackberry), scattered *Prosopis velutina* (mesquite), and relatively continuous herbaceous cover that was probably *Bouteloua rothrockii* (Rothrock grama) or annual grama species (*B. aristidoides* or *B. barbata*). (Photographer unknown, 111.1.1922.12).

B. (September 1935). *Isocoma tenuisecta* (burroweed) and *Cylindropuntia fulgida* (chainfruit cholla) have become established. There is a general decline of grass. (W. Cribbs, 111.1.1935.09).

C. (October 1947). The *I. tenuisecta* and *C. fulgida* have begun to decline, while *P. velutina* is becoming established on both sides of the exclosure fence. (S. C. Martin, 111.1.1947.10).

D. (August 1958). The *I. tenuisecta* and *C. fulgida* died, and the left half of the small *P. velutina* (foreground) died during the drought of the mid-1950s. (Photographer unknown, 111.1.1958.08).

E. (March 1969). The third eruption of *I. tenuisecta* is fully developed. (Photographer unknown, 111.1.1969.03).

F. (July 1970). There was a rapid decline of *I. tenuisecta* in the one year since 1969. (Photographer unknown, 111.1.1970.07).

G. (March 1990). The nonnative grass *Eragrostis lehmanniana* (Lehmann lovegrass) is very abundant, and beneath that grass canopy are the *I. tenuisecta* established in the late 1970s. (M. P. McClaran, 111.1.1990.03).

H. (March 2007). There was a decline of *I. tenuisecta* and the nonnative *Eragrostis lehmanniana* (Lehmann lovegrass) since 1990. (M. P. McClaran, 111.1.2007.03).

Figure 12.4. Box Canyon drainage, Santa Rita Experimental Range, southern Arizona, USA.
A. (1902). Ground repeat photography location 222 looking east into Box Canyon drainage, at 1,191-meter elevation. In 1902, *Prosopis velutina* (mesquite) was abundant in the drainage and a few plants scattered in open grasslands above the drainage. (D. Griffiths, 222.1.1902).
B. (September 1951). An increase of *P. velutina* in open grasslands above drainage by September 1951 is visible as dark objects above drainage escarpment. (J. Bohning, 222.1.1951.09).
C. (April 2007). There was little change in *P. velutina* abundance since 1951, except the death of a large *P. velutina* at bottom center that was a large plant in 1902. (M. P. McClaran, 222.1.2007.03).

Greater seed dispersal and cessation of fire are the primary reasons for the *P. velutina* increase, and the introduction of large numbers of cattle contributed to both (Griffiths 1910, Wooten 1916). Cattle will consume large amounts of *P. velutina* seed that can remain viable after digestion (Glendening and Paulsen 1955, Cox et al. 1993). The native kangaroo rats (*Dipodomys* spp.) bury seeds at optimum depth for germination (Cox et al. 1993), but their dispersal is less than 40 meters (Reynolds 1954). *P. velutina* seedlings were spared fatality from recurring fires because cattle consumed the grass fuel (Humphrey 1958, Bahre 1991). Fire is more likely to kill *P. ve-lutina* plants less than 1 centimeter basal diameter (Glendening and Paulsen 1955; Cable 1961, 1965), but larger plants are very likely to survive fires and resprout from surviving basal meristems.

Once established in former grasslands, *P. velutina* has not been reduced by rodent or cattle exclusion (fig. 12.3; Brown 1950, Glendening 1952, Havstad et al. 1999). Reintroducing fire has not eliminated these established stands (Cable 1967), even if four fires occur within 17 years (Geiger and McPherson 2005). Full recovery of *P. velutina* canopy cover occurs within 40 years after total plant removal (McClaran and Angell 2006), probably because seed

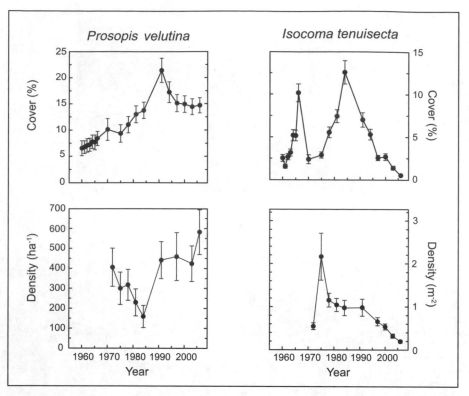

Figure 12.5. Prosopis velutina (mesquite) and *Isocoma tenuisecta* (burroweed) cover and density on 60 permanent transects, between 950- and 1,250-meter elevation, 1960–2007. Vertical lines represent one ± standard error of the mean.

sources and vectors are common in surrounding areas. A slowing rate of increase and possible stabilization of *P. velutina* suggests a carrying capacity near 25 percent canopy cover (Glendening 1952, Browning et al. 2008) because of inadequate water and nutrient resources (House et al. 2003, Sankaran et al. 2005).

The spatial anomalies of this *P. velutina* increase appear at both ends of the abundance spectrum. In large washes and arroyos, it was initially very abundant and remains so today (fig. 12.4), probably because of deep soils and favorable moisture conditions without frequent fires. Its abundance remains low on older geomorphic surfaces (more than 200,000 years BP) where soils have high clay and cobble content (fig. 12.6), probably because root penetration is limited by the fine soil texture (Wilson 1961, McAuliffe 1995) and because soil moisture is less available than on the sandy loam soils (Fravolini et al. 2005). Opportunities for seed dispersal and

the absence of fire are not different between these areas where *P. velutina* has increased or remained stable.

The general pattern of *P. velutina* increase on the SRER is common across the greater southwestern United States. It is widely documented in Arizona with repeat photography by Turner et al. (2003). A similar species (*P. glandulosa*) has increased in New Mexico and northern Mexico (Gibbens et al. 2005), but the associated development of shrub-centered sand dunes has not occurred on the SRER.

Isocoma tenuisecta

Isocoma tenuisecta is a short-lived (less than 40 years) shrub that can grow up to 1.3 meters tall (Humphrey 1937), and roots are common at depths greater than 1 meter but are scarce at 5- to 30-centimeter depths (Cable 1969). Growth is greatest in spring (Cable 1969), and seeds germinate in winter and spring

Figure 12.6. Madera Canyon alluvial fan, Santa Rita Experimental Range, southern Arizona, USA.
A. (June 1936). Ground repeat photography location 45 looking east across Madera Canyon alluvial fan, at 1,056-meter elevation. In 1936, grass cover was very sparse, and there were scattered *Fouquieria splendens* (ocotillo) and three *Prosopis velutina*
(mesquite) plants in the distance. (W. Cribbs, 45.2.1936.06).
B. (October 1984). Nonnative *Eragrostis lehmanniana* (Lehmann lovegrass) became established, and the abundance reflects wet
summer conditions in 1984. (R. F. Buttery, 45.2.1984.10).
C. (March 2007). The *E. lehmanniana* has declined. The same *Acacia greggii* (catclaw acacia) present at the bottom left in 1936
remained quite small in 2007, and *Calliandra eriophylla* (false mesquite) is a common low-growing shrub. Unlike most areas, *P.
velutina* has not increased, and there have not been eruptions of *Isocoma tenuisecta* (burroweed) or cacti; the anomaly may be
related to the higher concentration of clay in the soil at this location. (M. P. McClaran, 45.2.2007.03).

(Humphrey 1937). It is toxic and not commonly
eaten by livestock (Tschirley and Martin 1961).

Since 1902, there have been four cycles of *I. tenuisecta* increase and decline, each lasting about 15 to
20 years. The first cycle ended in 1914 (Wooten
1916), and the second cycle began in the 1930s and
significant declines occurred by the early 1950s (figs.
12.3 and 12.7; Canfield 1948, Humphrey 1937). The
third cycle occurred across much of the SRER below
1,200-meter elevation (figs. 12.3 and 12.5; Martin
and Cable 1974): cover increased from 2.4 percent to

more than 10 percent between 1957 and 1966 and
declined to less than 3 percent by 1970. The fourth
cycle began in the mid-1970s and ended in the early
1990s (figs. 12.3 and 12.5, Mashiri et al. 2008). The
cycles start with increased recruitment, followed by
self-thinning and increase in total cover, and then
both density and cover decline (fig. 12.5). The decline may be a function of its short longevity and occurrence of dry periods.

Cable (1967) found a strong correlation between
I. tenuisecta density and winter precipitation, and

Figure 12.7. Northern portion of Santa Rita Mountains Santa Rita Experimental Range, southern Arizona, USA.
A. (April 1905). Ground repeat photography location 231 looking east at the northern part of the Santa Rita Mountains, at 1,070-meter elevation. In 1905, abundant grass and dark patches of *Eschscholzia* spp. (poppies) are clearly visible on the near side of the fence, but intense grazing by cattle on the far side of the fence occurred because of a boundary dispute. (Photographer unknown, 231.2.1905.04).
B. (September 1941). Abundant perennial grass after the summer rains and members of the 1930s *Cylindropuntia fulgida* (cholla) eruption were evident. (Photographer unknown, 231.2.1941.09).
C. (June 1948). *Isocoma tenuisecta* (burroweed) plants were abundant, the cacti were dead or dying, and the decline of grass abundance was probably related to three consecutive dry summers. (M. Culley, 231.2.1948.06).
D. (April 2007). Numerous small *Prosopis velutina* (mesquite) and *Parkinsonia florida* (blue palo verde) became established and *Opuntia engelmanni* and *O. phaeacantha* (prickly pear cacti) were common. Unlike most areas, the nonnative *Eragrostis lehmanniana* (Lehmann lovegrass) has not established in this area; the anomaly may be related to the very deep, sandy, loam soil at this location. (M. P. McClaran, 231.2.2007.03).

there are compelling coincidences between wet winters (fig. 12.2) and the initiation of *I. tenuisecta* cycles. Three consecutive wet winters from 1929–30 to 1931–32 coincide with the beginning of the second cycle. The wet winter of 1957–58 may have initiated seedling establishment for the third cycle, and the very wet 1965–66 and 1967–68 winters may have contributed to an increase in plant size. Similarly, the wet 1972–73 winter and consecutive wet winters in 1977–78 and 1978–79 may have stimulated the establishment and subsequent growth during the fourth cycle.

Neighboring grasses and *P. velutina* have not influenced these cycles, possibly because the summer-growing grasses do not use much of the winter moisture (Cable 1969) and *P. velutina* cover is not great enough to limit *I. tenuisecta* recruitment (McClaran and Angell 2006). *Isocoma tenuisecta* abundance is relatively indifferent to manipulations of cattle grazing, rodents, and fire, and the persistence of any response will only last until a cycle ends or the next cycle starts (fig. 12.3; Brown 1950, Cable 1967, Martin 1983, Mashiri et al. 2008).

The spatial anomaly of these *I. tenuisecta* cycles is on the older geomorphic surfaces (greater than 200,000 years BP) where soils with high clay and cobble content may have limited root penetration (fig. 12.6, McAuliffe 1995).

Isocoma tenuisecta's cyclic pattern is similar to short-lived *Gutierrezia sarothrae* (snakeweed) in the Chihuahuan Desert grasslands (Torrel et al. 1992) and *Ambrosia deltoidea* (triangle-leaf bursage) in the Sonoran desertscrub (Goldberg and Turner 1986). Given the high interannual variability of winter precipitation (CV greater than 40 percent), these shrub cycles can be viewed as slightly longer-lived analogs

of the spring wildflower eruptions that occur after wet fall and winter conditions (Bowers 2005a).

Cacti

The primary cacti, *Opuntia engelmanni* and *O. phaeacantha* (prickly pear), and *Cylindropuntia fulgida* and *C. spinosior* (cholla), are relatively short-lived (less than 60 years, Bowers 2005b). Chollas have cylindrical sections and are taller (less than 2 meters) than broad (less than 1 meter), and prickly pear have flat, circular sections and are broader (less than 2 meters) than tall (less than 1 meter). Both can establish from seed and from fallen sections that develop roots when areoles are in contact with the soil, although vegetative reproduction is more common in the chollas (Bobich 2005, Bowers 2005b).

In 1902, cacti were most common below 1,000-meter elevations (Griffiths 1904), but at higher elevations (1,100 or more meters) there have been two

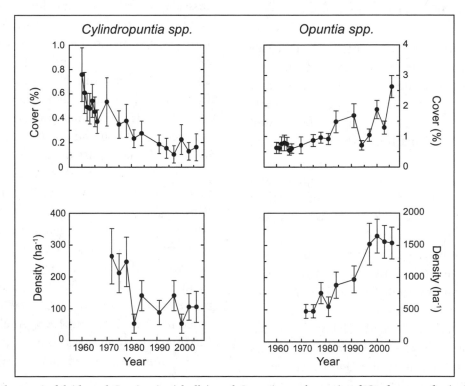

Figure 12.8. Cylindropuntia fulgida and *C. spinosior* (cholla), and *Opuntia engelmanni* and *O. phaeacantha* (prickly pear) cover and density on 60 permanent transects, between 950- and 1,250-meter elevation, 1960–2003. Vertical lines represent one ± standard error of the mean.

cholla eruptions and a steady increase of prickly pear since the 1930s (figs. 12.3, 12.7, and 12.8). In the first cholla eruption, areas supporting more than 840 plants per hectare doubled from 19 to 38 percent of the SRER (Humphrey and Mehrhoff 1958), and it declined by 1960. The second eruption occurred between 1972 and 1984 (Martin and Severson 1988), and it continues to decline. In contrast, prickly pear cover and density increased from 1970 to 1984 (Martin and Severson 1988), and it has continued to increase.

Frequent fires before 1880 may have confined the cacti to low elevations where fuel was limited, but the cessation of fires allowed these species to spread to higher-elevation grasslands (McLaughlin and Bowers 1982). Cholla and prickly pear densities recovered within 13 years after fire (Cable 1967). Cattle and rodent exclusion failed to alter the first eruption of cholla (Brown 1950, Glendening 1952), and the density of both cactus groups did not differ between yearlong and seasonal rotation of cattle grazing from 1972 to 1984 (Martin and Severson 1988).

These eruptions may follow wet winters (Bowers 2005b) similar to *I. tenuisecta* eruptions in the 1930s and 1960s. Cholla decline may be related to increasing levels of the bacterium *Erwinea carnegiea* that causes the fallen sections to desiccate before roots can be produced (Tschirley and Wagle 1964), as well as by increased competition from neighboring plants (Bobich 2005).

The spatial anomaly of no cactus eruptions occurs on the older geomorphic surfaces (greater than 200,000 years BP) where soils have high clay and cobble content (fig. 12.6), but restricted root penetration is not a likely cause. Cacti are shallow rooted, and therefore some other constraint must be operating because the shallow-rooted *Fouquieria splendens* (ocotillo) is common on this soil.

There are similar increases in prickly pear across southern Arizona (Turner et al. 2003, Bowers 2005b). However, the report of modest increases in cholla since 1970 (Turner et al. 2003) is contrary to the pattern on the SRER.

Perennial Grass

The common perennial grasses use the C_4 photosynthetic pathway, and their seeds germinate and plants grow most vigorously in the wet summer season (McClaran 1995). Plants are relatively short lived (5–10 years, Canfield 1957). Roots are most dense in the upper 15 centimeters of the soil, but some extend more than 60 centimeters deep (Blydenstein 1966, Cable 1969). Productivity is a function of both current and previous summer precipitation (Cable 1975), so there will be considerable intra- and inter-annual variation in grass abundance (figs. 12.3, 12.6, and 12.7).

Since 1902, there have been three major changes in grass abundance and composition: an increase from 1903 to 1915, the introduction and eventual dominance of the nonnative *Eragrostis lehmanniana* (Lehmann lovegrass) by 1990, and the decline of all grasses from 1995 to 2007. In 1902, after at least 20 years of uncontrolled cattle grazing, perennial grasses were largely absent below 1,350 meters in elevation, the annual *Bouteloua aristidoides* (six-weeks grama) was dominant below 1,100 meters, and only scattered *B. rothrockii* (Rothrock grama), *Muhlenbergia porteri* (bush muhly), and *B. eriopoda* (black grama) were present at higher elevations (Griffiths 1904). By 1915, perennial grasses were more common between 1,000 and 1,250 meters; *B. rothrockii* was the dominant grass on about 50 percent of the area, and *Aristida* spp. (threeawns) were dominant above 1,250 meters (figs. 12.3 and 12.7; Wooten 1916).

Nonnative *E. lehmanniana* became the most abundant (density and cover) perennial grass by 1984, and by 1991, it was more common than all native grasses combined (figs. 12.3, 12.6, and 12.9). It spread from about 50 (200 total hectares) areas seeded between 1945 and 1975 (Anable et al. 1992). Since 1995, native grasses continued declining, and *E. lehmanniana* declined sharply (figs. 12.3, 12.6, and 12.9).

The recovery of native grasses from 1903 to 1915

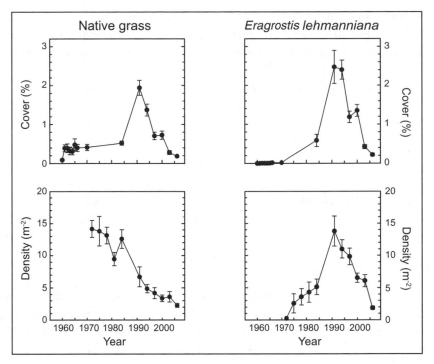

Figure 12.9. Native perennial grass and *Eragrostis lehmanniana* (Lehmann lovegrass) cover and density on 60 permanent transects, between 950- and 1,250-meter elevation, 1960–2003. Vertical lines represent ± one standard error of the mean.

was probably a response to cattle exclusion and six consecutive wet summers from 1906 through 1911 (fig. 12.2, Wooten 1916). A sequence of six wet summers has not occurred since. Grass response to cattle management is typically contingent on precipitation patterns. For example, during the very dry period from 1932 to 1949, total grass cover declined from 2.0 to 0.1 percent on grazed areas as well as all combinations of rodent and cattle exclusion (Glendening 1952). Gibbens and Beck (1989) placed greater emphasis on precipitation when interpreting similar patterns of grass abundance from 1915 to 1979 in the Chihuahuan Desert grassland.

Similarly, the spread and dominance of nonnative *E. lehmanniana* were largely independent of livestock grazing because density did not differ among areas with different grazing intensity or grazing exclusion (McClaran and Anable 1992, Angell and McClaran 2001). Four wet summers starting in 1981 may have fostered this spread (figs. 12.2 and 12.6).

Fire appears to stimulate seed germination by increasing the red light reaching the soil surface (Sumrall et al. 1991, Roundy et al. 1992), and *E. lehmanniana* plants may be less susceptible to fire-related mortality than native species (Cable 1965, Geiger and McPherson 2005).

The simultaneous decline of native grasses and increase of *E. lehmanniana* is the result of highly successful *E. lehmanniana* recruitment rather than the hastened death of adult native grasses (Angell and McClaran 2001). Recently, the simultaneous decline of all grasses may be the result of dry summer conditions since 2000 (fig. 12.2).

Prosopis velutina is more likely to decrease grass abundance at the higher and wetter elevations where *P. velutina* is most abundant on the SRER (Parker and Martin 1952, Cable 1971, Williams 1976, McClaran and Angell 2006). At lower elevations, the removal of *P. velutina* did not influence *E. lehmanniana* abundance (Kincaid et al. 1959, Cable 1971,

McClaran and Angell 2006), but older *P. velutina* are more likely to exclude *B. rothrockii* and support *Muhlenbergia porteri* beneath the canopy (McClaran and Angell 2007).

There is a spatial anomaly of no *E. lehmanniana* on very deep, sandy, loam soils (fig. 12.7), even though lovegrass is abundant on adjacent areas with less sandy soils.

Spatial Scale and Variation

Ground-Level and Oblique-Ground Comparison

Ground-level and oblique-ground images illustrate the same conversion of open grassland to shrub-savanna, and in combination they identify *P. velutina* as the main species contributing to that change (figs. 12.10 and 12.11). Among Griffiths's original photographs is one showing open grassland near Huérfano Butte (fig. 12.10) and a nearly panoramic series from the butte summit (fig. 12.11). The butte, at 1,220 meters, is about 40 meters above the base elevation, and its isolation from the mountain range provides an oblique view of the alluvial fan and drainages.

The ground-level series shows open grassland in 1902, numerous cholla cacti and scattered *P. velutina* in 1941, and dense stands of *P. velutina* and prickly pear cacti in 2007 (fig. 12.10). Close inspection of the 1902 image reveals dark patches of shrub-dominated vegetation in the distance, directly behind the horse-drawn buggy and at both sides of the image. The 1902 oblique-ground image from the summit shows similar open grassland vegetation with very widely scattered shrubs, and by 1997–2000, shrubs were much more common and large patches of open grasslands were rare (fig. 12.11).

The location of the ground-level photographs is near the center of the second from left panel of the panorama (fig. 12.11B,F). The oblique angle resolves the identity and distribution of the dark patches in the ground photograph. Those patches are *Celtis*

pallida (desert hackberry) shrubs in and along the widest drainage in the lower-center portion of that panorama panel. Most of these patches were still present in 2007.

The combination of ground-level and oblique-ground panoramic photographs shows that the larger landscape is represented by this single ground-level location and gives species-specific precision to the broader landscape perspective. It confirms that shrub-free open grasslands were common in 1900, and scattered shrubs, primarily *C. pallida*, were associated with drainages and were widely scattered in the grasslands, and *P. velutina* now dominates near the drainages and in the former open grasslands. However, the oblique-ground images cannot resolve the extent of the cactus increase and grass decline depicted in the ground-level photographs because those plants are not visible at such a coarse resolution.

Ground-Level, Aerial, and Satellite Comparisons

Sixty years of repeat ground-level and aerial photography show a similar pattern of shrub increase in former open grasslands, and the shorter, 13-year series of repeat satellite spectral data produced total shrub cover estimates very similar to aerial photography (fig. 12.12). Repeat ground-level photography location 111 (fig. 12.3) is at the center of the northwest edge of the crescent-shaped shrub in the center of the 150- by 150-meter area represented in the aerial photographs (fig. 12.12). Ground-level images look west-northwest from that point and represent the 270- to 300-degree azimuth zone from the center of the aerial and satellite images. The fence line around the ungrazed Rodent Station exclosure is visible in the upper right of the ground-level photographs and as a horizontal line through the middle of the aerial photographs. Aerial and satellite imageries in 1996 each estimate about 25 percent shrub cover and show a similar spatial arrangement of cover, with maximum in the bottom right corner and minimum in the top right.

Figure 12.10. Huérfano Butte, Santa Rita Experimental Range, southern Arizona, USA.
A. (April 1902). Ground repeat photography location 233 looking west at Huérfano Butte, at 1,159-meter elevation. In 1902, photographer David Griffiths's horse-drawn buggy was clearly visible in the open grassland, and scattered *Celtis pallida* (desert hackberry) plants were behind and on either side of the buggy. (D. Griffiths, 233.1.1902).
B. (September 1941). Members of the 1930s *Isocoma tenuisecta* (burroweed) and *Cylindropuntia fulgida /C. spinosior* (cholla) eruptions were visible in 1941 as well as new *Prosopis velutina* plants that were already quite large. (Photographer unknown, 233.1.1941.09).
C. (March 2007). More *P. velutina* were visible, and *Opuntia engelmanni* and *O. phaeacantha* (prickly pear) had replaced cholla as the dominant cacti. (M. P. McClaran, 233.2.2007.03).

The combination of ground and aerial photography shows the extent of shrub cover increase since 1922 and identifies the species-specific contributions to shrub cover dynamics. A large *C. pallida* and widely scattered large *P. velutina* are visible in the 1922 ground-level photograph, and correspond with dark patches in the 1936 aerial photograph. Analysis of the aerial images suggests an increase in shrub cover from 11 percent in 1936 to 25 percent in 1996. This change is not discernible with the ground-level images alone because the distance to far objects is compressed in the two-dimensional representation.

Apparently, an individual *P. velutina* canopy must be about 4 square meters to be detected in the aerial photographs (Browning et al. 2009). For example, the *P. velutina* plant in the left-center foreground of the ground-level photograph (fig. 12.3C) was small (less than 4 square meter canopy) but well established by 1947, but that plant is not visible with the naked eye in the aerial photograph until 1977 (fig. 12.12C).

The stagnant or declining shrub cover estimated by the aerial and satellite-based analyses may reflect the decline of *I. tenuisecta* shown in the ground-level

Figure 12.11. View from Huérfano Butte, Santa Rita Experimental Range, southern Arizona. Oblique-ground repeat photography from locations 332 and 333 at the summit of Huérfano Butte (1,220-meter elevation). The panorama begins looking northeast (332 [A,E]) and ends looking south (333 [D,H]). The ground-level repeat photography location 233 is in the bottom right center of (B,F). In 1902 (A) and 1904 (B–D), the landscape was open grasslands with *Celtis pallida* (desert hackberry) plants concentrated near the drainages and scattered in the grassland. By 2007 (E–H), *Prosopis velutina* (mesquite) had increased and converted the open grassland to a savanna, but *C. pallida* had not increased, and the plants present in 1900 were about the same size in 2000. The broad-scale conversion of open grassland to *P. velutina* savanna is consistent with that observed at ground-level location 233. The shadow of David Griffiths is cast on the fore-ground rocks in (B), and the shadow of Mitchel McClaran is cast similarly in (F).

(A) 1902, D. Griffiths, 332.2.1902; (B) 1904, D. Griffiths, 333.1.1904.09; (C) 1904, D. Griffiths, 333.2.1904.09; (D) 1904, D. Griffiths, 333.3.1904.09; (E) 2007, M.P. McClaran, 332.2.2007.10.2; (F) 2007, M.P. McClaran, 333.1.2007.10; (G) 2007, M.P. McClaran, 333.2.2007.10; (H) 2007, M.P. McClaran, 333.3.2007.10.

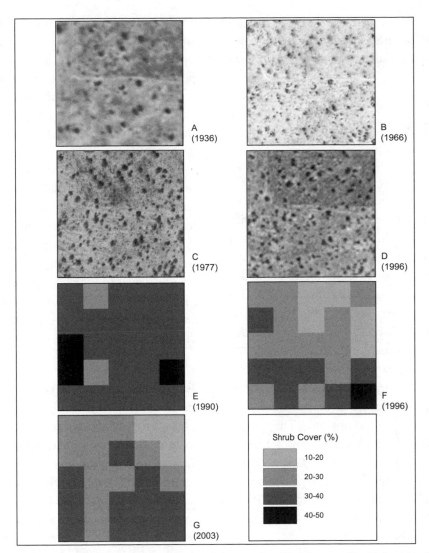

Figure 12.12. Repeat aerial photography (A–D) and satellite multispectral Landsat images (E–G) for 150- by 150-meter area centered at ground-level repeat photography location 111 (see fig. 12.3). The ground-level view looks west-northwest (270- to 300-degree azimuth) from the northwest edge of the crescent-shaped shrub in the middle of the aerial photographs. The fence line around the ungrazed Rodent Station exclosure is the horizontal line through the middle of the aerial photographs. In the aerial photography, estimates of shrub cover increased from 11.2 percent in February 1936 (A), to 17.7 percent in August 1966 (B), 17.1 percent in October 1977 (C), and 24.9 percent in June 1996 (D). In the Landsat images, estimates of shrub cover are represented in 30- by 30-meter pixels, and the overall shrub cover (± standard deviation) declined from 35 ± 5 percent in 1990 (E), to 26 ± 7 percent in 1996 (F), and 28 ± 7 percent in 2003 (G).

photography. In the aerial photography, shrub cover remained approximately 17 percent between 1966 and 1977, and the satellite spectral analysis estimated a decline from 35 to 25 percent cover between 1990 and 1996 (fig. 12.12). The abundant *I. tenuisecta* in 1966 and 1990 may have contributed to total shrub cover in addition to *P. velutina* and *C. pallida*, but when *I. tenuisecta* declined by 1977 and again by 1996, it no longer contributed to shrub cover. For aerial photography–based estimates of cover, this could occur if *I. tenuisecta* patches were greater than or equal to 4 square meters or *I. tenuisecta* plants extended the perceived size of *P. velutina* or *C. pallida* canopies because isolated *I. tenuisecta* plants are too small to be detected. In contrast, the multispectral data that are used for the satellite-based estimates are more likely to sense even smaller *I. tenuisecta* plants.

Discussion and Conclusions

The widely distributed and frequently repeated ground-level photography collection for the SRER provides a detailed, species-specific description of temporal vegetation dynamics and spatial variability in the desert grassland since 1902. In combination with vegetation measurements and precipitation data, these photographs were frequent enough to describe the rapid initial rate of *P. velutina* increase and a slowing of that increase more recently; three of the four *I. tenuisecta* cycles; cactus eruptions and declines; and changes in grass composition and abundance. The frequent repeat photography (every 1 to 10 years) supports the interpretation that winter precipitation had a positive influence on initiating and sustaining *I. tenuisecta* eruptions. The widely distributed array of ground-level locations identified the spatial anomalies in these dynamics. Since 1902, *P. velutina* did not increase in large drainages, where it was common prior to 1902; nor has *P. velutina*, *I. tenuisecta*, or cactus increased on older soils with high clay and cobble content.

A broader spatial representation of these dynamics was available from oblique-ground, aerial, and satellite perspectives, but as expected it was more difficult to identify species at these coarser resolutions. However, in combination with ground-level repeat photography, we described how *P. velutina* increased across the broad landscape, but scattered *C. pallida* plants were common in the grasslands before the *P. velutina* increase and most of those *C. pallida* plants are still alive.

Ground photographs provide information for interpreting the declines in shrub cover represented in aerial and satellite imagery. The ground-level photographs suggest that crashes in *I. tenuisecta* populations may contribute to those declines in shrub

cover rather than short-term fluctuations in the larger-sized *P. velutina* and *C. pallida*. The combination of sources supports a more sophisticated interpretation of shrub cover dynamics than one based exclusively on large shrubs. We refer to this combination of sources as a "retrospective ground truthing" of historic aerial photographs and satellite data.

Therefore, we overcame the challenge of representing the broader spatial perspective by amassing a widely distributed ground network of repeat photography locations, and by analyzing coarser resolution aerial and satellite images. More importantly, we found that combining repeat photography from ground-level and elevated positions is the optimal solution for understanding the temporal dynamics and spatial variability of vegetation dynamics in the desert grassland.

Acknowledgments

This effort is dedicated to all the scientists who contributed to the research legacy on the Santa Rita. The Santa Rita Experimental Range Digital Database at ag.arizona.edu/srer provided some of the data sets used in this chapter. The US Department of Agriculture, Forest Service, Rocky Mountain Research Station, and University of Arizona provided funding for the digitization of these data. All photographs are used courtesy of the Santa Rita Experimental Range.

Literature Cited

Anable, M. E., M. P. McClaran, and G. B. Ruyle. 1992. Spread of introduced Lehmann lovegrass *Eragrostis lehmanniana* Nees in southern Arizona, U.S.A. *Biological Conservation* 61:181–188.

Angell, D. L., and M. P. McClaran. 2001. Long-term influences of livestock management and a non-native grass on grass dynamics on the desert grassland. *Journal of Arid Environments* 49:507–520.

Anys, A., D. Bannari, D. He, and D. Morin. 1994. Texture analysis for the mapping of urban areas using airborne

MEIS-II images. In *Proceedings of the First International Airborne Remote Sensing Conference and Exhibition*, ed. S. B. Serpico, 3:231–245. Ann Arbor: Environmental Research Institute of Michigan.

Asner, G. P., S. Archer, R. F. Hughes, R. J. Ansley, and C. A. Wessman. 2003. Net changes in regional woody vegetation cover and carbon storage in Texas drylands, 1937–1999. *Global Change Biology* 9:1–20.

Asner, G. P., and D. B. Lobell. 2000. A biogeophysical approach for automated SWIR unmixing of soils and vegetation. *Remote Sensing of Environment* 74:99–112.

Bahre, C. J. 1991. *A legacy of change: Historic impact on vegetation in the Arizona borderlands*. Tucson: University of Arizona Press.

Bahre, C. J., and M. L. Shelton. 1996. Rangeland destruction: Cattle and drought in southeastern Arizona at the turn of the century. *Journal of the Southwest* 38:1–22.

Blydenstein, J. 1966. Root systems of four desert grassland species on grazed and protected sites. *Journal of Range Management* 19:93–95.

Bobich, E. G. 2005. Vegetative reproduction, population structure, and morphology of *Cylindropuntia fulgida* var. *mamillata* in a desert grassland. *International Journal of Plant Science* 166:97–104.

Bowers, J. E. 2005a. El Niño and displays of spring-flowering annuals in the Mojave and Sonoran deserts. *Journal of the Torrey Botanical Society* 132:38–49.

Bowers, J. E. 2005b. Influence of climatic variability on local population dynamics of a Sonoran Desert platyopuntia. *Journal of Arid Environments* 61:193–210.

Breckenfeld, D. J., and D. Robinett. 2003. Soil and ecological sites on the Santa Rita Experimental Range. In *Santa Rita Experimental Range: 100 years (1903 to 2003) of accomplishments and contributions*, ed. M. P. McClaran, P. F. Ffolliott, and C. B. Edminster, 157–165. USDA Forest Service Proceedings RMRS-P-30. Rocky Mountain Research Station, Fort Collins, CO.

Brown, A. L. 1950. Shrub invasion of a southern Arizona desert grassland. *Journal of Range Management* 3:172–177.

Browning, D. M., S. R. Archer, G. P. Asner, M. P. McClaran, and C. A. Wessman. 2008. Woody plants in grasslands: Post-encroachment stand dynamics. *Ecological Applications* 18:928–944.

Browning, D. M., S. R. Archer, A. T. Byrne. 2009. Field validation of 1930s aerial photography: What are we missing? *Journal of Arid Environments* 73:844–853.

Cable, D. R. 1961. Small velvet mesquite seedlings survive burning. *Journal of Range Management* 14:160–161.

Cable, D. R. 1965. Damage to mesquite, Lehmann lovegrass, and black grama by a hot June fire. *Journal of Range Management* 18:326–329.

Cable, D. R. 1967. Fire effects on semidesert grasses and shrubs. *Journal of Range Management* 20:170–176.

Cable, D. R. 1969. Competition in the semidesert grass-shrub type as influenced by root systems, growth habits, and soil moisture extraction. *Ecology* 50:27–38.

Cable, D. R. 1971. Lehmann lovegrass on the Santa Rita Experimental Range, 1937–1968. *Journal of Range Management* 24:17–21.

Cable, D. R. 1975. Influence of precipitation on perennial grass production in the semidesert southwest. *Ecology* 56:981–986.

Cable, D. R. 1977. Seasonal use of soil water by native velvet mesquite. *Journal of Range Management* 30:4–11.

Cable, D. R., and S. C. Martin. 1975. *Vegetation responses to grazing, rainfall, site condition, and mesquite control on a semidesert range*. USDA Forest Service Research Paper RM-149. Rocky Mountain Forest and Range Experiment Station, Fort Collins, CO.

Canfield, R. H. 1948. Perennial grass composition as an indicator of condition of southwestern mixed grass ranges. *Ecology* 29:190–204.

Canfield, R. H. 1957. Reproduction and life span of some perennial grasses of southern Arizona. *Journal of Range Management* 10:199–203.

Cohen, J. 1960. A coefficient of agreement for nominal scales. *Educational and Psychological Measurement* 20:37–46.

Congalton, R. G., and K. Green. 1999. *Assessing the accuracy of remotely sensed data: Principles and practices*. Boca Raton, FL: Lewis Publishers.

Cox, J. R., A. de Alba-Avila, R. W. Rice, and J. N. Cox. 1993. Biological and physical factors influencing *Acacia constricta* and *Prosopis velutina* establishment in the Sonoran Desert. *Journal of Range Management* 46:43–48.

ERDAS, I. 2002. *ERDAS Field Guide*. Atlanta, GA: Leica Geosystems, Incorporated.

Fravolini, A., K. R. Hultine, E. Brugnoli, R. Gazal, N. B. English, and D. G. Williams. 2005. Precipitation pulse use by an invasive woody legume: The role of soil texture and pulse size. *Oecologia* 144:618–627.

Geiger, E. L., and G. R. McPherson. 2005. Response of semi-desert grasslands invaded by non-native grasses to altered fire regimes. *Journal of Biogeography* 32:895–902.

Gibbens, R. P., and R. F. Beck. 1989. Changes in grass basal area and forb densities over a 64-year period on grassland types of the Jornada Experimental Range. *Journal of Range Management* 41:186–192.

Gibbens, R. P., R. P. McNeely, K. M. Havstad, R. F. Beck, and B. Nolen. 2005. Vegetation changes in the Jornada Basin from 1858 to 1998. *Journal of Arid Environments* 61:651–668.

Glendening, G. E. 1952. Some quantitative data on the increase of mesquite and cactus on a desert grassland range in southern Arizona. *Ecology* 33:319–328.

Glendening, G. E., and H. A. Paulsen. 1955. *Reproduction and establishment of velvet mesquite as related to invasion of semidesert grasslands.* US Department of Agriculture, Technical Bulletin 1127.

Goldberg, D. E., and R. M. Turner. 1986. Vegetation change and plant demography in permanent plots in the Sonoran Desert. *Ecology* 67:695–712.

Griffiths, D. 1904. *Range investigations in southern Arizona.* US Department of Agriculture, Bureau of Plant Industry, Bulletin 67.

Griffiths, D. 1910. *A protected stock range in Arizona.* US Department of Agriculture, Bureau of Plant Industry, Bulletin 177.

Hastings, J. R., and R. M. Turner. 1965. *The changing mile: An ecological study of vegetation change with time in the lower mile of an arid and semiarid region.* Tucson: University of Arizona Press.

Havstad, K. M., R. P. Gibbens, C. A. Knorr, and L. W. Murray. 1999. Long-term influences of shrub removal and lagomorph exclusion on Chihuahuan Desert vegetation dynamics. *Journal of Arid Environments* 42:155–166.

House, J. I., S. Archer, D. D. Breshears, R. J. Scholes, and NCEAS Tree-Grass Interaction Participants. 2003. Conundrums in mixed woody-herbaceous plant systems. *Journal of Biogeography* 30:1763–1777.

Huang, C., S. E. Marsh, M. P. McClaran, and S. R. Archer. 2007. Postfire stand structure in a semiarid savanna: Cross-scale challenges estimating biomass. *Ecological Applications* 17:1899–1910.

Humphrey, R. R. 1937. Ecology of burroweed. *Ecology* 18:1–9.

Humphrey, R. R. 1958. The desert grassland: A history of vegetational change and analysis of causes. *Botanical Review* 24:193–252.

Humphrey, R. R., and L. A. Mehrhoff. 1958. Vegetation changes on a southern Arizona grassland range. *Ecology* 39:720–726.

Kincaid, D. R., G. A. Holt, P. D. Dalton, and J. S. Tixier. 1959. The spread of Lehmann lovegrass as affected by mesquite and native perennial grasses. *Ecology* 40:738–742.

Martin, S. C. 1970. Longevity of velvet mesquite seed in the soil. *Journal of Range Management* 23:69–70.

Martin, S. C. 1983. Responses of semidesert grasses and shrubs to fall burning. *Journal of Range Management* 36:604–610.

Martin, S. C., and D. R. Cable. 1974. *Managing semidesert grass-shrub ranges: Vegetation responses to precipitation, grazing, soil texture, and mesquite control.* USDA Forest Service Technical Bulletin RM-1480. Rocky Mountain Forest and Range Experiment Station, Fort Collins, CO.

Martin, S. C., and K. E. Severson. 1988. Vegetation responses to the Santa Rita grazing system. *Journal of Range Management* 41:291–295.

Martin, S. C., and D. E. Ward. 1970. Rotating access to water to improve semidesert cattle range near water. *Journal of Range Management* 23:22–26.

Mashiri, F. E., M. P. McClaran, and J. S. Fehmi. 2008. Long-term vegetation change related to grazing systems, precipitation and mesquite cover. *Rangeland Ecology and Management* 61:368–379.

McAuliffe, J. R. 1995. Landscape evolution, soil formation, and Arizona's desert grasslands. In *The Desert Grassland*, ed. M. P. McClaran and T. R. Van Devender, 100–129. Tucson: University of Arizona Press.

McClaran, M. P. 1995. Desert grasslands and grasses. In *The Desert Grassland*, ed. M. P. McClaran and T. R. Van Devender, 1–30. Tucson: University of Arizona Press.

McClaran, M. P. 2003. A century of vegetation change on the Santa Rita Experimental Range. In *Santa Rita Experimental Range: 100 years (1903 to 2003) of accomplishments and contributions*, ed. M. P. McClaran, P. F. Ffolliott, and C. B. Edminster, 16–33. USDA Forest Service Proceedings RMRS-P-30. Rocky Mountain Research Station, Fort Collins, CO.

McClaran, M. P., and M. E. Anable. 1992. Spread of introduced Lehmann lovegrass along a grazing intensity gradient. *Journal of Applied Ecology* 29:92–98.

McClaran, M. P., and D. L. Angell. 2006. Long-term vegetation response to mesquite removal in desert grassland. *Journal of Arid Environments* 66:686–697.

McClaran, M. P., and D. L. Angell. 2007. Mesquite and grass relationships at two spatial scales. *Plant Ecology* 191:119–126.

McClaran, M. P., D. L. Angell, and C. Wissler. 2002. *Santa Rita Experimental Range Digital Database user's guide.* USDA Forest Service General Technical Report RMRS-GTR-100. Rocky Mountain Research Station, Fort Collins, CO.

McClaran, M. P., P. F. Ffolliott, C. B. Edminster, ed. 2003. *Santa Rita Experimental Range: 100 years (1903 to 2003) of accomplishments and contributions.* USDA Forest Service Proceedings RMRS-P-30. Rocky Mountain Research Station, Fort Collins, CO.

McLaughlin, S. P., and J. E. Bowers. 1982. Effects of wildfire on a Sonoran Desert plant community. *Ecology* 63:246–248.

Medina, A. L. 1996. *The Santa Rita Experimental Range: History and annotated bibliography (1903–1988).* USDA Forest Service General Technical Report RM-GTR-276. Rocky Mountain Forest and Range Experiment Station, Fort Collins, CO.

Medina, A. L. 2003. Historical and recent flora of the Santa Rita Experimental Range. In *Santa Rita Experimental Range: 100 years (1903 to 2003) of accomplishments and contributions*, ed. M. P. McClaran, P. F. Ffolliott, and C. B. Edminster, 141–148. USDA Forest Service Proceedings RMRS-P-30. Rocky Mountain Research Station, Fort Collins, CO.

Nellis, M. D., and J. M. Briggs. 1989. The effect of spatial scale on Konza landscape classification using textural analysis. *Landscape Ecology* 2:93–100.

Parker, K. W., and S. C. Martin. 1952. *The mesquite problem on southern Arizona ranges.* US Department of Agriculture, Circular 908.

Reynolds, H. G. 1954. Some interrelations of the Merriam kangaroo rat to velvet mesquite. *Journal of Range Management* 7:176–180.

Roundy, B. A., R. B. Taylorson, and L. B. Sumrall. 1992. Germination responses of Lehmann lovegrass to light. *Journal of Range Management* 45:81–84.

Ruyle, G. B. 2003. Rangeland livestock production: Developing the concept of sustainability on the Santa Rita Experimental Range. In *Santa Rita Experimental Range: 100 years (1903 to 2003) of accomplishments and contributions*, ed. M. P. McClaran, P. F. Ffolliott, and C. B. Edminster, 34–47. USDA Forest Service Proceedings RMRS-P-30. Rocky Mountain Research Station, Fort Collins, CO.

Sankaran, M., N. P. Hana, R. J. Scholes, J. Ratnam, D. J. Augustine, B. S. Cade, J. Gignoux, S. I. Higgins, J. Le Roux, F. Ludwig, J. Ardo, F. Banyikwa, A. Bronn, G. Bucini, K. K. Caylor, M. B. Coughenour, A. Diouf, W. Ekaya, C. J. Feral, E. C. February, P. G. H. Frost, P. Hiernaux, H. Hrabar, K. L. Metzger, H. H. T. Prins, S. Ringrose, W. Sea, J. Tews, J. Worden, and N. Zambatis. 2005. Determinants of woody cover in African savannas. *Nature* 438:846–849.

Scholes, R. J., and S. R. Archer. 1997. Tree–grass interactions in savannas. *Annual Review of Ecology and Systematics* 28:517–544.

Sumrall, L. B., B. A. Roundy, J. R. Cox, and V. K. Winkel. 1991. Influence of canopy removal by burning or clipping on emergence of *Eragrostis lehmanniana* seedlings. *International Journal of Wildland Fire* 1:5–40.

Thornber, J. J. 1910. *The grazing ranges of Arizona.* University of Arizona, Agricultural Experiment Station, Bulletin 65, Tucson.

Torrel, L. A., K. C. McDaniel, and K. Williams. 1992. Estimating the life of short-lived, cyclic weeds with Markov processes. *Weed Technology* 6:62–67.

Tschirley, F. H., and S. C. Martin. 1961. *Burroweed on southern Arizona range lands.* University of Arizona, Agricultural Experiment Station, Technical Bulletin 146, Tucson.

Tschirley, F. H., and R. F. Wagle. 1964. Growth rate and population dynamics of jumping cholla (*Opuntia fulgida* Engelm.). *Journal of the Arizona Academy of Sciences* 3:67–71.

Turner, R. M., R. H. Webb, J. E. Bowers, and J. R. Hastings. 2003. *The changing mile revisited: An ecological study of vegetation change with time in the lower mile of an arid and semiarid region.* Tucson: University of Arizona Press.

van Auken, O. W. 2000. Shrub invasions of North American semiarid grasslands. *Annual Review of Ecology and Systematics* 31:197–215.

Wessman, C. A., S. R. Archer, L. C. Johnson, and G. P. Asner. 2004. Woodland expansion in US grasslands: Assessing land-cover change and biogeochemical impacts. In *Land change science: Observing, monitoring and understanding trajectories of change on the Earth's surface*, ed. G. Gutman, A. C. Janetos, C. O. Justice, E. F. Moran, J. F. Mustard, R. R. Rindfuss, D. Skole, B. L. Turner Jr., and M. A. Cochrane, 185–208. Dordrecht, the Netherlands: Kluwer Academic Publishers.

Williams, P. T. 1976. *Grass production changes with mesquite* (Prosopis juliflora) *reinvasion in southern Arizona*. MS thesis, University of Arizona, Tucson.

Wilson, D. G. 1961. *Characteristics of a southern Arizona desert grassland soil related to mesquite invasion*. PhD dissertation, Texas A&M University, College Station.

Wooten, E. O. 1916. *Carrying capacity of grazing ranges in southern Arizona*. US Department of Agriculture Bulletin 367.

Chapter 13

Disturbance and Vegetation Dynamics in the Southern Andean Region of Chile and Argentina

Thomas T. Veblen

During the final quarter of the twentieth century, plant ecology experienced a major shift away from paradigms that stressed progressive development toward some sort of stasis or equilibrium condition, in favor of conceptual frameworks that allow for repeated disturbance and climatic variation as major drivers of vegetation dynamics (Glenn-Lewin et al. 1992, Wu and Loucks 1995). Consistent with this trend in theory, the ecosystem-based land management paradigm explicitly recognizes the dynamic character of ecosystems (Christensen et al. 1996). For resource managers, it is important to know the range of ecological processes and conditions that have characterized particular ecosystems over specified time periods and under varying degrees of human influences. Formerly, there was a widespread expectation of a "balance of nature" that was reflected in concepts that stressed stability, such as the climax concept or homeostatic self-regulation of ecosystem properties (Glenn-Lewin et al. 1992). Today, ecosystem change is regarded as the norm, and periods of relatively rapid versus slow change should be expected and accommodated in management practices.

In 1975, I initiated research on the regeneration dynamics of the *Nothofagus* forests of south-central Chile (ca. 39–41° S) that was aimed at providing the basic ecological understanding necessary for the development of sustainable management schemes of these native forests. Our work revealed the critically important and previously unrecognized roles played by natural disturbances in these forests (Veblen et al. 1981). In the mid-1980s, we expanded this line of research eastward across the midlatitude Andes into Argentina, where the research foci included topics similar to those examined in Chile, such as treefall gap dynamics (Veblen 1985), as well as broad-scale vegetation changes potentially related to climate variation and/or land-use changes (Veblen and Lorenz 1988). Since the late 1980s, we expanded research in northwestern Patagonia on the effects of land-use (primarily introduced animals and fire exclusion) and climatic influences on fire regimes and vegetation changes (Veblen et al. 1992a, Kitzberger et al. 1997, Raffaele and Veblen 1998, Mermoz et al. 2005).

During the 30-plus years of conducting research in the southern Andes, historical landscape photographs have been a frequent source of research questions, inspiration, and testing of hypotheses of landscape changes. In this chapter, I present numerous repeated photographs and their interpretations relevant to some of the major themes of this long-standing research program. First, I briefly describe the setting of this region, and then present some of the major landscape changes at least partially captured in repeated photographs.

Setting

The areas of interest to this chapter are the Lake District (ca. 39–42° S) of south-central Chile and northwestern Patagonia, Argentina (ca. 39–43° S; fig. 13.1). At about 39° S along a west-to-east transect, the physiography consists of the Coastal Cordillera and the Central Depression in Chile, the Andes, and the Patagonian Plains in Argentina (Veblen 2007). On both the Chilean and Argentine sides of the Andes, Quaternary glaciers scoured large lakes in the Andean foothills. Northward of about 43° S on the Chilean side, the Andes are paralleled by the Coastal Cordillera that rarely exceed 1,000 meters in elevation. The two mountain systems are separated by an intervening Central Depression of structural origin that is filled with Quaternary glacial, fluvioglacial, aeolian, and alluvial deposits and volcanic ash. South of about 43° S, the Andes rise directly from a coastal maze of islands and steep fiords. East of the Andean summits are the precordilleran foothills and the vast Patagonian Plains.

The dominant climate mechanisms of the southern Andes are the persistent midlatitude westerlies, the seasonally shifting subtropical anticyclone of the southeastern Pacific region, and the topographic influences of the coastal and Andean mountains (Veblen 2007). Orographic uplift of the westerly airflow results in mean annual precipitations of 3,000 to over 5,000 millimeters on the windward slopes of mountains from 40° S southward. At higher elevations (e.g., above 900 meters at 40° S), winter precipitation is mostly in the form of snow. The Andes produce a strong precipitation shadow to their lee; for example, over a west-to-east distance of only 50 kilometers at about 40° S, mean annual precipitation declines from greater than 3,000 millimeters to less than 800 millimeters.

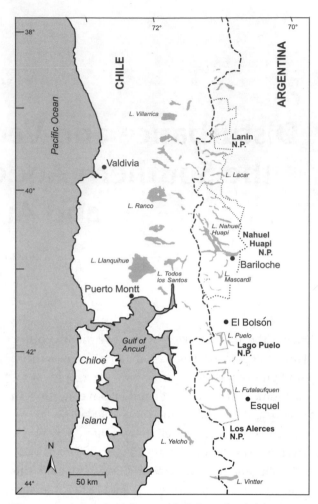

Figure 13.1. Map of south-central Chile and northwestern Patagonia, Argentina.
Repeat photographs included in the south-central Chile case study include locations from just to the southeast of Lago Villarrica to Lago Todos Los Santos. Dotted lines are the boundaries of Argentine national parks and encompass most of the area discussed in the northwestern Patagonia case study.

The *Nothofagus* forest region south of 37°45′ S considered here includes the temperate rain forests mostly west of the Andes and cool temperate *Nothofagus* forests and woodlands mostly east of the Andes (Veblen et al. 1996a). The Valdivian rain forest extends from 37°45′ to 43°20′ S, and the North Patagonian rain forest occurs from 43°20′ to 47°30′ S. The Valdivian rain forest typical of the Chilean Lake district and adjacent parts of Argentina is characterized mainly by evergreen broadleaved trees and evergreen conifers but also includes the deciduous *Nothofagus obliqua* and *N. nervosa* (synonym = *N. alpina*) in its northern, drier extent. In contrast, the North Patagonian rain forest is nearly purely evergreen, consisting of mainly broadleaved trees with a small conifer component. The giant evergreen conifer *Fitzroya cupressoides* occurs in both the southern part of the Valdivian rain forest and the North Patagonian rain forest. Common dominants of the North Patagonian rain forest are the evergreen *Nothofagus* (*N. dombeyi*, *N. nitida*, and/or *N. betuloides*), which typically occur in stands associated with fewer than five or six other angiosperm or conifer tree species. The deciduous *N. pumilio* and *N. antarctica* also occur within both rain forest districts in areas transitional to subalpine forests (*N. pumilio*) or on poorly drained sites (*N. antarctica*). *Chusquea* spp. (bamboos) dominate the understories of all the wetter forest types. At midlatitudes (ca. 40° S) the elevation gradient in forest composition from low to midelevations is floristically similar to the north–south pattern of Valdivian to North Patagonian rain forests (e.g., around 700 meters in the Chilean Andes). Subalpine forests (above 1,000 meters) are dominated by the deciduous *N. pumilio*.

On the eastern side of the Andes, parallel to the strong west-to-east decline in precipitation, is a gradient from temperate rain forests, through cool temperate *Nothofagus* forests and xeric woodlands to the Patagonian steppe of bunchgrasses and shrubs. Cool temperate *Nothofagus* forests and woodlands occur from 37°30′ S southward to 55° S on Tierra del Fuego, and include subalpine Andean forests as well as drier forests eastward in the rain shadow of the Andes (Veblen et al. 1996a). These forests and woodlands occur most extensively on the Argentinean side of the Andes, and at midlatitudes they include stands dominated by the evergreen *N. dombeyi* at mesic midelevation sites, and by deciduous *N. pumilio* and *N. antarctica* stands at high elevation or xeric sites. The evergreen conifer *Austrocedrus chi-*

lensis (about 37°30′ to 44° S) codominates with *N. dombeyi* in relatively mesic forests, and then forms pure stands eastward at more xeric sites in northwestern Patagonia.

Tectonic Disturbances and the Dynamics of *Nothofagus* Forests in South-Central Chile

A General Model of *Nothofagus* Regeneration Dynamics

In south-central Chile, *Nothofagus* species occur in old stands (i.e., more than 250 years old) in association with other tree genera from low elevations through the upper montane zone (up to about 1,000 meters). Commonly in old stands, *N. dombeyi*, *N. obliqua*, and *N. nervosa* occur primarily as trees greater than 50 centimeters in diameter at breast height (dbh) and more than 200 years old. Although *Nothofagus* seedlings and saplings less than 5 centimeters dbh may be present, especially of *N. nervosa*, the age and size structures of old stands indicate that the *Nothofagus* dominants are not regenerating in situ (Veblen and Ashton 1978, Veblen et al. 1981). In contrast, trees in other genera (e.g., *Aextoxicon*, *Laureliopsis*, *Saxegothaea*, *Amomyrtus*, *Luma*, and *Dasyphyllum*) occur in old stands as both young and old individuals and have age structures indicative of self-maintaining populations. This stand structure indicates that the shade-intolerant *Nothofagus* dominants are gradually being replaced successionally by shade-tolerant species. Although this structure and its interpretation were first presented for the forests of the province of Valdivia at 39–40° S (Veblen and Ashton 1978, Veblen et al. 1981), similar structures and interpretations have been widely documented in old *Nothofagus*-dominated mixed-species forests throughout south-central Chile, including the Island of Chiloé (Donoso et al. 1984, Armesto and Figueroa 1987, Armesto and Fuentes 1988, Pollmann 2003). Despite this prevalent stand structure suggesting

Nothofagus is undergoing successional replacement, large areas of old stands lacking *Nothofagus* are not present in most of the region, with the exception of some coastal areas (Veblen et al. 1981).

The apparent paradox of the typical stand structure indicating seral status for old *Nothofagus*-dominated rain forests versus the lack of widespread areas dominated by exclusively shade-tolerant species (i.e., a putative climax-species composition) stimulated the development in the 1970s of a general hypothesis about the dynamics of *Nothofagus*-dominated mixed-species rain forests. We suggested that in the Valdivian Andes the regeneration of *Nothofagus* species is largely dependent on coarse-scale disturbance (especially of tectonic origin, such as earthquake-triggered landslides, associated flooding, and volcanic disturbances); furthermore, the disturbance-free interval between these infrequent disturbances is typically too short to permit the complete replacement of the shade-intolerant species by shade-tolerant trees (Veblen and Ashton 1978). Although our initial focus was on *Nothofagus* regeneration triggered by mass movements and flood depositions related to earthquakes (Veblen et al. 1981), later we documented the importance of wind storms and large treefalls in creating infrequent opportunities for regeneration of *Nothofagus* in old stands (Veblen 1985, Veblen et al. 1996a). Although the history of fire is still not thoroughly studied in this region, fire scars from some of the wettest forests of south-central Chile indicate at least some role for infrequent, severe fire in these forests prior to permanent European settlement in the mid-1800s (Lara et al. 2003). Our interpretation was developed specifically for the Valdivian Andes but has been extended by ourselves and others to a broader region of *Nothofagus*-dominated, mixed-species forests (Armesto and Fuentes 1988, Donoso et al. 1984, Pollmann and Veblen 2004).

These and similar studies from other *Nothofagus* forests in the southwest Pacific region (Australia, New Zealand, New Guinea, and New Caledonia) led to a general model of how the successional status of shade-intolerant *Nothofagus* is related to site conditions and disturbance (Veblen et al. 1996a, 1996b). This general model postulates that at edaphically and climatically favorable sites where shade-tolerant tree species also grow, successful regeneration of *Nothofagus* is largely dependent on coarse-scale disturbance; at suboptimal sites where tree species richness is low, *Nothofagus* regeneration can occur after fine-scale treefalls as well as coarse-scale disturbance (Veblen 1989). Predictions from this general model have guided numerous studies of forest dynamics in the *Nothofagus* region of southern Chile where variations have been found according to the species of *Nothofagus*, associated tree species, and site conditions determined by climate and soil (Innes 1992, Pollmann 2003, Gutiérrez et al. 2004).

Coarse-Scale Disturbance and *Nothofagus* Forest Dynamics in South-Central Chile

The 1960 Valdivian earthquake was one of the strongest giant earthquakes recorded in modern times (Cisternas et al. 2005). The resulting landscape features formed throughout south-central Chile were key to the formulation of the hypothesis in the 1970s that the dynamics of *Nothofagus* forests are strongly related to tectonic disturbances (Veblen and Ashton 1978). From 21 May to 26 May 1960, this region was stricken by 11 earthquakes, each measuring at least 6.0 and one 8.5 on the Richter scale (Wright and Mella 1963). These earthquakes triggered thousands of debris avalanches, landslides, and mudflows in the Andes. This region is particularly susceptible to mass movements because of slopes oversteepened by glacial erosion, unstable volcanic ash covering these slopes, and the relatively frequent occurrence of strong earthquakes. Large percentages of the land surface—often 50 percent of the surface area over slopes covering tens of square kilometers—were largely denuded of their forest covers in 1960 (plate 13.1).

Earthquakes of similar magnitude to the 1960 event struck this region three times during the previous four centuries: in AD 1575, 1737, and 1837 (Veblen and Ashton 1978). Based on the tsunami effects and estuarine inundation, the AD 1575 and 1960 earthquakes are considered *giant* earthquakes that have occurred along the southwest coast of South America on average at 250- to 300-year intervals over the last two millennia (Cisternas et al. 2005). In association with the earlier earthquakes, large mass movements in the Andean region were also reported (Fonck 1896). For example, four months after the 16 December 1575 earthquake struck the coastal city of Valdivia, a large flood destroyed much of this Spanish outpost, which in turn motivated an expedition inland to investigate the cause of the flooding (Veblen and Ashton 1978). The Spanish expedition found that the flood resulted from the breaching of a dam formed by a debris flow that had blocked the egress of Lago Riñihue into the San Pedro River that eventually flows through the city of Valdivia (Veblen and Ashton 1978). The same sequence of events almost repeated in 1960 when a 30 million cubic meter mudflow again blocked the outlet of Lago Riñihue, except a channel was excavated by Chilean authorities to reestablish flow and prevent another catastrophic flood. Flows of enormous volumes of debris from mass movements into stream valleys and lakes in 1960 drowned extensive areas of forest (plate 13.2). They also created subsequent flood deposits in stream valleys that were abundantly colonized by establishment of *Nothofagus* spp. (Veblen and Ashton 1978).

In the Valdivian Andes, sampling of the vegetation over an elevation range from 350 to 800 meters on sites disturbed by mass movements in 1960 showed abundant tree establishment on denuded as well as depositional sites (Veblen and Ashton 1978). Toward the lower end of this elevation range, the common tree species on disturbed sites were *Nothofagus obliqua*, *Eucryphia cordifolia*, and *Weinmannia trichosperma*. Above 500 meters, *N. dombeyi* was the most abundant colonizing species. On the most favorable sites (i.e., finely textured depositional substrates), dense young thickets of *Nothofagus* species had developed by 1975. Although seed trees of shade-tolerant tree species such as *Laureliopsis*, *Saxegothaea*, and *Dasyphyllum* were common in the ad-

jacent forests, their seedlings were absent or rare on the bare surfaces created in 1960 (Veblen and Ashton 1978). Establishment of *Nothofagus* species on recently denuded substrates lacking any soil development has also been widely documented in other areas of the southern Andes and appears to be facilitated by ectomycorrhizal symbionts (Veblen et al. 1996a). Nitrogen fixers include root-nodule bacteria on *Discaria* and *Coriaria* species and blue-green algae (*Nostoc* spp.) on the giant herb *Gunnera chilensis*, and are important in contributing biomass and nutrients to bare sites (fig. 13.2).

Surfaces disturbed by mass movements in the Andes in 1960 exhibit a wide range of microsite variation depending on the degree of removal of the overlying volcanic ash and the nature of the exposed underlying rocks (plate 13.3). The typical volcanic lithology of the region consists of alternating layers of highly porous pumiceous lapilli and finer andesitic ash, sometimes including fossil soils buried by ash (Wright and Mella 1963). The weathered andesite is rich in allophone, which can absorb large quantities of moisture and is subsequently expressed when shaken under pressure, producing a phenomenon similar to liquefaction. Thus, even on relatively low-angle slopes, extensive mudflows occurred in 1960 from weathered andesite. However, where patches of volcanic ash remained intact on otherwise open sites, moisture and nutrient conditions are relatively favorable for the establishment of vascular plants, including tree species. At such sites that are adjacent to forests and where the typical dominant of the post-disturbance vegetation is *Nothofagus*, areas of exposed bedrock are much less favorable for plant establishment, but slow encroachment of plants onto these sites is also evident 47 years after the 1960 earthquake (plates 13.3 and 13.4). Many sites that initially were completely denuded of vegetation, 47 years later support a complete cover of young forest (fig. 13.2).

Vegetation sampling conducted in 1975 demonstrated that denuded and depositional sites disturbed by the 1960 mass movements were colonized by *Nothofagus* species (Veblen and Ashton 1978),

Figure 13.2. Lago Todos Los Santos, south-central Chile.
A. (January 1966). Landslides at Lago Todos Los Santos were triggered by the 1960 earthquake. This photograph shows that the site was completely denuded in 1960 and that a large fraction of the surface consisted of bedrock exposed by the sliding of soil and volcanic ash. (Photographer unknown, courtesy of Carlos Vargas; archived as T. T. Veblen, 1005).
B. (March 1979). The most conspicuous plant in 1979 is the giant herb *Gunnera chilensis* with leaf diameters greater than 1 meter. *G. chilensis* is host to a nitrogen-fixing blue-green alga (*Nostoc* spp.). (T. T. Veblen, 1006).
C. (December 1993). Although trees were too small to be conspicuous in 1979, by 1993 the site had become dominated primarily by *Nothofagus dombeyi* and other tree and shrub species. (T. T. Veblen, 1007).

and more than 30 years later photographs document continued dominance of these sites by the same tree species. The dependence of *Nothofagus* spp. on coarse-scale disturbance, however, required fine-scale investigations of stand structures across a wide range of habitat types (Veblen et al. 1981, Veblen 1985). Fine-scale examinations of the structures of old *Nothofagus* forests revealed that the vegetation response to small gaps created by treefalls is dominated by shade-tolerant tree species and also by *Chusquea* bamboos (fig. 13.3). In mixed-species rain forests, the regeneration of *Nothofagus* is rarely successful in the gaps (often larger than 0.1 hectare) created by treefalls because such gaps are preempted by abundant juveniles of shade-tolerant tree species and *Chusquea* that are already established in the understory at the time of the gap creation. This is particularly the case for *N. dombeyi* and *N. obliqua*, which are less shade-tolerant than *N. nervosa*; the latter species has a somewhat higher probability of successfully regenerating beneath a small canopy

gap (Veblen et al. 1981). In the case of large blowdowns that open the forest canopy over large areas, the overall advantage is still in favor of the advanced regeneration of shade-tolerant species but these larger events may also permit some regeneration of *Nothofagus* (Veblen 1985). Studies of forest structure in the Valdivian Andes and similar forests in south-central Chile (Donoso et al. 1984, Armesto and Fuentes 1988, Pollmann 2003, Pollmann and Veblen 2004) have shown that in the absence of coarse-scale disturbance (i.e., in the presence of only small treefalls), the regeneration of *N. dombeyi*, *N. obliqua*, and *N. nervosa* is insufficient to account for their typical abundance in the canopy of old forests.

Fire and Vegetation Changes in Northwestern Patagonia

Since the mid-1980s in northwestern Patagonia (Argentina), we have examined the history of fire and

A

B

C

D

Plate 1.1. Stake 1, foothills of the Santa Catalina Mountains, Arizona, USA.

A. (April 1926). This photograph documented what at the time was called the "tallest saguaro" (*Carnegiea gigantea*). This northerly view is northeast of Tucson, Arizona, just off the Mount Lemmon Highway. (D. T. MacDougal, A1-23).

B. (6 July 1960). Hastings and Turner took this image as the first stake number in the then unnamed Desert Laboratory Repeat Photography Collection, using a Graflex medium format (120 roll film) camera. The tall *Carnegiea gigantea* is dead, although another persists at right; the increase in short shrubs likely reflects a decrease in livestock grazing. Hastings secured this image using both black-and-white film, which remains of good quality, and this version on Kodak Ektachrome color transparency roll film. As is the case of most Ektachrome shot between 1946 and 1976, the yellow and cyan dyes in the film have faded, giving the image a magenta cast. (J. R. Hastings).

C. (11 February 1995). The view was replicated using a 4- by 5-inch Crown Graphic camera equipped with a modern 135 millimeter lens, the current standard for the Desert Laboratory Collection. In addition to this color image (shot with both color positive and negative film), the view was taken with two sheets of TMax 100 black-and-white film. (D. P. Oldershaw).

D. (19 June 2007). Again, this view was matched using a 4- by 5-inch Crown Graphic camera. Numerous species present in 1995 remain, including *Phoradendron* sp. (dwarf mistletoe), which persists in the low *Acacia greggii* (catclaw acacia) at left center; *Cylindropuntia fulgida* (jumping cholla) on the left has increased, while *Carnegiea gigantea* has declined. (R. H. Webb, Stake 1).

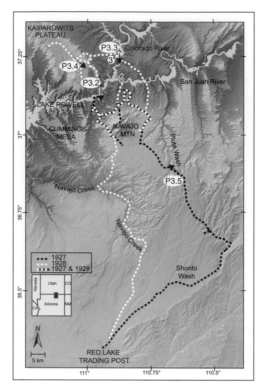

Plate 3.1. This color-coded digital-elevation model (DEM) shows parts of southern Utah and northern Arizona. The 1927 and 1928 routes of the Kluckhohn party are approximately shown. The arrows P3.2–P3.5 locate the actual camera stations and virtual photographs shown in plates 3.2 to 3.5, respectively; the tail of the arrow is at the photographic site, and the direction of the arrow indicates the orientation of the virtual and actual cameras.

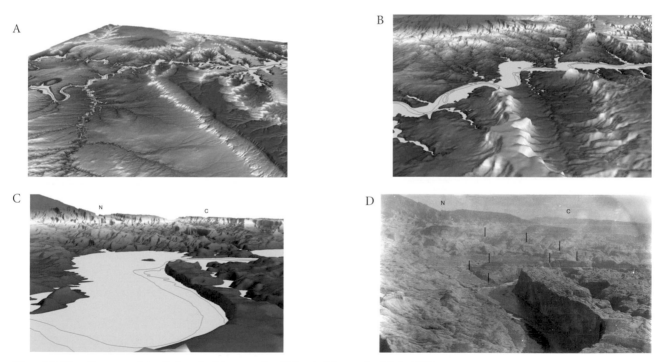

Plate 3.2. Glen Canyon of the Colorado River north of Navajo Mountain, Utah, USA.
A. This view of the DEM shown in plate 3.1 is south-southeast along the northeast edge of the Kaiparowits Plateau, with Navajo Mountain in the distance. Light blue denotes Lake Powell at full-pool level; dark blue lines show the location of the Colorado River before construction of Glen Canyon Dam in 1963. The camera station we are seeking is in the midground.
B. This view is closer to the photographic site (site P3.2 in plate 3.1). The view position is in the air and is too high.
C. This view shows that we have landed slightly behind the camera station so that we can identify it on the DEM. *N* and *C* denote the west flank of Navajo Mountain and Cummings Mesa, respectively.
D. (1928). The original photograph, taken in 1928, shows the Colorado River flowing away from the camera position. Oak Creek enters the river corridor in the midground; vertical lines in the midground point to more distant tributary positions. (J. J. Hanks, NAU.PH.2005.3.2.1.5).

A

B

C

Plate 3.3. The Colorado River upstream from Hole in the Rock, Utah, USA.

A. (1928). The original photograph is annotated "Navajo Mountain in the background. Walls of Colorado canyon in foreground" and provides little help in locating this camera station. (J. J. Hanks, NAU.PH.2005.3.2.15.2).

B. Our virtual image from the DEM is at the approximate location and direction of A. The detail of the foreground outcrop in A is too fine to be portrayed in the DEM. Significant differences are apparent above lake level on the right side of the image.

C. (18 September 2003). Most of the changes in the view are related to the filling of Lake Powell, which began in 1963 and peaked in 1986. The east buttress of the wall across the river has collapsed, probably related to saturation-failure. The lake level is at 1,098 meters, with the "bathtub ring" of calcium carbonate above it to the full-pool level. (D. P. Oldershaw, Stake 4721).

A

B

C

D

Plate 3.4. Kaiparowits Plateau, Utah, USA.

A. (1928). The original image was labeled "Looking down into the valley where our camp was pitched" and was taken at an unidentified location on the Kaiparowits Plateau. Navajo Mountain, which appears in the hazy distance, provides a clue that the view is toward the south. (J. J. Hanks, NAU.PH.2005.3.2.19.21).

B. This view from the DEM shown in plate 3.1, at a smaller scale and anchored to a position inferred from the text and sketch maps in Kluckhohn (1933), shows only general topography.

C. This view, from the same position as B, shows the combination of the DEM with a digital orthophotograph extracted from the digital orthophoto quarter quadrangle (DOQQ) of the Kaiparowits Plateau. The DOQ was rectified from a 1983 aerial photograph and is draped over the topography using common registration points. Now, shrub-filled bottomlands are easily distinguished from woodlands on the hillslopes.

D. (13 July 2003). This view is taken from an overlook of Trail Canyon, one historical access point to the southern Kaiparowits Plateau. (R. H. Webb, Stake 4662).

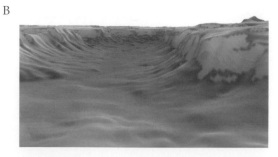

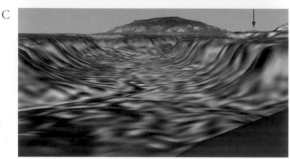

Plate 3.5. Piute Wash in Piute Canyon, Arizona, USA.

A. (July 1927). This downstream view of Piute Wash, a tributary that crosses the Arizona–Utah border as it drains northward into the San Juan River, shows a recently widened arroyo with a nearly trapezoidal cross section and no low terraces or riparian vegetation. Navajo Mountain, a major geographic feature in Utah, appears in the distance. The location where the wash was crossed was not specifically identified in the original caption, providing few clues about the location of the camera station in this extensive canyon system. (J. J. Hanks, NAU.PH.2005.3.1.6.20).

B. This DEM shows a topographic feature at upper right skyline that guided us to the camera station on Piute Wash. This feature is not defined in the 2005 or 2006 implementations of Google Earth, which uses a 90-meter DEM. Navajo Mountain is not visible because it is too far away and would have required a substantial increase in the size of the DEM to include in our analysis, but it is not necessary to locate the camera station.

C. This Google Earth image (2008) was obtained by virtually landing on the known camera station and orienting toward Navajo Mountain. This image accurately depicts the view from the camera station, in contrast to the 2005 imagery from Google Earth, because the shape of the distinctive mountain on the right skyline is correct, and the top of Navajo Begay (arrow) appears between the mountain at right and Navajo Mountain.

D. (26 September 2005). This camera station is in an extremely remote area and documents how the Kluckhohn expedition accessed this canyon to gain water for animals and move northward to Navajo Mountain. The obvious increase in pinyon and juniper in the foreground prevented an exact match, which might have been possible if the rocks at lower right in the 1927 photograph could be seen. The channel of Piute Wash is much smaller, and floodplain development is indicated by the dense growth of mostly Russian olive between the arroyo banks. The arroyo at this site appears to be in the process of filling without additional widening. (R. H. Webb, Stake 4835).

Plate 4.1. The "Time Reveal" window from the Third View project's interactive DVD (Klett et al. 2004). The information taken at one time period may be juxtaposed through the window on top of the same location at a different time period. In this case, the window shows a house that is part of the rephotograph made in 1997 at Green River, Wyoming, USA, as it appears in the window superimposed on the scene as originally photographed by Timothy O'Sullivan in 1867. Different time periods may be selected in either the reveal window or base photograph below. (B. Wolfe).

Plate 4.2. The map of the western United States used by the Third View DVD's "Journey" section is a geographically based navigation device (Klett et al. 2004). Rolling over the enlarged colored dots triggers a photograph to appear, and clicking on the dot brings the user to that site location.

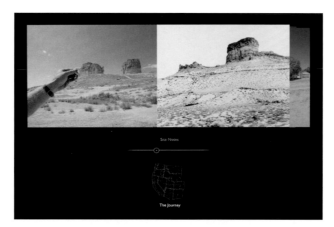

Plate 4.3. After selecting a geographical location on the western states map, the user is brought to another screen providing a scrolling lineup of images that when selected connect to rephotographs, still photographs, video, panoramas, text, and audio files.

Plate 4.4. A rephotography tutorial from the Third View DVD may be found among the options at the Pyramid Lake, Nevada, site. The text describes how to align and make a test rephotograph, and the viewer moves the position of the camera shown at the right along three possible directions as shown by the axis to line up the scene until it matches the original image on the left. As the camera moves, the simulated view also changes. When a test shot is taken with a click of the computer mouse, the tutorial demonstrates the choice of vantage point may be verified by comparing measurements between the original and the test print.

Plate 4.5. This panorama, "Above Lake Tenaya, connecting views from Edward Weston to Eadweard Muybridge, 2002," connects views by Edward Weston, 1936 (the tree at far left) to Eadweard Muybridge, 1872 (the polished boulders on the right) over approximately 300 meters of terrain, using images taken in 2002. This kind of panorama represents an option when multiple vantage points surround features within popular landscapes. (E. Weston, Center for Creative Photography; E. Muybridge, Bancroft Library; M. Klett and B. Wolfe).

Plate 4.6. (2002). We entitled this view "Four views from four times and one shoreline, Lake Tenaya," taken at Yosemite National Park, California. The left inset was taken by Eadweard Muybridge (1872); center top: Ansel Adams (ca. 1942); inset center bottom: Edward Weston (1937). The three historical photographs were made within 6 meters of one another over a period of about 70 years. The panorama combines all photographs into a visual composite of time layers analogous to the lithographic layers of a stratigraphic rock. One way to approach rephotography of image-dense landscapes is to map the vantage points into real space. In the scene, photographer Byron Wolfe swats at the high country mosquitoes mentioned in Weston's historical account of the scene. (E. Muybridge, Mount Hoffman, Sierra Nevada Mountains. From Lake Tenaya. Number 48, Bancroft Library; A. Adams, Tenaya Lake, Mount Conness, Yosemite National Park, Center for Creative Photography; E. Weston, Lake Tenaya, Center for Creative Photography; M. Klett and B. Wolfe).

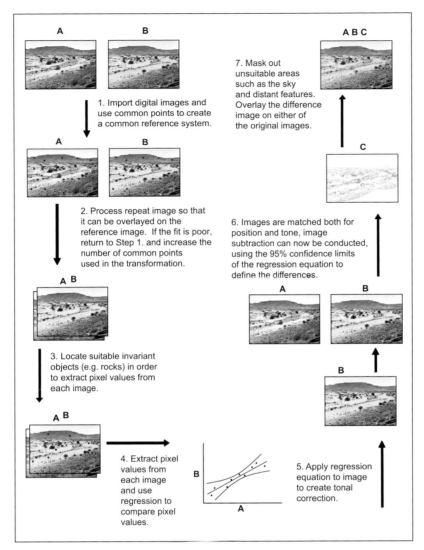

Plate 5.1. Schematic diagram of the basic steps followed first to match, then to correct, and then to subtract repeat landscape photographs.

Plate 5.2. A composite image from camera station 22 showing the vegetation that has decreased (orange-shaded area) or increased (green-shaded area). The foreground rocky slope and distant mountain are not included in the analysis. The results of the photographic analysis should be interpreted together with the more detailed vegetation survey (tables 5.1 and 5.2).

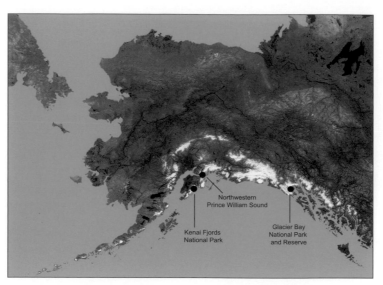

Plate 6.1. Alaska map (AVHRR mosaic compiled by Michael Fleming, USGS) showing the location of the areas where detailed repeat photography investigations were conducted.

Plate 6.2. Barnard Glacier, Alaska, USA.

A. (June 1937). Oblique aerial photograph of Barnard Glacier, St. Elias Mountains, Alaska, taken toward the northeast, showing the upper Barnard Glacier. The numerous regularly spaced, subparallel, medial moraines have contributed to the popularity of this photograph. (B. Washburn, 1355, NSIDC).

B. (28 July 2006). In the more than 69 years since Washburn's 1937 photograph, the margins of the glacier have thinned more than 30 meters, and the surface of the glacier has been lowered and has developed several meandering streams with entrenched channels. A well-developed trim line is visible on the west side of the glacier. This photograph was made from an altitude of about 3,500 meters; Washburn's 1937 photograph was from an altitude of over 5,000 meters. (Molnia-2006-07-28-545).

Plate 6.3. Muir Glacier and Muir Inlet, Alaska, USA.

A. (28 July 1976). This photograph of Muir Glacier and Muir Inlet is taken toward the northwest and shows the nearly 100-meter-high tidewater terminus of the glacier with a face capped by angular séracs. A thinning unnamed tributary flows into Muir Glacier (upper center of the photograph). Numerous icebergs are evidence of recent calving. Two gray aplite dikes penetrate the bedrock in the foreground. Each hosts a growth of green marine algae. Frequent calving creates waves, keeping the bedrock surface wet and cool and contributing to the growth of the algae. (Molnia-1976-07-28-75).

B. (7 September 2003). Note the disappearance of Muir Glacier and its tributary, and the absence of all floating ice. In the 27 years between photographs, Muir Glacier has retreated more than 6 kilometers and ceased to have a tidewater terminus. The cessation of calving has led to loss of habitat for the green algae. Evidence of the establishment of terrestrial vegetation is *Alnus* sp. (alder bush) located at the upper left of the photograph and the green tint on the land surface on the fiord's opposite side. (Molnia-2003-09-08-57).

Plate 6.4. Muir Glacier and Muir Inlet, Alaska, USA.

A. (2 September 1892). Photograph of Muir Glacier and Muir Inlet, taken toward the northeast, showing the over 100-meter-high, more than 4-kilometer-wide glacier's tidewater terminus, capped by angular séracs. Icebergs, evidence of recent calving, are floating in Muir Inlet. Mount Wright is located right of center. Note the absence of vegetation. (H. F. Reid 417, NSIDC).

B. (11 August 2005). In the 113 years between photographs, Muir Glacier has retreated more than 50 kilometers and ceased to have a tidewater terminus, which is indicated by the lack of floating ice. Note the abundant vegetation on many slopes throughout the photograph. (Molnia-2005-08-11-12).

A

B

Plate 6.5. Muir Glacier and Muir Inlet, Alaska, USA.

A. (Mid-1880s–mid-1890s). Photograph of Muir Glacier and Muir Inlet, taken toward the northwest, showing the over 100-meter-high tidewater terminus of the glacier. Icebergs, some more than 2 meters in diameter, are grounded on the adjacent tidal flat. The tidal flat is probably ice cored and underlain by glacier ice. Tides are greater than 3 meters. A photographer and seven tourists explore the icebergs. Note their formal attire, including the bustle on the woman on the right and the bowler hat and morning coat on the photographer. Many late-nineteenth-century photographs of Muir Glacier were made by tourists. This photograph is attributed to G. D. Hazard, possibly a camera-carrying passenger on a steamer that visited Glacier Bay. Several bedrock ridge tops poke above the glacier on the left side of the photograph. (G. D. Hazard, 7807, Glacier Bay National Park and Preserve Archive).

B. (11 August 2005). In the approximately 115 years between photographs, Muir Glacier has retreated more than 50 kilometers and is out of the field of view. Riggs Glacier, formerly a tributary to Muir Glacier, is in the distance. It is approximately 40 kilometers from the late-nineteenth-century photo point and about 10 kilometers from Muir Glacier's 2005 location. The beach in the foreground is covered by a cobble and pebble deposit. Many of the cobbles are covered by barnacles. Note the vegetation on the west side of the inlet. (Molnia-2005-08-11-52).

A

B

Plate 6.6. Muir Glacier and Muir Inlet, Alaska, USA.

A. (Late nineteenth century). Postcard showing Muir Glacier and Muir Inlet, with view to the northeast. The glacier's approximately 100-meter-high tidewater terminus is capped by angular séracs. Adams Glacier, which flows out of Adams Inlet, located in the gap between the essentially snow-free mountains in the center of the picture, forms the eastern part of the terminus. Icebergs, some over 5 meters in diameter, are grounded in the foreground. Photographs used on many early postcards were airbrushed to simplify the area's complex topography. No vegetation is visible. (Winter and Pond postcard C141, author's collection).

B. (11 August 2005). In the 105+ years between photographs, Muir Glacier has retreated more than 50 kilometers and is out of the field of view. Note the vegetation on the slopes of Muir and Adams inlets. (Molnia-2005-08-11-126).

A

B

Plate 6.7. Reid Glacier and Reid Inlet, Alaska, USA.

A. (10 June 1899). Photograph of Reid Glacier and Reid Inlet taken toward the northwest, showing Reid Glacier's approximately 60-meter-high retreating tidewater terminus, located adjacent to the mouth of Reid Inlet. The foreground hillside is covered by a few centimeters of snow. No trees are present in the field of view. Icebergs of various sizes float in front of the glacier. A large piece of grounded ice is located between the hillside and the glacier. It probably recently separated from the retreating glacier and is stranded adjacent to the shoreline. Concentric ripples suggest that a large calving event has recently occurred. (G. K. Gilbert, 258, USGS Photographic Library).

B. (6 September 2003). In the 104 years between photographs, Reid Glacier has retreated about 3 kilometers and is just visible at the head of the fiord. The hillside is now covered with dense vegetation, including both conifers and deciduous trees. Vegetation rooted in glacier till covers much of the lower slopes of the inlet. Species present include *Alnus* (alder), *Salix* (willow), and *Populus* (cottonwood). The foreground spit is part of Reid Glacier's recessional moraine, deposited at the mouth of its fiord during the early twentieth century. (Molnia-2003-09-06-45).

A

B

Plate 6.8. Carroll Glacier and Queen Inlet, Alaska, USA.

A. (6 August 1906). North-looking photograph of Carroll Glacier and Queen Inlet taken from Triangle Island. Shown is Carroll Glacier's approximately 40-meter-high debris-free, slowly retreating tidewater terminus. The glacier spans the width of Queen Inlet. No vegetation or icebergs are visible. An early-twentieth-century Coast and Geodetic Survey nautical chart shows water depths adjacent to the glacier exceed 125 meters. (C. W. Wright, 333, USGS Photographic Library).

B. (7 September 2003). In the 97 years between the dates of the photographs, Carroll Glacier's terminus has become stagnant and debris covered. The glacier has significantly thinned and retreated several hundred meters. The head of Queen Inlet has been filled by more than 125 meters of sediment. The water in the foreground is a meandering braided stream. When photographed, the stream was in flood as a result of a glacial lake outburst. The ice strewn on the sediment plain was transported by these floodwaters. Note the diverse vegetation on Triangle Island. (Molnia-2003-09-07-35).

Plate 6.9. Carroll Glacier and Queen Inlet, Alaska, USA.

A. (7 August 1906). Photograph of Carroll Glacier and Queen Inlet taken from a talus-alluvial cone, located at an elevation of about 200 meters on the east side of Queen Inlet. Taken a day after plate 6.8A by the same photographer, the view looks toward the northwest and shows Carroll Glacier's approximately 40-meter-high debris-free, slowly retreating tidewater terminus. This view provides a significant look "up-glacier" and presents a better perspective of the near vertical face of the glacier. While a few small tundra plants are present, no trees are visible. (C. W. Wright, 335, USGS Photographic Library).

B. (21 June 2004). In the intervening 98 years, Carroll Glacier has become a stagnant, debris-covered glacier that has significantly thinned and retreated several hundred meters from its 1906 position. Maximum sediment thicknesses on the glacier exceed 6 meters. About 1 cubic kilometer of sediment has filled the head of Queen Inlet to just above sea level. From this perspective, the meandering stream that flows along the west wall of the valley (plate 6.8B) cannot be seen. Note the vegetation, especially *Alnus* and conifers, that has become established. The isolated trees on the floor of the fiord are *Alnus*. (Molnia-2004-06-21-76).

Plate 6.10. Harvard Glacier, Harvard Arm, College Fiord, Alaska, USA.

A. (1 July 1909). North-looking photograph showing the western part of the terminus of the advancing Harvard Glacier, taken from a location on the east side of the Harvard Arm. Harvard Glacier, which covers half of the width of the photograph, sits at the head of the fiord with Radcliff Glacier, its largest tributary comprising nearly all of the glacier ice visible at tidewater. Baltimore Glacier, a retreating hanging glacier, is at the left. Note the abandoned, elevated lateral moraine adjacent to its north margin. A smaller, nameless cirque glacier is located to the right side of Baltimore Glacier. No vegetation is visible. The 3,000+ meter peak on the left side of the photo is unnamed. (U. S. Grant, 208, USGS Photographic Library).

B. (3 September 2000). During the 91 years between photographs, the terminus of Harvard Glacier has advanced more than a kilometer and thickened by at least 150 meters. The continuing advance has completely obscured the view of Radcliff Glacier. Note that the hanging glacier located on the mountain slope north of and above Radcliff Glacier shows no change. However, Baltimore Glacier and the nameless cirque glacier both have continued to retreat and thin. *Alnus* sp. has become established on many hill slopes but is difficult to see from the photo location. Barren zones have developed around the perimeters of Baltimore Glacier and the unnamed small hanging glacier located immediately to its north. (Molnia-2000-09-03-127).

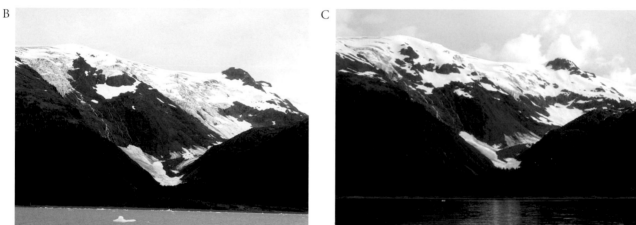

Plate 6.11. Toboggan Glacier, Harriman Fiord, Alaska, USA.

A. (20 August 1905). Northeast-looking photograph showing the retreating terminus of Toboggan Glacier, taken from a location about half a kilometer offshore. The two arcuate ridges, capped by several hummocky mounds that are located on the surface of the outwash sediments, are recessional moraines, dating from the late nineteenth century. By 1905, the terminus had thinned to about 50 percent of its former LIA maximum thickness. The northernmost tributary still makes contact with Toboggan Glacier. Note the large bedrock barren zone on both margins of the glacier. Little vegetation existed on the fiord-facing hill slopes. (S. Paige, 731, USGS Photographic Library).

B. (4 September 2000). After 95 years, the glacier is still thinning and retreating. Its terminus, a thin tongue of ice, can be seen surrounded by a mass of debris. The glacier has thinned as much as 150 meters and retreated more than 500 meters. Snow sits on the valley floor adjacent to where the northernmost hanging glacier tributary previously joined Toboggan Glacier. This tributary no longer makes contact, having retreated more than 600 meters up the valley wall. The former barren zone is now covered by vegetation. (Molnia-2000-09-04-78).

C. (22 August 2008). After 103 years, the glacier continues to thin and retreat. Terminus ice is no longer visible. Both former hanging glacier tributaries continue to retreat. (Molnia-2008-08-22-4663).

Plate 6.12. Muir and Riggs glaciers, Muir Inlet, Alaska, USA.

A. (13 August 1941). Northeast-looking photograph was made from photo station 4, established in 1941 by William O. Field, on the southeastern side of White Thunder Ridge showing the lower reaches of Muir Glacier, then a large tidewater calving valley glacier, and its tributary Riggs Glacier. The séracs in the lower right-hand corner of the photograph mark Muir Glacier's terminus. The total ice thickness is more than 700 meters. Muir Glacier had been retreating since the mid-eighteenth century, with maximum retreat exceeding 50 kilometers. In places, more than a kilometer thickness of ice had been lost. Note the absence of vegetation and the bare bedrock faces present on both sides of the glacier (W. O. Field, 41-64, NSIDC and Glacier Bay National Park and Preserve Archive).

B. (4 August 1950). This, the first of two repeat photographs, documents significant changes that have occurred during the nine years between photographs A and B. Although Muir Glacier has retreated more than 3 kilometers and thinned more than 100 meters, exposing Muir Inlet, it remains connected with tributary Riggs Glacier. White Thunder Ridge remains devoid of vegetation. In places, erosion has removed some of the surface till. (W. O. Field, F50-R29, Glacier Bay National Park and Preserve Archive).

C. (31 August 2004). The second repeat photograph documents significant changes that have occurred during the 63 years between photographs A and C, and during the 54 years between photographs B and C. Muir Glacier has retreated out of the field of view and is now more than 7 kilometers northwest. Riggs Glacier has retreated as much as 600 meters and thinned more than 250 meters. Note the dense vegetation that has developed on White Thunder Ridge. Also note the correlation between Muir Glacier's 1941 thickness and the trimline on the left side of this photograph. (Molnia-2004-08-31-28).

Plate 6.13. Plateau Glacier, Alaska, USA.

A. (9 September 1961). Northwest-looking photograph made from a photo station established by Field's colleague M. T. Millet, showing the lower reaches of the tidewater calving Plateau Glacier, then a valley glacier with parts of its terminus being land based on either side of the fiord. Note the séracs rising about 35 meters above tidewater. The total ice thickness is over 200 meters. Note the absence of vegetation in the boulder-till-covered foreground. Note the black, linear medial moraine on the surface of a tributary to Plateau Glacier that descends from Mount Wordie. Two people are located in the center of the photograph. (M. T. Millet, M-61-P51, Glacier Bay National Park and Preserve Archive).

B. (14 September 2003). This Ron Karpilo photograph documents changes that occurred during the ensuing 42 years. Plateau Glacier has retreated out of the field of view. The tributary glacier that formerly supported the medial moraine has retreated nearly 3 kilometers, thinned by more than 300 meters, and left an area of debris-covered ice in the path of its retreat. The dense vegetation covering much of the foreground area includes *Alnus*, *Salix*, *Populus*, and *Picea* (spruce). It was so dense that the two geologists standing at the shoreline on the right side of the peninsula were unable to reach the point of the headland occupied by the two individuals in the 1961 photograph. (Karpilo-2003-273).

Plate 6.14. Muir Glacier and Muir Inlet, Alaska, USA.

A. (28 August 1980). This ship-deck-based photograph of Muir Glacier and Muir Inlet is taken toward the north-northwest and shows the nearly 50-meter-high retreating tidewater terminus of the glacier. Note the icebergs, especially in the smoother, arcuate ship's wake in the lower right side of the photograph. The location of Muir's terminus is less than a kilometer from the landward end of Muir Inlet. (Molnia-1978-08-18-46).

B. (7 September 2003). In the 23 years between photographs, Muir Glacier has retreated more than 2 kilometers and ceased to have a tidewater terminus. Note that retreat has left several ice-cored morainal mounds between the shoreline and the terminus. Since 1980, Muir Glacier has thinned over 100 meters, permitting a view of the previously unseen approximately 1,500-meter-high mountain, located in the center of the photograph. A reexamination of the 1980 photograph shows that the then-snow-covered summit of this mountain was visible but that it blended in with adjacent clouds. No evidence of vegetation is seen anywhere in the photograph. (Molnia-2003-09-08-78).

Plate 6.15. Yale Glacier, Alaska, USA.

A. (June 1937). North-looking oblique aerial photograph of the retreating, calving, tidewater terminus of Yale Glacier. In 1937, Yale Glacier's terminus was located at about the same position that it occupied when visited by the Harriman Alaska Expedition in 1899. Icebergs issue from several embayments in the approximately 45-meter-high glacier terminus. Several current and former tributary glaciers descend the fiord's east wall. The two closest to the terminus have lost contact with Yale Glacier. (B. Washburn, 122, National Snow and Ice Data Center).

B. (28 July 2006). During the intervening 69 years, Yale Glacier has retreated as much as 6 kilometers, with most of the retreat occurring post-1957. The tidewater part of the terminus is less than half of its 1937 width. Yale Glacier has thinned in places by more than 250 meters. All of the eastern tributaries have lost contact with Yale Glacier. An island and a large glacially sculpted bedrock ridge have emerged from beneath the retreating glacier. Retreat of the land-based western portion of the terminus has kept pace with the retreat of the eastern tidewater portion of the glacier. A well-developed trimline is visible on the west side of the glacier. This 2006 photograph was made from an altitude of approximately 3,500 meters, while Washburn's 1937 photograph was made from an altitude of more than 5,000 meters. (Molnia-2006-07-28-345).

Plate 6.16. Bear Glacier, Alaska, USA.

A. (2 September 2002). North-looking oblique aerial photograph of the retreating, calving terminus of Bear Glacier, located at the head of a large ice-marginal lake, informally named Bear Lake. Prior to 1950, the entire basin of Bear Lake was filled by Bear Glacier's piedmont lobe. By 1961, a small lake occupying less than 10 percent of the basin had formed. By 1984, the lake had nearly doubled in size. In the 18 years between 1984 and September 2002, the lake quadrupled in size. The glacier's triangular-shaped terminus has retreated approximately 2 kilometers from its 1984 position. The large tabular icebergs and the low-relief, low-gradient terminus indicate that the terminus has thinned so much that much of it is afloat. Floating glacier termini typically retreat rapidly, calving large tabular icebergs. (Molnia-2002-09-02-574).

B. (6 August 2005). During the 35 months between photographs A and B, intensive passive calving has resulted in the triangular-shaped terminus (plate 6.17A) retreating over 3 kilometers and the large triangular lobe disappearing. The glacier has thinned by about 10 meters. (Molnia-2005-08-06-264).

C. (13 August 2007). From 2005 to 2007, the terminus continued to retreat through passive calving, although at a slower rate than previously. A few large, tabular icebergs are remnants of even larger icebergs present in 2005. Others are the result of continuing passive calving of the glacier's western terminus. In the 24 months between images B and C, the western margin of the glacier retreated more than 400 meters. (Molnia-2007-08-13-326).

A

B

Plate 6.17. Bear Glacier, Alaska, USA.

A. (Summer view, probably mid-1920s). This northeast-looking photograph of Bear Glacier is taken from a ridge in Bulldog Cove, near Bear Glacier Point. The photograph is from a postcard labeled "Harding Glaciers, Resurrection Bay, Alaska." The water in the foreground is part of lower Resurrection Bay. The name Harding Glacier or Glaciers was never officially adopted. In 1923, President Warren G. Harding visited Alaska, including a trip to Seward and nearby Resurrection Bay. The name Harding Icefield was officially approved in 1950 for the upland accumulation area that feeds Bear Glacier and a number of other glaciers of the Kenai Fjords. In the 1920s, Bear Glacier's piedmont lobe filled most of the lake basin shown in plate 6.16A. Note that a few small shrubs or trees are present in front of the glacier. (Unknown photographer, Kenai Fjords National Park archive).

B. (12 August 2005). In the approximately 80 years between photographs, Bear Glacier's piedmont lobe has retreated out of the field of view. Large icebergs, floating in the ice-marginal lake that fills the basin formerly occupied by the piedmont lobe, are the only visible ice. Isolated patches of snow are present at a few higher-elevation locations. Note the dense vegetation that has developed around the margin of the lake and on most of the lower-gradient slopes. (Molnia-2005-08-12-86).

A

B

Plate 6.18. Pedersen Glacier, Alaska, USA.

A. (Summer view, pre-1915). North-looking photograph of Pedersen Glacier, taken from the west shoreline of Aialik Bay. The water in the foreground is part of an ice-marginal lake located adjacent to Aialik Bay. When photographed, Pedersen Glacier was calving icebergs into the lake from a séracs-capped terminus that was 20 to 40 meters high. No vegetation is visible. (Photograph attributed to Reverend Pedersen, Kenai Fjords National Park archive).

B. (10 August 2005). In the 90 or so years between photographs, most of the lake has filled with sediment and now supports several varieties of grasses, shrubs, and aquatic plants. Dead trees are remnants of a mid-twentieth-century forest that was drowned by more than 3 meters of downwarping of the coast during the 1964 Alaskan earthquake. Pedersen Glacier's terminus has retreated more than 2 kilometers. The tributary located high above Pedersen Glacier separated from it sometime after 1950. Note the stands of trees that have developed between the sediment-filled wetland and the glacier. (Molnia-2004-08-13-76).

A  B

Plate 7.1. Shepard Glacier, Glacier National Park, Montana, USA.
A. (1913). Well-defined boundaries and crevasses are apparent in this photo of Shepard Glacier when its mass filled the cirque in 1913. (W. C. Alden, USGS Photographic Library).
B. (2005). The thick, crevassed ice floes of historic Shepard Glacier have been diminished to less than 0.1 square kilometer in area by 2005. According to the criteria set by the USGS Repeat Photography Project, Shepard Glacier is now considered to be too small to be defined as a glacier. (B. Reardon).

A  B

Plate 7.2. Boulder Glacier from Chapman Peak, Glacier National Park, Montana, USA.
A. (circa 1910). Around the time that GNP was established as a national park in 1910, Boulder Glacier had substantial mass with a lobe extending beyond Boulder Pass. (M. Elrod).
B. (2007). A climb to Chapman Peak reveals that the thick glacier once covering the Boulder Pass region has virtually disappeared. (D. B. Fagre and G. Pederson).

A  B  C  D

Plate 7.3. Grinnell Glacier from the summit of Mt. Gould, Glacier National Park, Montana, USA.
A. (1938). This early photograph from the summit of Mt. Gould shows the ice filling the cirque where, prior to 1938, the ice-surface elevation was high enough to connect with the upper band of ice. (T. J. Hileman).
B. (1981). Proglacial Upper Grinnell Lake becomes an obvious feature as Grinnell Glacier recedes. (C. Key).
C. (1998). The glacier's forward movement is evident through the progress of the dark, triangular rock pile that has moved from the center of the ice in the 1981 photo and now protrudes into the lake. (D. B. Fagre).
D. (2009). These oblique views of Grinnell Glacier show the decrease in the glacier's area as well as the reduction in the height of the glacier along the cirque wall. (Lindsey Bengtson).

Plate 7.4. Northeast portion of Sperry Glacier, Glacier National Park, Montana, USA.

A. (1913). Sperry Glacier extends its northeastern edge, with substantial depth, to the rocky cliffs in the background. Sparsely vegetated moraines are evident at the foot of the glacier. (W. C. Alden).

B. (2008). In addition to recording glacial recession, numerous sets of repeat photographs have recorded other landscape changes indicative of climate change, such as this photograph depicting the establishment of conifers and other vegetation on the moraines of Sperry Glacier. (L. A. McKeon).

Plate 9.1. Mount Mansfield, Vermont, USA. (ca. 1857). Local artists recorded the mostly denuded Vermont landscape. Charles Heyde painted this view of Mount Mansfield and the Browns River in northwestern Vermont. He faithfully recorded stumps in the streamside field and deep gullying that resulted from undercutting the toe of the adjacent hillslope. See Jonas (chapter 21) for more examples of using artwork to study landscape change. (C. Heyde, LS09924, The Robert Hull Fleming Museum, University of Vermont, Gift of Mrs. Guy Bailey, 1944.3).

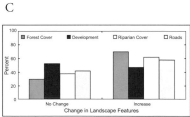

Plate 9.2. 1927 flood at Montpelier, Vermont, USA.

A. (1927). Rephotography of oblique imagery taken just after the 1927 flood in Montpelier, Vermont, shows specific examples (A and B) and the broader pattern of landscape change between 1927 and 2004. (E. S. Mann, LS01429_000).

B. (2004). Green arrow shows area that has reforested. Red arrow shows overbank deposits now covered by a high school. Blue arrow shows commercial development of once-wooded site. Note the lack of bridges (they have been washed out) in the older image. (E. S. Mann, LS01429_001).

C. Characterization of images (*n* = 67 pairs) showing percentage in which characteristics studied (forest cover, development, riparian cover, and roads) either remained similar (no change) or increased.

Plate 9.3. Woolen mill at Winooski, Vermont, USA.

A. (1927). Imagery used to calculate Manning's *n* for a bedrock reach of the Winooski River. Image of woolen mill in Winooski, Vermont, during high water of 1927 flood (3,340 cubic meters per second). (H. Kurilchyk, LS01132_000).

B. (2009). The same mill, which is now an apartment complex, from a slightly different angle, in summer 2009. Note the considerably lower level of water (about 35 cubic meters per second). The level of the windows was used to match the imagery and determine river stage. The dam is present in both images but washed over in A. (C. Zephir, LS26033_001).

Plate 9.4. The 1927 flood in Vermont, USA.

A. (12 November 1927). The earliest widespread aerial photography of Vermont was taken by the Army Air Corps within days of the devastating 1927 flood. This image was taken eight days after the flood. Extensive, light-colored sandy splay deposits are visible. (Photographer unknown, LS01418_000).

B. (19 May 2004). There are many differences between the images, but the most striking is the magnitude of channel change, specifically the narrowing of the channel near the bridge (same abutments marked by arrows in both images) and the reappearance of the island, farther north (circled). (E. S. Mann, LS01418_001).

Plate 11.1. Near Steinkopf, Namaqualand, South Africa.

A. (1970). These two senescent *Aloe dichotoma* individuals are part of a larger population near Steinkopf, Namaqualand. (T. Oliver, courtesy of Ted Oliver).

B. (18 March 2005). The small leaf clusters present at the tips of the slender branches in 1970 have all died in the individual on the left. The individual on the right will soon follow suit and shed some of its branches as it collapses and dies. (R. F. Rohde, 364).

Plate 11.2. Cornell's Kop, South Africa.

A. (ca. 1950). Three large *Aloe pillansii* individuals are on a north-facing slope on Cornell's Kop. In 1950, the large *Aloe pillansii* (individual no. 60) in the foreground was 7.32 meters tall. (P. van Heerde Sr., courtesy of P. van Heerde Jr.).

B. (1 April 2004). When rephotographed in 2004, the large *Aloe pillansii* (or individual no. 60) was 8 meters tall, giving an annual growth rate of 0.013 meter per year. The individual on the right (no. 63) was 3.94 meters in the original photograph and was recorded at 4.40 meters tall in 2004, giving an annual growth rate of 0.009 meter per year. (M. T. Hoffman, 277).

Plate 11.3. Cornell's Kop, South Africa.

A. (September 1953). *Aloe pillansii* (individual no. 19) is on a northeast-facing slope on Cornell's Kop. In 1953, this individual was 2.75 meters tall. (H. Hall, 6-6-20, Harry Hall Archives).

B. (6 September 2003). When rephotographed 50 years later, individual no. 19 had grown to be 3.55 meters tall, giving an annual growth rate of 0.016 meter per year. The collapse of the general population in the background is also evident in this repeat photograph. (R. F. Rohde, 228).

Plate 13.1. Pampa de Pilmaiquén, Valdivia, Chile.

A. (March 1977). Landslides at Pampa de Pilmaiquén were triggered by the 1960 giant Valdivian earthquake. The old forest on the slopes at the time of the 1960 earthquake was dominated by *Nothofagus dombeyi, Laureliopsis philippiana, Dasyphyllum dia-canthoides*, and *Saxegothaea conspicua*, but the trees that have colonized the denuded sites are exclusively *N. dombeyi*. Tree establishment has been abundant within about 50 meters of the edges of the old forests, but colonization of sites more distant from forest edges is proceeding slowly, presumably due to lack of seed sources. The foreground shows shrubby *Nothofagus antarctica*, which was burned in a fire that probably occurred in the early 1970s and subsequently resprouted. At the foot of the slope is a narrow belt of *N. pumilio*, which typically occurs at higher elevations, but in this large cold-air drainage basin the usual elevation gradient of vegetation is inverted so that the subalpine *N. pumilio* and *N. antarctica* species are found in the frost pocket, and the montane *N. dombeyi* occurs on the slopes where cold air does not pond. The early-1970s fire burned through the mixture of grasses and *N. antarctica* in the valley bottom but did not spread into the adjacent stand of tall *N. pumilio* at the edge of the valley bottom. (T. T. Veblen, 1001).

B. (March 2006). The 2006 photograph shows continued colonization of the denuded slopes by *Nothofagus dombeyi*. The foreground shows regrowth of the postfire resprouts of *N. antarctica* as well as a line of planted introduced pines. (T. T. Veblen, 2001).

Plate 13.2. Lago Pellaifa, Valdivia, Chile.

A. (October 1965). Landslides at Lago Pellaifa were triggered by the 1960 earthquake. In the bottom left of the scene are standing dead trees that were drowned by the rise in the lake level caused by debris flows into the lake in 1960. (T. T. Veblen, 1002).

B. (January 1977). This scene taken 12 years later shows continued growth and establishment of *Nothofagus* species on the landslides. (T. T. Veblen, 1003).

C. (January 1984). The 1984 photograph shows further vegetation establishment on the sites denuded by the landslides. (T. T. Veblen, 1004).

D. (March 2006). This photograph taken 46 years after the exposure of volcanic ash and bedrock by landslides shows continued growth and establishment of *Nothofagus* species and other trees on sites exposed in 1960. (T. T. Veblen, 2002).

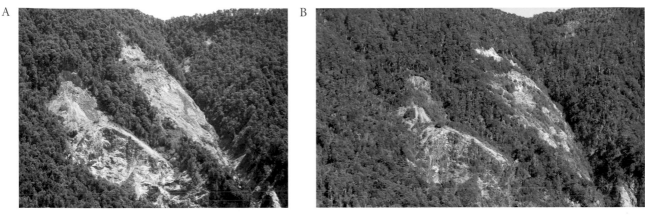

Plate 13.3. Lago Pellaifa, Valdivia, Chile.
A. (January 1977). These landslides were triggered by the 1960 earthquake. (T. T. Veblen, 1008).
B. (March 2006). Colonization of the disturbed sites is primarily by *N. dombeyi* and shrubs. (T. T. Veblen, 2003).

Plate 13.4. Near Anticura, Osorno, Chile.
A. (February 1978). Landslides near Anticura, which is northeast of Lago Todos Los Santos (fig. 13.1), were triggered by the 1960 earthquake. Among the woody species that have colonized the denuded sites is *N. dombeyi*. Lighter patches in this view are thickets of *Chusquea* bamboo, which probably dominated these sites following a previous fire. *Chusquea* proliferates after fire because of its vigorous vegetative reproduction, but it is slow to disperse to the denuded landslide sites because of its infrequent flowering and seeding. (T. T. Veblen 1009).
B. (March 2007). This photograph shows that 47 years after exposure of the landslide sites, *N. dombeyi* and other woody species have nearly covered most of the areas exposed in 1960. (T. T. Veblen 2004).

Plate 14.1. Mount Tsibet, Ethiopia.
A. (April 1868). When these photographs of the southern slopes of Mount Tsibet (southern Tigray, 3,928 meters) were taken, only a small forest remained around the church on the crestline at left. (Royal Engineers of the British Navy, "Panoramic View of Mountain at Bolago," MA-1868-RE-26 and MA-1868-RE-27, King's Own Royal Regiment Museum, Lancaster).
B. (April 2008). The *Juniperus* forest has expanded significantly, which even rendered repeat photography difficult because trees blocked the view from the camera station. This is the first repeat photographic evidence confirming earlier authors' claims that the "40% forest cover in Ethiopia by 1900" is nothing but a myth. (J. Nyssen, MA-1868-RE-26-R2-2008-NY and MA-1868-RE-27-R2-2008-NY).

A   B

Plate 14.2. Mount Tsibet, Ethiopia.

A. (1975). The northern slopes of Mount Tsibet in 1975 were similar to the status of its southern slopes in 1868 (plate 14.1A), with basically only a church forest remaining. (R. N. Munro, MA-1975-MU-218).

B. (2006). Large parts of the slopes have been converted into exclosures. The changes portrayed here represent the average changes of vegetation cover in the study area over the last 30 years. (J. Nyssen, MA-1975-MU-218-R1-2006-NY).

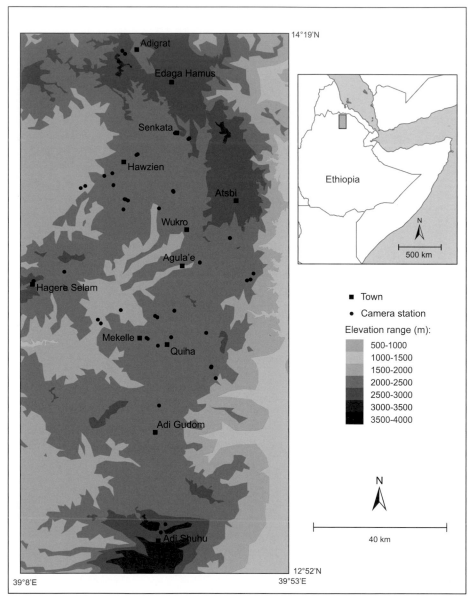

Plate 14.3. Location of repeat photography sites and plates in this chapter.

Plate 14.4. Debre Damo site, Ethiopia.

A. (1868). A rendering of the Debre Damo site, in which the artist linked up some of the mountains and widened the valley bottom. (R. Holmes, AD-1868-RE-13, King's Own Royal Regiment Museum, Lancaster).

B. (2007). The drawing, as well as the photograph, were produced in the second part of the dry season. Evidence that the 2007 photographer used the same position as the 1868 artist is provided by the fact that several elements from the drawing are found back in the current landscape. (J. Naudts, AD-1868-RE-13-R1-2007-NA).

Plate 14.5. Makhano, Ethiopia.

A. (1975). When this image was taken, the ancient agricultural terraces that utilize sandstone benches in Makhano near Senkata were largely free of woody vegetation. (R. N. Munro, SE-1975-MU-16).

B. (2006). The terraces have been rehabilitated, and vegetation has established on the risers. Many farmsteads have shelter belts, but the key feature is the lack of change in the church woodland (center right of the picture). This set of repeat photographs acted as a control with which to demonstrate that changes on the surrounding plains and slopes are not related to increased rainfall after a drought but to land husbandry. This contributed to the rejection of the hypothesis that the improvement is caused by higher rainfall; if such were the cause, the woodland would have improved too, because people do not cut down trees in such areas. (J. Nyssen, SE-1975-MU-16-R1-2006-NY).

Plate 14.6. Debre Tsion, Ethiopia.

A. (1975). A major fault crosses this landscape at Debre Tsion near Hawzen, Ethiopia, creating this scene with uplifted Adigrat sandstone scarp land at the middle and back, and younger limestone in the foreground. (R. N. Munro, HA-1975-MU-58).

B. (2006). From 1975 to 2006, there is a considerable improvement in growth of shrubs on slopes and of *Eucalyptus* (eucalyptus) trees around villages. There is limited streambed incision. (J. Nyssen, HA-1975-MU-58-R1-2006-NY).

Plate 14.7. Dugum, Ethiopia.

A. (1975). This image shows farmland on gently sloping sandy soils in Geralta (south of Hawzen). At the back are Mesozoic sandstone cliffs that cap Precambrian tillites, with less steep slopes that are largely used as rangeland. (R. N. Munro, HA-1975-MU-20).

B. (2006). Established contour bunds (foreground) and the regrowth of *Acacia* around farmsteads and on skyline slopes attest to the success of land management. The camel caravan in the midground provides scale. (J. Nyssen, HA-1975-MU-20-R1-2006-NY).

Plate 14.8. Northern slopes of Amba Alaje, Ethiopia.

A. (1975). As Amba Alaje (mountain tip at second plan, 3,440 meters) consists of Tertiary trap basalts, which yield fertile soils, even steep slopes were cultivated. Intense erosion took place leading to outcropping bedrock. (R. N. Munro, MA-1975-MU-228).

B. (2006). Photo interpretations by six experts revealed that sheet-and-rill erosion was reduced by approximately half in the viewed landscape. Whereas the disappearance of the riparian forest will definitely have major consequences on bank erosion and downstream flooding, as well as on the ecosystem in general, the slopes are better protected with woody vegetation. (J. Nyssen, MA-1975-MU-228-R1-2006-NY).

Plate 14.9. Gogua, Ethiopia.

A. (1975). The vegetation in the foreground is part of the protected church forest of Gogua (north of Wukro). (R. N. Munro, WU-1975-MU-47).

B. (2006). The foreground vegetation is largely unchanged. Background slopes on granite tors have been depleted of *Euphorbia candelabra* vegetation. The granite boulder in the lower left corner of the 1975 photograph was no longer present. (J. Nyssen, WU-1975-MU-47-R1-2006-NY).

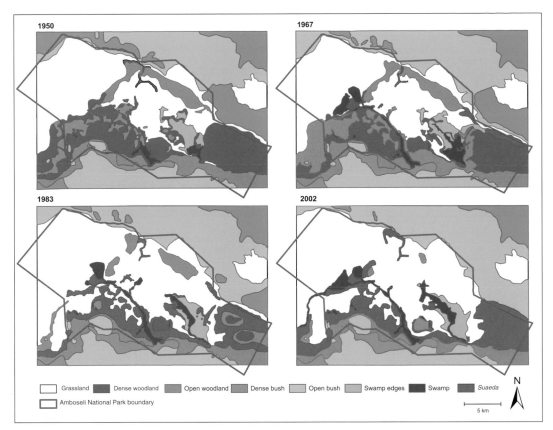

Plate 16.1. GIS maps of Amboseli Basin and National Park and environs showing the distribution of eight major habitats in 1950, 1967, 1983, and 2002. Note in particular the collapse of the woodlands and expansion of grasslands, *Suaeda-Salvadora* scrub, and swamps.

Plate 18.1. Beveridge photograph of Arnisdale, Scotland, UK.

A. (ca. 1898). Erskine Beveridge's photo shows the beginnings of the new, improved Arnisdale village, financed by the landlord, and is included here as historical background to an era of improvements in housing and social conditions that is often evident in Robert Adam's other township photos taken more than 30 years later. Note the numerous fishing boats on the shore and the small fields behind the houses. The wooded area and buildings from center right are part of the Arnisdale Farm, soon to become a sporting estate with its signature "big house." (E. Beveridge, SC748658, © RCAHMS [Erskine Beveridge Collection]. Licensor www.rcahms.gov.uk).

B. (28 September 2006). The new row of houses was completed and the old "black houses" closer to the shore were demolished shortly after Beveridge's photograph was taken. The most noteworthy change in these repeat images is the increase of woodland, due almost entirely to cessation of firewood cutting and the effects of reduced grazing pressure on the inbye fields. (R. F. Rohde, Sco 53b).

A

B

Plate 18.2. Adam's photograph of Arnisdale, Scotland, UK.

A. (20 August 1939). Adam's photograph of Arnisdale, taken from a position left of Beveridge's camera station in plate 18.1A, shows how extensively it was cultivated in 1939. (R. M. Adam, RMA-H7115).

B. (28 August 2006). The most noticeable changes evident in 2006 are the amenity trees near the row of houses, while the trees in the arable croft land have grown more as a result of agricultural abandonment in the last 30 or 40 years. Other signs of agricultural neglect are evident in the *Pteridium aquilinum* (bracken) and *Juncus effusus* (rushes) that have invaded the arable ground and *Ulex europaeus* (gorse) on the steeper slopes. Comparison of plates 18.1 and 18.2 underscores that these changes are recent. (R. F. Rohde, Sco 10b).

A

B

Plate 18.3. Lower Tarskavaig village, Scotland, UK.

A. (24 September 1931). This view of lower Tarskavaig village, from just outside the head dike of this crofting township, shows extensively cultivated fields during the late harvest. (R. M. Adam, RMA-H2769).

B. (23 September 2006). The effects of changing agricultural technology and improved housing are evident. Within the frame of the repeat photograph, there are now five occupied modern houses and one new house under construction while several of the old thatched or corrugated roofed houses are either derelict or used as barns and sheds. Today there are no haystacks, corn stooks, and strips of potatoes; earlier harvesting of the grass for big bale silage has left a lush green sward for early winter grazing. Underlying these changes is a remarkable stability in the social fabric of the crofting community reflected in the persistence of basic settlement and field patterns. (R. F. Rohde, Sco 6b).

A

B

Plate 18.4. Bornesketaig township, Scotland, UK.
A. (27 September 1936). This view of Bornesketaig township shows partially harvested hayfields and small patches of potatoes with "black houses" behind. (R. M. Adam, RMA-H5269).
B. (1 September 2006). Surrounded by new modern bungalows, the "black houses" inhabited in 1936 but abandoned soon afterward have now been restored—one as a heritage museum. The fields in Adam's photo are now fallow and used solely for grazing sheep. (R. F. Rohde, Sco 46b).

A

B

Plate 18.5. Flodigarry township, Scotland, UK.
A. (27 September 1936). This view of Flodigarry township shows a recently built "white house" with large stacks of peat to the right and winter fodder stacked in the yard behind. All the arable land in the township has been cropped for oats and hay. (R. M. Adam, RMA-H5276).
B. (1 September 2006). Modern bungalows have replaced the improved "white houses." Over half of the fields are now fallow and invaded by *Pteridium aquilinum* and *Juncus effusus*, and native woodlands are on the increase. (R. F. Rohde, Sco 47b).

A

B

Plate 19.1. Village of Newtonmore, Scotland, UK.
A. (ca. 1890s). A view looking northeast along the main street in the village of Newtonmore, in what essentially served as the village center. A large *Acer* sp. (sycamore) tree is on the left; it appears in many other vintage views of the village. Signs on the buildings at the right foreground advertise a "Jeweller" and "car-hire"; beyond them is a small roadside cottage. (Probably a James Valentine and Co. photograph, published by D. A. Anderson as a postcard).
B. (2005). The *Acer* has greatly increased in size. The buildings in the right foreground have been removed, while those on the left have changed little in form. The village center has shifted about half a kilometer away. (P. R. Moore).

Plate 19.2. Cairn Gorm car park, Scotland, UK.

A. (ca. 1964). The Cairn Gorm car park in Coire Cas, probably photographed in the month of July. Signs of excavation and construction of the Cairn Gorm ski area are evident on the slopes at left. (J. Brooks, 1964).

B. (August 2007). Prominent within the view is a new building to house the workings of a funicular railway, opened in 2001. The funicular train can be seen approaching on raised rails. Rigorous environmental specifications for the earthworks and construction and sympathetic coloring of buildings and infrastructure have helped to minimize the visual impacts of this significant imposition on the landscape. (P. R. Moore).

Plate 20.1. Market at Gracias, Lempira, Honduras.

A. (August 1957). In the pottery section of the Gracias market, Lenca women sell pots carried on their backs 15 kilometers from their village of La Campa. Gracias was founded by the Spanish in 1526, and its colonial fort, Castillo San Cristóbal, is visible in the background. The hillsides surrounding the fort are bare of vegetation except for grasses. (R. C. West, H1, 18-5).

B. (July 2001). Women from La Campa still sell their pottery, though they came to town in pickup trucks and there are fewer of them. The men now appear more *ladino* in their dress, and mountain bikes, a pickup truck, an electronics store, and an illuminated Pepsi sign indicate the modernization that is happening. The *castillo* is now a historic attraction and public park, covered in landscaped trees, mostly *Pinus* and *Quercus*. (J. O. Bass, HOND. PP4b).

Plate 20.2. La Campa, Lempira, Honduras.

A. (August 1957). La Campa is a Lenca village about 15 kilometers south of Gracias. The village church is at center. Most dwellings have tile roofs, and the distant montana forest has been almost completely cleared for cultivation. (R. C. West, H1, 18-4).

B. (August 2001). The most obvious changes are the addition of a new road running through town south toward San Manuel de Colohete, and an increase in vegetation in and on the slopes around the town (see also Southworth and Tucker 2001). The distant forest has apparently changed. The new trees in town are a mix of natives and imports, including *Eucalyptus*. Outside town, most of the increase appears to be in pines, likely the result of an array of socioeconomic changes (see Nagendra et al. 2004). (J. O. Bass, HOND. PP37b).

A

B

Plate 20.3. Marcala, La Paz, Honduras.

A. (July 1957). This covered bridge over a stream flowing through Marcala, La Paz, is typical of bridges seen elsewhere in Central and South America. The area to the left of the adobe wall was a *finca* (farm) that grew citrus and perhaps coffee. Marcala, a commercial center, has been an important coffee-growing region for some time (see Bass 2006b). Alcerro-Castro claimed that in 1893 Marcala became the first Honduran town to export coffee to Germany (1989: 41). (R. C. West, H, 1-8).

B. (2001). The most obvious changes are in commercialization. The town now includes this once marginal street in its commercial district, with the accompanying stores, advertising, brands indicating global–local linkages, changes in resource access, and changes in material culture. The farm behind the left wall is a large municipal market. The bamboo and trees appear to be larger, and there is a greater variety of vegetation, partially due to an increase in trees in domestic landscapes and those planted for landscaping. (J. O. Bass, HOND. PP18b).

A

B

Plate 21.1. Leroux Island, Little Colorado River, Arizona, USA.

A. (1851). Richard H. Kern's "Leroux Island, Little Colorado River" appears as plate 8 from Sitgreaves (1853). This site is not mentioned in Sitgreaves's journal, but its location appears on a map in his account. The men on the rock and the bush to the left of it (which appears to be a stylized *Juniperus* [juniper]) were apparently added by the New York lithographer. (R. H. Kern, from Sitgreaves [1853]).

B. (4 November 2001). The author and professor Andrew Wallace on Kern's rock with Leroux Island in the distance. Foreground obstructions necessitated a camera position further back than Kern's viewpoint. Besides the perennial grasses, which add a yellow tone to the photograph, the trees are *Juniperus osteosperma* (juniper), and *Gutierrezia sarothrae* (snakeweed), a common overgrazing indicator in this region, appears at lower right. (T. Jonas and R. A. Tompkins).

Plate 21.2. Canyon of the Gila River, Arizona, USA.

A. (ca. 1853). Seth Eastman's watercolor, "Great Cañon, River Gila," was based on a field sketch taken by an unknown participant in the US and Mexican Boundary Survey in 1851. The scene depicts the "jumping off place" of the survey, where the survey had to be interrupted due to this impassable canyon known today as the Needle's Eye. The view looks west, or downstream, on the river. The artist was never at the site, which explains why he erroneously showed the water flowing toward the viewer. This scene was described as being downstream from the San Francisco River, which has led many to believe it is located in the Gila Box, southwest of Clifton, Arizona; in fact, it is a few kilometers west of Coolidge Dam, downstream from the San Carlos River, which was known as the San Francisco in 1851. (S. Eastman, Museum of Art, Rhode Island School of Design. Gift of the RISD Library, reproduction photography by Del Bogart).

B. (1 November 2003). The match demonstrates that the original view was along the Gila River downstream from Coolidge Dam. Riparian vegetation, mostly nonnative *Tamarix*, has increased, but large rocks in the river channel remain remarkably similar to the rocks depicted in 1851. The accuracy of Eastman's watercolor suggests that the original field sketch was made with the help of a camera lucida. (T. Jonas).

Plate 22.1. Kodak House check dam, Mesa Verde National Park, Colorado, USA .

A. (1891). On the cliff-top directly above Kodak House is a drainage where the inhabitants built a check dam to impound water for irrigation; there are many check dams in Mesa Verde National Park. When Nordenskiöld took this photo, the dam had been breached on the left side, many of the rocks had moved, and a large dead tree had fallen across the right side of the dam. (G. Nordenskiöld, N182).

B. (1987). The match shows that most of the rocks remain, and, surprisingly, the fallen tree and two of its branches can still be seen! This site is exposed and subject to runoff from the drainage and the adjacent slopes. Note the increase in vegetation density and size in 1987. (B. Howard, B182).

C. (30 June 2006). The devastating effect of a 2000 wildfire is shown in this rematch. Only a small, charred piece of the fallen tree remains. Although nearly all of the vegetation burned, some regrowth has occurred in the ensuing six years. Lack of vegetation increased the velocity of runoff, which moved several rocks. With less ground cover present, more check dams are faintly visible upstream from the one in the 1891 and 1987 photographs. (D. Hamilton, 6-30-2002).

Figure 13.3. San Pablo de Tregua, Valdivia, Chile.

A. (January 1977). A treefall gap in an old-growth forest at San Pablo de Tregua was created by the fall of a 45-meter-tall *N. dombeyi* in the austral winter of 1976 in a previously installed permanent forest plot. The arrow points to the stump of a codominant tree snapped by the fall of the *N. dombeyi*. (T. T. Veblen, 1010).

B. (January 1979). Two years later, the understory response to the gap was dominated by accelerated growth of juveniles of previously established trees (especially *Laureliopsis philippiana*) and the bamboo *Chusquea culeou*. (T. T. Veblen, 1011).

C. (January 1984). Seven years after the creation of the gap, the understory response continues to consist mostly of accelerated growth of previously established stems of *Laureliopsis philippiana* and *Chusquea culeou*. Sampling of the site in 1984 indicated a lack of successful establishment of *N. dombeyi* seedlings. (T. T. Veblen, 1012).

its ecological consequences along the striking vegetation gradient from Andean rain forests to the steppe at 39 to 42° S. Fire and changes in fire regimes play a major role in structuring this landscape, as we have previously illustrated using repeat photography (Veblen and Lorenz 1988). Research conducted largely since the early 1990s has revealed some of the complexity of climatic and human influences on fire regimes as well as the key roles played by introduced wild animals and livestock in vegetation responses to fire (Veblen et al. 2003). Here, repeat photography is used to further examine the roles of both natural and anthropogenic fire on structuring the vegetation of this landscape, the interactions between landscape patterns and fire behavior, the ecological consequences of fire exclusion, and the changes in the potential for fire related to land-use practices.

The Role of Fire across the Rain Forest to Steppe Gradient

Along the gradient from rain forest to steppe in northwestern Patagonia, differences in fuel types and their responses to interannual climatic variation strongly differentiate historical fire regimes (Veblen et al. 1992a, Kitzberger et al. 1997). At the western end of the gradient, fires in rain forests of *Fitzroya cupressoides* and *Nothofagus dombeyi*, ignited either by lightning or humans, can spread only during infrequent dry years (Veblen et al. 1999). In the wettest and most remote forests of the region, fire scars on *F. cupressoides* and postfire cohort age structures of *N. dombeyi* indicate that at intervals of a century or more, severe crown fires historically shaped forest structure (Veblen et al. unpublished data). In these rain forests and also in the mesic forests dominated by *N. dombeyi* and *Austrocedrus chilensis*, fuel quantity is clearly not limiting to fire spread; instead fire occurrence is limited by a combination of ignition sources and desiccating weather.

The occurrence of lightning in this landscape is highly variable at an annual scale. Records from Argentine national parks in this region (fig. 13.1) from 1937 to 2006 record some individual years with no lightning-ignited fires and others with many lightning-ignited fires (Administración de Parques Nacionales, 2006, unpublished data). Thus the fire regime of this region cannot be considered to be "ignition-saturated" by lightning, at least not on an annual scale. On the other hand, at a supra-annual to decadal time scale, lack of lightning is not limiting to fire ignition. Since the mid-1970s, this region has recorded a sharp increase in the number of lightning-ignited fires that is associated with a trend toward warmer summer temperatures (Kitzberger and Veblen 2003). Park records indicate that both humans and lightning have been the sources of ignition for extensive fires often burning greater than 2,000 hectares in single events, even in the wettest forests (Administración de Parques Nacionales, 2006, unpublished data).

In the mesic forests dominated by *N. dombeyi* and *A. chilensis*, extensive fires occurred during the second half of the nineteenth century and the first decade of the twentieth century; these fires coincided with both severe droughts and intentional burning by European colonists (Veblen and Lorenz 1988, Kitzberger et al. 1997). Photographs taken by early scientific explorers in the 1890s and early 1900s show widespread, recently burned forests either dominated purely by *N. dombeyi* or mixed with *Austrocedrus* (figs. 13.4 and 13.5). These fires occurred mostly in the 1890s to 1914 period when Euro-American settlers attempted to clear forest and replace it with cattle pasture (Moreno 1897, Willis 1914). For the provinces of Neuquén, Río Negro, and Chubut, Rothkugel (1916) mapped 692,000 hectares (37 percent) of the total forest area as having burned during the European settlement period from the 1890s to 1914. However, as noted by Willis (1914), years of widespread fire also coincided with extreme droughts, which have been corroborated by dendrochronological dating of fires and tree-ring reconstructions of climate conditions (Kitzberger et al. 1997, Veblen et al. 1999). In much of the mesic forest

A

B

Figure 13.4. Lago Moreno and Lago Nahuel Huapi, Nahuel Huapi National Park, Argentina.
A. (1896). This westward view of Lago Moreno (left center), Lago Trebol (foreground), and the southwestern part of Lago Nahuel Huapi (distant background) is from the Campanario scenic overlook. Most of the distant burnt forests are dominated by *N. dombeyi*. The burnt forests in the foreground are mixed stands of *N. dombeyi* and *Austrocedrus chilensis*. (From Moreno 1897).
B. (March 2007). In 2007, the area of forest recently burned in 1896 is now dominated by *N. dombeyi* and *Austrocedrus chilensis* that established soon after the fires of the late nineteenth century. The large Llao-llao Hotel is visible in the center right of the 2007 photograph. (T. T. Veblen, 2101).

A

B

Figure 13.5. Península San Pedro and Isla Huemul, Nahuel Huapi National Park, Argentina. A. (1902). This eastward view of Península San Pedro and Isla Huemul, taken from the Campanario scenic overlook (see fig. 13.4), shows foreground forests that had been recently burned near the end of the nineteenth century. (From Holdich 1904). B. (March 2007). The burned areas have recovered to mixed-dominance of *Austrocedrus chilensis* and *N. dombeyi*. (T. T. Veblen, 2102).

zone burned from about 1890 to 1914, today there are even-aged stands of *N. dombeyi* and *A. chilensis* (Veblen et al. 1992a). Despite the success of postfire regeneration of these two dominant tree species over large areas, regeneration sometimes fails or is delayed due to lack of available seed, postfire climate conditions, and the browsing of introduced mammals (Veblen et al. 1992b, Kitzberger and Veblen 1999, Tercero-Bucardo et al. 2007). As will be discussed here, this potential for the failure of postfire tree regeneration creates the possibility of positive feedbacks between fire and vegetation responses that increase the likelihood of subsequent fire.

In dry *Austrocedrus* woodlands, at the opposite end of the vegetation gradient, fuel and climate controls on fire spread differ importantly from the pattern in mesic and rain forests. The *Austrocedrus* woodlands occur at the ecotone with the steppe, which is dominated by bunchgrasses and small shrubs. At relatively dry sites, quantity and connectivity of fine fuels (mostly grasses) are seen to limit the spread of modern fires, such as the extensive fires that burned in the eastern part of Nahuel Huapi National Park in 1996 (Kitzberger et al. 2005a). In open woodlands of *Austrocedrus*, historical fires burned as relatively low-severity surface fires, permitting the survival of scattered, large-diameter trees. Fire-scar data from *Austrocedrus* show that in this habitat type,

years of widespread fire both coincide with drought and follow moist years by one to several years (Kitzberger et al. 1997). Thus years of above-average moisture and increased growth of fine fuels favor widespread fires at a lag of one to several years by increasing the quantity and continuity of fine fuels.

Interactions between Landscape Patterns and Fire Behavior

Presence of sharp boundaries between tall forests and adjacent, recently burned tall shrublands led to the hypothesis that structural and site differences in these vegetation types result in differences in their flammability (Veblen and Lorenz 1988). Midslopes in the foothills, especially on drier north-facing aspects, are typically characterized by tall shrublands dominated by the shrubby tree *Nothofagus antarctica*, the shrubs *Diostea juncea*, *Schinus patagonicus*, *Embothrium coccineum*, and *Chusquea culeou*. Toward higher elevations, midslope shrublands are bordered by tall subalpine forests of *N. pumilio*, and below by tall mesic forests of *N. dombeyi* and *A. chilensis*. Sharp boundaries between recently burned midslope shrublands and subalpine *N. pumilio* forest indicate that past fires burned extensively through the shrublands but then stopped or only encroached slightly into the subalpine forest (fig. 13.6).

A

B

Figure 13.6. Pampa del Toro and the Cordón Blanco southeast of Nahuel Huapi National Park, Argentina.
A. (1912). This eastward view of Pampa del Toro and the Cordón Blanco shows a sharp boundary between recently burned midslope shrublands dominated by *Nothofagus antarctica* and the bamboo *Chusquea culeou* and higher-elevation forests dominated by *N. pumilio*. (From Willis 1914).
B. (March 2007). More than 90 years later, the sharp fire boundary between the midslope shrublands and the subalpine forest is still evident. Fires that burn frequently through the highly flammable midslope shrublands often stop at the edge of the more fire resistant tall forests of *N. pumilio*. The road that crosses the foreground is the main highway from Bariloche to El Bolsón. (T. T. Veblen, 2103).

The idea, originally derived from repeat photography, that differences in the flammability of adjacent vegetation types of contrasting structures may allow fire to maintain sharp vegetation boundaries, stimulated further research on the interactions of vegetation types and fire behavior. Analyses of fire–vegetation relationships in a 780,000 hectare study area from 1985 to 1999 showed that shrublands, dominated by *N. antarctica*, *Chusquea*, and other woody resprouters, are proportionally more affected by fire than adjacent tall mesic forests (Mermoz et al. 2005). Tall shrublands, dominated mainly by *N. antarctica* and *Chusquea* bamboos, show a positive association with burnt areas over this large area, and analyses at a fine spatial scale show that once shrubland patches are ignited, they tend to burn completely until a topographic break or a less flammable vegetation type is encountered. In contrast, tall forests dominated by *N. dombeyi* or *N. pumilio* burn less extensively than expected. Patches of the subalpine *N. pumilio* forests often serve as natural fire breaks, except under the most severe drought conditions when they also burn (Mermoz et al. 2005).

These analyses of fire–vegetation relationships based on modern fire records are consistent with tree-ring studies that show higher fire frequencies in areas encompassing tall shrublands in comparison with adjacent tall forests (Veblen et al. 1992a).

Several factors may contribute to the higher flammability of the shrublands (Veblen et al. 2003): (1) high density of multistemmed small trees and shrubs; (2) rapid fuel recovery due to resprouting; (3) heavy loads of flammable epiphytes (e.g., *Usnea* lichens) and vines (e.g., *Mutisia* spp.) that provide fuel ladders; (4) rapid accumulation of dead biomass from partial crown dieback of the shrub and tree species and from annual foliage replacement of vines; (5) location on steep, northerly midslopes where fuels are prone to desiccation; and (6) dominance of shrub species with high contents of volatile chemicals. In contrast, tall *Nothofagus* forests have coarser, moister fuels, and lack vertical fuel continuity between the understory and canopy.

Contrasts in the susceptibility of shrublands and mesic forests to fire create important feedbacks that facilitate long-term shifts in the relative abundance of these landscape components (Veblen et al. 2003, Kitzberger et al. 2005b). The life-history characteristics of the dominant tree and shrub species and the effects of introduced herbivores in this landscape play key roles in determining postfire vegetation patterns. As a broad generalization, over the past century the area of resprouting tall shrubs and small trees has expanded at the expense of tall tree species that are all obligate seed reproducers. At many sites, in the long-term absence of fire, these shrublands can be successionally replaced by forests and can be considered quickly recovering, fire-prone systems (Veblen et al. 2003, Kitzberger et al. 2005b). In contrast, the tall forests are dominated by tree species (mainly *N. dombeyi, N. pumilio,* and *Austrocedrus chilensis*) that depend exclusively on reproduction by seed, which is a slower and less certain mechanism of postfire recovery. Following fire, these obligate seed producers may regenerate to form dense postfire populations if seed trees remain and climate and other environmental conditions are favorable for seedling survival (Villalba and Veblen 1997). However, their regeneration may be slow or fail entirely due to exceptionally severe fires, fire-induced edaphic changes, lack of seed sources, postfire herbivory, and/or unfavorable postfire weather (Veblen et al. 2003, Kitzberger et al. 2005a, Tercero-Bucardo et al. 2007).

Postfire herbivory can be important to long-term conversion of tall forests to shrublands because the resprouting shrubs resist browsing better than the seed-producing tall tree species (Raffaele and Veblen 1998, 2001; Kitzberger et al. 2005b). Fire followed by browsing favors the replacement of tall forests by tall shrublands that appear to be self-replacing because of the positive feedback between fire and shrub cover. Measurement of vegetation responses to recent fires under experimentally varied animal impacts and microclimate clearly shows that under drier conditions, with or without presence of livestock, postfire tree regeneration fails (Tercero-Bucardo et al. 2007). These recent studies of postfire vegetation responses, when combined with historical information from the beginning of the twentieth century, strongly imply that a combination of fire and animal browsing has resulted in broad-scale vegetation changes, creating a more flammable landscape. In an extensive resource survey conducted in 1912–1913 covering the Andean area from 39° to 44° S, large areas of forest and tall shrubland were described and mapped as recently burned (Willis 1914). This widespread burning of tall forests resulted in changes in vegetation cover (i.e., transition to tall shrublands), which further increased the overall flammability of this landscape as shown by analyses of fire–vegetation relationships (Mermoz et al. 2005). The episode of forest burning from 1890 to 1914 that coincided with widespread livestock browsing of burned sites would have increased the connectivity of fire-prone shrublands and favored the spread of subsequent fires. Thus, during subsequent episodes of burning, such as in the droughts of the early 1940s and late 1990s, the higher proportion of tall shrublands in the landscape mosaic facilitated more extensive burning.

Ecological Consequences of Fire Exclusion

The tree-ring record of fires in the *Austrocedrus* woodlands near the ecotone with the steppe documents a dramatic decline in fire occurrence since the early 1900s (Kitzberger et al. 1997, Veblen et al. 1999). Several factors contribute to this decline in formerly frequent surface fires: (1) fewer intentionally set fires by Native Americans, (2) fuel reductions due to increased grazing by livestock, and (3) active fire suppression. The Native American populations occupying this habitat were nomadic hunters who depended mainly on guanaco (*Lama guanicoe*) and secondarily on large flightless birds (*Pterocnemia pennata*) and deer (*Hippocamelus bisulcus*) (Fonck 1900). Eighteenth- and nineteenth-century explorers describe the use of fire as a Native American hunting tool to drive game in the forest/steppe ecotone (Machoni 1732, Musters 1871). As early as the seventeenth century, Native Americans here had adopted the horse and other European livestock

(Machoni 1732, Fonck 1900), so that intentional burning may also have been motivated by a desire to improve forage for domestic as well as game animals (Veblen and Lorenz 1988). The Argentine-Indian war of 1879–1883 drastically reduced the Native American population, and the surviving population had access to only a small fraction of their traditional hunting lands (Moreno 1897). Thus formerly intentional burning by Native Americans declined at the end of the nineteenth century. At the same time, in conjunction with European colonization, livestock raising massively expanded, which would have reduced the quantity and connectivity of grass fuels. Later, with the establishment of national parks in the 1930s, active fire suppression in the area of *Austrocedrus* woodlands, which are easily accessible by road, would have further contributed to the decline in fire occurrence.

During the twentieth-century period of reduced fire frequency in the *Austrocedrus* woodlands, there has been a regionally extensive increase in tree-dominated cover types at the expense of cover types dominated by grasses and small shrubs (Kitzberger and Veblen 1999). Historical photographs from the 1890s show woody encroachment, especially by *Austrocedrus*, but also by other trees and tall shrubs such as *Lomatia hirsuta*, *Maytenus boaria*, and *Schinus patagonicus* (figs. 13.7 and 13.8). At these sites of woody encroachment, there is no field or photographic evidence of a former woodland cover in the form of spars or stumps. Thus former open woodlands that previously supported moderately frequent surface fires have converted to denser stands of trees. Tree ages over a large region of *Austrocedrus* woodlands from 39°30′ to 43° S indicate that most trees were established after the late-nineteenth-century decline in fire frequency (Veblen and Lorenz 1988, Villalba and Veblen 1997). Although the timing of this tree encroachment reflects supra-annual to decadal periods of more favorable climate (Villalba and Veblen 1997), the survival of newly established trees would not have been possible without the decline in fire frequency. Woody encroachment into the steppe and increased stand densities of formerly open woodlands of *Austrocedrus* have created more contiguous woody fuels so that crown fires are now possible at sites that formerly supported only surface fires.

A

B

Figure 13.7. Dedo de Dios, Nahuel Huapi National Park, Argentina.
A. (1896). This southwesterly view looking toward the Dedo de Dios from 4 kilometers south of the confluence of the Río Limay and Río Traful shows relatively open slopes with sparse trees. During the nineteenth century, relatively frequent surface fires killed tree seedlings, which maintained the open character of the woody vegetation. (From Moreno 1897).
B. (December 1985). Following a decline in fire frequency during the twentieth century, *Austrocedrus chilensis* has expanded dramatically into the formerly open steppe in the background and midground. In the midground and foreground, the small trees and shrubs *Schinus patagonicus*, *Discaria articulata*, and *Berberis buxifolia* have become more abundant. (T. T. Veblen, 1101).

A

B

Figure 13.8. Río Limay, Nahuel Huapi National Park, Argentina.

A. (1912). This northeasterly view along the Río Limay, 4.5 kilometers upstream from the confluence with the Río Traful, shows steppe vegetation. The lack of logs or burnt spars indicates that forests did not exist on these slopes earlier in the nineteenth century. Fire frequency was relatively high on these slopes in the nineteenth century but declined dramatically during the twentieth century. (From Willis 1914).

B. (December 1985). Comparison of the two photographs shows substantial woody encroachment by *Austrocedrus chilensis* into areas formerly dominated by small shrubs and bunchgrasses. Construction of a reservoir downstream on the Río Limay in the late 1970s resulted in the flooding of trees along the river (T. T. Veblen, 1102).

Changes in the Potential for Fire Related to Land-Use Practices

Overall fire risk in the dry habitat near the forest and steppe ecotone has increased, not only due to woody encroachment by native trees and shrubs but also because of widespread planting of exotic conifers as well as increased residential and recreational use of this landscape. In the national parks of northwestern Patagonia, since the 1930s, small-scale experimental plantations of exotic conifers have been established, including *Sequoiadendron giganteum*, *Sequoia sempervirens*, *Picea* spp., and *Pinus* spp. (Dimitri 1972). Beginning in the 1970s, large areas in the provinces of Neuquén, Río Negro, and Chubut were planted to *Pinus ponderosa* (accounting for 90 percent of the area planted), *Pinus contorta*, and *Pseudotsuga menziesii*. By 1998, there were 50,000 hectares of planted conifers in these provinces, including roughly 10,000 hectares planted within the national reserve portions of national parks (Schlicter and Laclau 1998). The potential area suitable for afforestation with conifers, mainly *Pinus ponderosa*, has been estimated at 700,000 to 2,000,000 hectares, which is larger than

the current area occupied by the native conifer *Austrocedrus* (Schlicter and Laclau 1998).

The transformation of large areas of relatively open steppe with small patches of tall shrubs and trees is a dominant feature of the 2007 Patagonian landscape compared with the 1980s (figs. 13.9 and 13.10). Although the effects of plantations on native biodiversity have been little studied in this region, preliminary studies indicate a decline in plant and animal diversity in plantations (Raffaele and Schlicter 2000, Paritsis and Aizen 2008). Plantations also favor the increase of exotic plant and animal species, and the local extirpation of rare and specialist species (Paritsis and Aizen 2008). In addition to the detrimental effects of these plantations on native biodiversity, there is the potential for negative consequences for soil erosion depending on silvicultural technique; there is also concern about the impact on the quality and quantity of hydrologic resources and the cascading effects on wetlands and riparian habitats (Schlicter and Laclau 1998). Furthermore, the creation of large areas of highly combustible, densely planted conifers has radically increased the hazard of

Figure 13.9. Confluence of the Río Limay and Río Traful, Nahuel Huapi National Park, Argentina.
A. (1902). This southerly view, at the confluence of the Río Limay and Río Traful, shows an old wooden bridge crossing the Río Traful. (From Anchorena 1902).
B. (March 1994). The bridge is now gone, and the higher waters in the view are the result of a reservoir constructed on the Río Limay in the late 1970s. Comparison of this view with A shows encroachment of *Austrocedrus chilensis*, especially on the slope at the far right and in the rocky area in the center left. (T. T. Veblen, 1103).
C. (March 2007). The 2007 photograph records the growth of exotic pines planted on those same slopes in the 1980s, creating a relatively continuous cover of flammable crown fuels. (T. T. Veblen, 2104).

Figure 13.10. Río Meliquina, Nahuel Huapi National Park, Argentina.
A. (1902). This view, which is west along the Río Meliquina, shows the river meandering through an open steppe. (From Anchorena 1902).
B. (March 1998). Much of the midground that had been open steppe in 1902 has been invaded by *Austrocedrus chilensis*. In both the mid- and foreground exotic pines have been planted over most of the surface, creating continuous woody fuels. (T. T. Veblen 1104).

severe crown fires. Given the regularity of annual summer droughts, as well as the historical susceptibility of this region to extreme droughts at multidecadal intervals (Kitzberger and Veblen 2003), the high fire risk to these plantations is well established. Less well appreciated is the strong warming trend in this region since the mid-1970s (Villalba et al. 2003), and the prediction of reduced precipitation over the next century (Vera et al. 2006). These climate trends, plus the documented recent increase in lightning-ignited fires in association with warmer summer temperatures associated with subtropical air masses (Kitzberger and Veblen 2003), imply an increasing risk of severe crown fires in the region planted to exotic conifers. Observations of fire behavior in recent years (e.g., the 1996 Challhuaco fire in Nahuel Huapi National Park) indicate that pine plantations increase both fire spread and severity, and current climate trends are favoring increased fire occurrence.

In addition to increased fire hazard related to enhancement of woody fuels due either to native woody encroachment or to plantations of exotic conifers, the drier habitats near the forest/steppe ecotone are also experiencing increased risk of fire ignition due to greater human presence and release of some areas from grazing. These habitats are the primary location of residential expansion as both exurban and suburban development (fig. 13.11). The mesic forest districts, especially along the shores of the large lakes, have been the locations of important tourist and residential development since the 1940s, but, especially since the 1980s, development has expanded eastward into the area of the forest/steppe ecotone. Developments in this habitat typically replace livestock raising, so that grass fuels have increased under reduced pressure from livestock. Proximity to Bariloche, the major urban area of northwestern Patagonia, has been shown to result in a significant increase in fire occurrence in comparison to areas less frequented by humans. For example, although the wildland–urban interface within 10 kilometers of Bariloche accounted for only 2 percent of a larger study area analyzed for fire occurrence between 1985 and 1999, it accounted for 36.5 percent of the total burned area (Mermoz et al. 2005). Tourism and recreation continue to grow rapidly, and so does the related risk of human-set fires.

Discussion and Conclusions

Repeat photography has supplemented a variety of ecological methods and has contributed significantly to our understanding of disturbance ecology

A  B

Figure 13.11. Nahuel Huapi National Park, Argentina.
A. (1902). This view is southeasterly from the Campanario scenic overlook and shows part of Lago Nahuel Huapi to the left and Lago Gutiérrez to the right. The crowns of recently burnt *N. dombeyi* and *Austrocedrus chilensis* are visible throughout the foreground and midground. (From Holdich 1904).
B. (March 2007). The urban and suburban areas of Bariloche are visible in the midground and background. Exurban development in the foreground is surrounded by postfire forests. Fire risk is high in this wildland–urban interface due to both the fire-prone nature of vegetation and the high potential for fire ignition by humans. (T. T. Veblen, 2105).

and vegetation dynamics in the midlatitude Andes of Chile and Argentina. Repeat photography has been particularly useful in communicating our research findings on landscape changes to public and professional audiences. It has also been instrumental in generating specific research questions that have subsequently been investigated through a combination of experimental and retrospective methods.

The case study of the ecological impacts of the 1960 giant earthquake on the forested landscape of the Chilean Lake District illustrates the profound and long-lasting effects of large, infrequent disturbances. Research on the dynamics of *Nothofagus*-dominated mixed-species forests in relation to the tectonic disturbances of 1960 revealed that over large areas these forests are not in species compositional equilibrium. Instead, the dominants of these forests are relatively old populations that established after previous coarse-scale disturbances similar to the earthquake-triggered mass movements of 1960, but also including blowdown, fire, flooding, and volcanic ash deposition. The continued presence of the shade-intolerant *Nothofagus* spp. in mixed assemblages with shade-tolerant tree species (e.g., *Laureliopsis*, *Saxegothaea*, *Dasyphyllum*) reflects the longevity (greater than 400 years) of these trees relative to the interval between coarse-scale disturbances. South-central Chile was previously affected by severe tectonic disturbances in 1575, 1737, and 1837 that created denuded and depositional sites, and therefore opportunities for establishment of *Nothofagus* that continue to dominate old stands. The findings for *Nothofagus* forests in south-central Chile are similar for forests in many parts of the world in showing that current tree population structures do not represent steady-states or compositional equilibrium (i.e., a climax in traditional terminology), and that repeated, large-scale disturbances have shaped their structures (Attiwell 1994).

Strong parallels between disturbance ecology and *Nothofagus* regeneration have been found in the tropical Southwest Pacific, Australia, and New Zealand, which have led to the development of a general model of *Nothofagus* regeneration dynamics (Veblen et al. 1996b). This model stresses the interaction of site characteristics and coarse-scale disturbances in determining the successional status of *Nothofagus*. Throughout its worldwide distribution, *Nothofagus* often occurs as monotypic stands on sites too harsh for other tree species that may be superior competitors. Where *Nothofagus* occurs on sites favorable to superior tree competitors, it can often coexist with them due to its ability to colonize and grow rapidly on sites severely disturbed by coarse-scale disturbances.

In northwestern Patagonia, the case study on landscape changes over approximately the past century reveals a dominant role for fire and interactions, sometimes subtle, with human activities such as the introduction of plants and animals. Repeat photography has been particularly useful in documenting the occurrence of widespread fires during the nineteenth century in mesic and wet forests that can be linked to both climatic variation and intentional burning by European colonists. Likewise, repeat photography documented dramatic woody encroachment in xeric woodland environments during the twentieth-century period of reduced fire occurrence.

Photographic evidence of differences in flammability of different vegetation types was key in developing a hypothesis of positive feedbacks between fire and subsequent fuel types that accounts for a broad-scale tendency toward conversion from fire-resistant tall forests to fire-prone tall shrublands. In the past century, this conversion has tended to occur in a stepwise fashion of short periods of drought-induced fire; it has been exacerbated by inhibitory effects of introduced animals and subsequent drought on postfire tree regeneration. Under current warming and drying trends that are strongly evident in regional climatic records since the mid-1970s (Daniels and Veblen 2000, Villalba et al. 2003), this stepwise process of forest to shrubland conversion is likely to accelerate. Repeat photography showing widespread planting of exotic conifers in the forest/steppe ecotone that has increased the risk of crown fires in areas of formerly heterogeneous and discontinuous woody fuels further presages an increasing role for fire in the future of this landscape.

Repeat photography in the studies of the ecological roles of tectonic disturbances in south-central Chile and of fire in the vegetation dynamics of northwestern Patagonia is particularly useful in evaluating the precedents and consequences of relatively infrequent large-scale ecological events. Clearly, the average interval between large-scale tectonic events in south-central Chile is beyond the typical human life span. Appreciating the role of these infrequent events is of vital importance to understanding the potential for natural hazards to humans in this region, as well as their role in shaping renewable resources such as soils and vegetation. Research findings about the role of coarse-scale disturbance in the dynamics of *Nothofagus* forests in this region have directly informed traditional silvicultural management of these and similar *Nothofagus* forests throughout the southern Andean region (Donoso and Lara 1999). These findings are also essential for more broadly conceived, ecosystem-based management aimed at sustaining both commodity production and ecosystem services (Christensen et al. 1996). As a practical way of managing landscapes to preserve biodiversity, understanding of natural variability in ecosystem conditions and processes provides a coarse-filter strategy for dealing with sustainability of diverse and often unknown species requirements.

Likewise, the study of the history of fire and ecological consequences of changes in fire regimes in northwestern Patagonia has provided a basis for evaluating and informing policy discussions in relation to recent years of widespread fires. Although intervals between severe fires recurring at the same site or between recurring years of widespread multiple fires over a large region of northwestern Patagonia may be on the order of only a decade or two, humans often find it difficult to appreciate the consequences of events that are outside of the recent memory of one or two decades. Such may have been the case when major drought-induced fires in northwestern Patagonia in 1996 and 1998 were perceived as unprecedented, yet there was strong historical evidence that much larger areas were burned in this region in 1943–1944 and the 1890s–1910s (Veblen et al. 2003). For communicating the ecological consequences of these large-scale events separated by intervals of several decades or more, repeat photography is a powerful tool.

Acknowledgments

Research was supported by the Council on Research and Creative Work of the University of Colorado, the National Science Foundation (Award 0117366), and the National Geographic Society (Awards 7155-01 and 7988-06). I thank many collaborators who facilitated this and related projects through assistance in obtaining photographs and finding field sites. Among others they include R. Alvarez, H. Amado, M. Besley, A. Dezotti, M. González, A. Holz, T. Kitzberger, D. Lorenz, C. Martín, M. Mermoz, J. Paritsis, E. Raffaele, G. Rep, and J. Sibold.

Literature Cited

Anchorena, A. 1902. Descripción geográfica de la Patagonia y valles Andinos. Buenos Aires: Compania Sudamericana.

Armesto, J. J., and J. Figueroa. 1987. Stand structure and dynamics in the temperate rain forests of Chiloé Archipelago, Chile. *Journal of Biogeography* 14:367–376.

Armesto, J. J., and E. R. Fuentes. 1988. Tree species regeneration in a mid-elevation, temperate rain forest in Isla de Chiloé, Chile. *Vegetatio* 74:151–159.

Attiwell, P. M. 1994. The disturbance of forest ecosystems: The ecological basis for conservative management. *Forest Ecology and Management* 63:247–309.

Christensen, N. L., A. M. Bartuska, J. H. Brown, S. Carpenter, C. D'Antonio, R. Francis, J. F. Franklin, J. A. MacMahon, R. F. Noss, D. J. Parsons, C. H. Peterson, M. G. Turner, and R. G. Woodmansee. 1996. The report of the Ecological Society of America committee on the scientific basis for ecosystem management. *Ecological Applications* 6:665–691.

Cisternas, M., B. F. Atwater, F. Torrejón, Y. Sawai,

G. Machuca, M. Lago, A. Eipert, C. Youlton, I. Salgado, T. Kamataki, M. Shishikura, C. P. Rajendran, J. K. Malik, Y. Rizal, and M. Husni. 2005. Predecessors of the giant 1960 Chile earthquake. *Nature* 437:404–407.

Daniels, L. D., and T. T. Veblen. 2000. ENSO effects on temperature and precipitation of the Patagonian-Andean region: Implications for biogeography. *Physical Geography* 21:223–243.

Dimitri, M. J. 1972. *La región de los bosques Andinos Patagónicos. Sinopsis general.* Buenos Aires: Instituto Nacional de Tecnología Agropecuario.

Donoso, C., R. Grez, B. Escobar, and P. Real. 1984. Estructura y dinámica de bosques del tipo forestal siempreverde en un sector de Chiloé Insular. *Bosque* 5:82–104.

Donoso, C., and A. Lara. 1999. *Silvicultura de los bosques nativos de Chile.* Santiago: Editorial Universitaria.

Fonck, F. 1896. *Libro de los diarios de Fray Francisco Menéndez.* Valparaiso: C. F. Niemeyer.

Fonck, F. 1900. *Viajes de Fray Francisco Menéndez a Nahuelhuapi.* Valparaiso: C. F. Niemeyer.

Glenn-Lewin, D. C., R. K. Peet, and T. T. Veblen, eds. 1992. *Plant succession: Theory and prediction.* Population and community biology series. London: Chapman and Hall.

Gutiérrez, A. G., J. J. Armesto, and J. C. Aravena. 2004. Disturbance and regeneration dynamics of an old-growth north Patagonian rain forest in Chiloé Island, Chile. *Journal of Ecology* 92:598–608.

Holdich, T. H. 1904. *The countries of the King's Award.* London: Hurst and Blackett.

Innes, J. L. 1992. Structure of evergreen temperate rain forest on the Taitao Peninsula, southern Chile. *Journal of Biogeography* 19:555–562.

Kitzberger, T., E. Raffaele, K. Heinemann, and J. Mazzarino. 2005a. Effects of fire severity in a north Patagonian subalpine forest. *Journal of Vegetation Science* 16:5–12.

Kitzberger, T., E. Raffaele, and T. T. Veblen. 2005b. Variable community responses to herbivory in fire-altered landscapes of northern Patagonia, Argentina. *African Journal of Range and Forage Science* 22:85–91.

Kitzberger, T., and T. T. Veblen. 1999. Fire-induced changes in northern Patagonian landscapes. *Landscape Ecology* 14:1–15.

Kitzberger, T., and T. T. Veblen. 2003. Influences of climate on fire in northern Patagonia, Argentina. In *Fire and*

climatic change in temperate ecosystems of the western Americas, ed. T. T. Veblen, W. Baker, G. Montenegro, and T. W. Swetnam, 290–315. Ecological Studies 160. New York: Springer Verlag.

Kitzberger, T., T. T. Veblen, and R. Villalba. 1997. Climatic influences on fire regimes along a rain forest-to-xeric woodland gradient in northern Patagonia, Argentina. *Journal of Biogeography* 24:35–47.

Lara, A., A. Wolodarsky-Franke, J. C. Aravena, M. Cortés, S. Fraver, and F. Silla. 2003. Fire regimes and forest dynamics in the Lake Region of south-central Chile. In *Fire and climatic change in temperate ecosystems of the western Americas,* ed. T. T. Veblen, W. Baker, G. Montenegro, and T. W. Swetnam, 322–342. Ecological Studies 160. New York: Springer Verlag.

Machoni, A. 1732. *Las Siete Estrellas de la Mano de Jesús, Tratado Histórico.* Córdoba: Colegio de Asumpción.

Mermoz, M., T. Kitzberger, and T. Veblen. 2005. Landscape influences on occurrence and spread of wildfires in Patagonian forests and shrublands. *Ecology* 86:2705–2715.

Moreno, F. P. 1897. Reconocimiento de la región Andina de la República Argentina. Apuntes preliminares sobre una excursión a los Territorios de Neuquén, Río Negro, Chubut y Santa Cruz. *Revista del Museo de La Plata* 8:1–180.

Musters, G. C. 1871. *At home with Patagonians: A year's wandering over untrodden ground from the Straits of Magellan to the Río Negro.* London: John Murray.

Paritsis, J., and M. A. Aizen. 2008. Effects of exotic conifer plantations on the biodiversity of understory plants, epigeal beetles and birds in *Nothofagus dombeyi* forests. *Forest Ecology and Management* 255:1575–1583.

Pollmann, W. 2003. Stand structure and dendroecology of an old-growth *Nothofagus* forest in Conguillio National Park, south Chile. *Forest Ecology and Management* 176:87–103.

Pollmann, W., and T. T. Veblen. 2004. *Nothofagus* regeneration dynamics in south-central Chile: A test of a general model. *Ecological Monographs* 74:615–634.

Raffaele, E., and T. Schlicter. 2000. Efectos de las plantaciones de pino ponderosa sobre la heterogeneidad de micrositios en estepas del noroeste Patagónico. *Ecología Austral* 10:49–155.

Raffaele, E., and T. T. Veblen. 1998. Facilitation by nurse shrubs on resprouting behavior in a post-fire shrubland in northern Patagonia, Argentina. *Journal of Vegetation Science* 9:693–698.

Raffaele, E., and T. T. Veblen. 2001. Effects of cattle grazing on early postfire regeneration of matorral in northwest Patagonia, Argentina. *Natural Areas Journal* 21:243–249.

Rothkugel, M. 1916. *Los bosques Patagónicos*. Buenos Aires: Ministerio de Agricultura.

Schlicter, T., and P. Laclau. 1998. Ecotono estepa-bosque y plantaciones en la Patagonia norte. *Ecologia Austral* 8:285–296.

Tercero-Bucardo, N., T. Kitzberger, T. T. Veblen, and E. Raffaele. 2007. A field experiment on climatic and herbivore impacts on post-fire tree regeneration in northwestern Patagonia. *Journal of Ecology* 95:771–779.

Veblen, T. T. 1985. Forest development in tree-fall gaps in the temperate rain forests of Chile. *National Geographic Research* 1:161–184.

Veblen, T. T. 1989. Tree regeneration responses to gaps along a transandean gradient. *Ecology* 70:543–545.

Veblen, T. T. 2007. Temperate forests of the southern Andean region. *The physical geography of South America*, ed. T. T. Veblen, A. Orme, and K. Young, 217–231. New York: Oxford University Press.

Veblen, T. T., and D. H. Ashton. 1978. Catastrophic influences on the vegetation of the Valdivian Andes. *Vegetatio* 36:149–167.

Veblen, T. T., C. Donoso, T. Kitzberger, and A. J. Rebertus. 1996a. Ecology of southern Chilean and Argentinean *Nothofagus* forests. In *The ecology and biogeography of* Nothofagus *forests*, ed. T. T. Veblen, R. S. Hill, and J. Read, 293–353. New Haven, CT: Yale University Press.

Veblen, T. T., C. Donoso, F. M. Schlegel, and B. Escobar. 1981. Forest dynamics in south-central Chile. *Journal of Biogeography* 8:211–247.

Veblen, T. T., R. S. Hill, and J. Read. 1996b. Epilogue: Commonalities and needs for future research. In *The ecology and biogeography of* Nothofagus *forests*, ed. T. T. Veblen, R. S. Hill, and J. Read, 387–397. New Haven, CT: Yale University Press.

Veblen, T. T., T. Kitzberger, E. Raffaele, and D. Lorenz. 2003. Fire history and vegetation changes in northern Patagonia, Argentina. In *Fire and climatic change in temperate ecosystems of the western Americas*, ed. T. T. Veblen, W. Baker, G. Montenegro, and T. W. Swetnam, 265–295. Ecological Studies 160. New York: Springer Verlag.

Veblen, T. T., T. Kitzberger, R. Villalba, and J. Donnegan. 1999. Fire history in northern Patagonia: The roles of humans and climatic variation. *Ecological Monographs* 69:47–67.

Veblen, T. T., T. Kitzberger, and A. Lara. 1992a. Disturbance and forest dynamics along a transect from Andean rain forest to Patagonian shrublands. *Journal of Vegetation Science* 3:507–520.

Veblen, T. T., and D. C. Lorenz. 1988. Recent vegetation changes along the forest/steppe ecotone in northern Patagonia. *Annals of the Association of American Geographers* 78:93–111.

Veblen, T. T., M. Mermoz, C. Martín, and T. Kitzberger. 1992b. Ecological impacts of introduced animals in Nahuel Huapi National Park, Argentina. *Conservation Biology* 6:71–83.

Vera, C., G. Silvestri, B. Liebmann, P. Gonzalez. 2006. Climate change scenarios for seasonal precipitation in South America. *Geophysical Research Letters* 33:L13707, doi:10.1029/2006GL025759.

Villalba, R., A. Lara, J. A. Boninsegna, M. Masiokas, S. Delgado, J. C. Aravena, F. A. Roig, A. Schemelter, A. Wolodarsky, and A. Ripalta. 2003. Large-scale temperature changes across the southern Andes: 20th-century variations in the context of the past 400 years. *Climatic Change* 59:177–232.

Villalba, R., and T. T. Veblen. 1997. Regional patterns of tree population age structures in northern Patagonia: Climatic and disturbance influences. *Journal of Ecology* 85:113–124.

Willis, B. 1914. *El Norte de la Patagonia*. Buenos Aires: Dirección de Parques Nacionales.

Wright, C., and A. Mella. 1963. Modifications to the soil pattern of south-central Chile resulting from seismic and associated phenomena during the period May to August 1960. *Bulletin of the Seismology Society of America* 53:1367–1402.

Wu, J., and O. Loucks. 1995. From balance of nature to hierarchical patch dynamics: A paradigm shift in ecology. *Quarterly Review of Biology* 70:439–466.

Chapter 14

Repeat Photography Challenges Received Wisdom on Land Degradation in the Northern Ethiopian Highlands

Jan Nyssen, Mitiku Haile, R. Neil Munro, Jean Poesen, A. T. Dick Grove, and Jozef Deckers

Deforestation and related soil erosion increase in the Ethiopian highlands are as old as the bases of colluvial deposits overlying palaeosols or burnt horizons, which have been dated around 2,450 ^{14}C years BP in parts of Wollo (Hurni 1985). The onset of deforestation is also indicated by the signal of ruderal species in pollen diagrams in Bale and Arsi, which occurred from 2,000 ^{14}C years BP to the present (Bonnefille and Hamilton 1986), or are marked by the end of freshwater tufa deposition in large rivers that have been dated earlier than 3,000 ^{14}C years BP (Moeyersons et al. 2006). However, modern population growth is assumed to have accelerated soil-erosion rates due to a progressive change in land cover with the main purpose of increasing food production within a subsistence farming system (Wøien 1995, Kebrom and Hedlund 2000). As land resources are pushed to their limits, ruptures in the fragile equilibrium contribute to catastrophes such as the 1984 famine (Nyssen et al. 2004a).

Resource exhaustion in Tigray (northern Ethiopia) is seen as part of a widely accepted paradigm, many variants of which exist, such as the following: In 1900 (or 1930 or 1965 . . .), 40 percent (or 45 percent or 30 percent . . .) of Ethiopia (or northern Ethiopia or Tigray or Eritrea . . .) was covered with forests. Due to population increase (or unsustain-

able exploitation), this forest cover has decreased to 1 percent (or 0.5 percent or 3 percent . . .). Comprehensive literature searches (Pankhurst 1995, Wøien 1995, McCann 1998, Ritler 2003, Boerma 2001 for Eritrea) show that no study ever produced such data and that there is no evidence for such facts in travelers' tales. In 1946, Logan estimated that only 5 percent of the highlands were forested. Taking into account the climatic potential for forest cover in Ethiopia, 40 percent would have been the approximate amount at the beginning of agricultural exploitation (Von Breitenbach 1961). Nevertheless, in many reports and even in scientific papers dealing with environmental degradation in Ethiopia or Eritrea, the commonplace still pops up: "During the Eritrean war for independence from Ethiopia (1961–1991), 30 percent forest cover of the country was reduced to less than 1 percent" (Asmeret 2001). This received wisdom may be expected to persist because it is present in virtually every geography manual at secondary school level in Ethiopia.

Landscape photographs, traced back as far as photographic records in this once poorly accessible country of Eastern Africa can be traced (i.e., the 1868 British military expedition to Abyssinia), show a landscape that, at first sight, is as barren as it is today (Nyssen et al. 2009). In many cases, it has even undergone significant positive changes over the last 140 years (plate 14.1).

At the current level of knowledge, most researchers will agree with the view of Brown (1973) that the highlands of Ethiopia had become progressively exhausted: there were high rates of soil erosion, widespread destruction of the natural vegetation, and little thought given to any rehabilitation of land or its cover. Runoff was rapid and destructive, and the diminishing returns on agricultural production were leading the country on a downward spiral of land degradation that necessitated urgent change. Such change could achieve results that would sustain the landscape, but only by rigorous application of modern concepts of land husbandry. The choice, though, lay with the people of Ethiopia.

As a response to such situations, huge efforts have been made in Tigray at a regional scale (100,000 square kilometers) to control soil erosion through the construction of stone bunds (0.5- to 1-meter-high erosion-control walls along the contour; Nyssen et al. 2007a) and the rehabilitation of steep slopes (Descheemaeker et al. 2006a, 2006b; plate 14.2). However, despite this record, some continue to insist on "the ever dwindling soil and forest resources of Tigray" (Fitsum et al. 1999, Amacher et al. 2004). Moreover, looking at it from a distance, some authors state that these conservation efforts do not lead to the desired effect or are even counterproductive (Keeley and Scoones 2000, Hengsdijk et al. 2005).

In this chapter, we assess changes in land degradation status in the Ethiopian highlands using information deduced from the comparison of 51 historical photographs taken in 1975 (Virgo and Munro 1978) with the current conditions. These comparisons are then linked to results from field research, conducted both on-farm and at catchment scale, over a period of more than 10 years.

Study Area

The study area lies on the western shoulder of the Rift Valley in the north of Ethiopia between 13° and 14° N, extending over an area of some 10,000 square kilometers (plate 14.3). Major lithologies are Precambrian metavolcanics and metasediments, Mesozoic sandstones and limestones, and Tertiary basalts (Nyssen et al. 2002). Erosion, in response to the Miocene and Plio-Pleistocene tectonic uplifts (on the order of 2,500 meters), resulted in the formation of tabular, stepped landforms between 2,000 and 2,800 meters that reflect the subhorizontal geological structure. Intervening mountain ranges rise locally to 3,000 meters; these high elevations result in a more temperate climate than would normally be associated with the latitude (Virgo and Munro 1978). Average yearly rainfall ranges between 500 and 900 millimeters, with a unimodal pattern, except in the southern part of the study area, where a second, smaller rainy season locally allows two crops within one year (Nyssen et al. 2005).

Since our research involved the use of time-lapsed photographs to analyze soil erosion phenomena, rainfall as well as rain seasonality of the Mekelle/Quiha meteorological station were examined for the period encompassed by the photographs. Soil erosion, especially under natural conditions, in a specific area, is dependent on annual and seasonal rain as expressed by Fournier's (1962) degradation coefficient (C_f, millimeters):

$$C_f = p^2/P$$

where p = monthly precipitation (millimeters) during the wettest month and P = yearly precipitation (millimeters). The period preceding 2006 was slightly wetter but also had a slightly higher degradation coefficient than the early 1970s (table 14.1).

The dominant land use in the study region is small-scale, rain-fed subsistence agriculture, for which the main constraints are inadequate soil water and excessive soil erosion (Virgo and Munro 1978). Since the 1980s, a land-tenure regime has been introduced leading to an approximate equalization in size of landholdings between households: "There is no single household or other kind of social group capable of concentrating land in large amounts" (Hendrie 1999).

Table 14.1. Average yearly precipitation (P) and degradation coefficient (C_f) for Mekelle, Ethiopia, in the periods preceding the taking of repeated photographs

	P (millimeters)		C_f (eq. 1)	
	1975	2006	1975	2006
Preceding year	451	610	70	162
5 preceding years	514	538	88	111
10 preceding years	532	578	95	115

Photo Monitoring Methods

Repeat Photography

Repeat photography is used for many purposes and can take on many different forms (Hall 2001; see Boyer et al., chapter 2; Klett, chapter 4; Hoffman and Todd, chapter 5). It has been used for landscape rephotography covering up to 100 years of change (Progulske and Sowell 1974, Skovlin and Thomas 1995, Grove and Rackham 2001, Rohde and Hilhorst 2001, Lätt 2004) or for sampling change in vegetation (Johnson 1984, Nader et al. 1995, Boerma 2006, Turner et al. 2003).

The analytical possibilities of repeat photography exceed by far those of other historical sources, such as narratives or drawings. For instance, the author of a well-known lithography of the Debre Damo site in Eastern Tigray in 1868 (plate 14.4), in an effort to romanticize his artwork, strongly departed from the existing landscape when representing topography. One may reasonably assume that such embellishments also affect the representation of vegetation, which renders historical comparisons impossible (see Jonas, chapter 21).

The methodology of ground-based photographic monitoring (Johnson 1984, Nievergelt 1998, Hall 2001, Lätt 2004) has been used for this analysis. As the name implies, repeat photography means retaking photographs from the same spot and of the same subject several times (Boyer et al., chapter 2); it requires precise repositioning of the camera and composition of the subject (Hall 2001), which in our case meant rephotographing a distant landscape.

Table 14.2. Number of analyzed landscapes by elevation class and lithology, the major agroecological determinants in the study area in Ethiopia

	Elevation (meters)			
Lithology	1,500–2,000	2,000–2,500	2,500–3,000	3,000–3,500
Basalt and dolerite	1	2	5	2
Limestone	2	20		
Sandstone	1	8	2	
Granite		4		
Precambrian	2	2		

Photographs of the environment in Tigray, taken in 1975 (Virgo and Munro 1978) were obtained. Fifty-one landscapes photographed in early 1975 (dry season) were revisited in the same season in 2006, and a new set of photographs was prepared. They cover the wide range of agroecologies present in the study area (plate 14.3, table 14.2; all repeat photographs are presented in Nyssen et al. 2007b). It can be assumed that the location of the interpreted landscapes is random. In no way could the original photographers foresee what specific landscape changes would take place over a period of 30 years, including two revolutions, a long civil war, and two famine epochs.

The relocation of the historical photographs was based on rough indications by the original photographers, knowledge of geomorphic features induced by various lithologies, and a dozen-year-long geomorphological research experience in the study area. The camera position was obtained by identification of unique landscape features such as mountain peaks, drainage lines, and their relative positions. Finally, the exact camera position and orientation were obtained by lining up near and distant objects in a triangulation system. Not all photographs, however, could be repeated: particular problems concerned the growth of nonnative *Eucalyptus* (eucalyptus) trees, which were an obstacle to rephotography, or absence of identifiable objects.

The photographs used in the analysis were made by Neil Munro (48 photographs, 1975), Ruth Trummer (3, 1972), Jan Nyssen (46, 2001 and 2006), Fikir

Alemayehu (1, 2005), Annelies Beel (1, 2005), and Jozef Naudts (3, 2006). The 1975 photographs were taken with a Praktica Exa-1a camera (35 millimeter slide film) and the 2006 by a digital Kodak Easyshare CX4230 camera.

For this study, the photographs were analyzed by Jozef Deckers, Dick Grove, Mitiku Haile, Neil Munro, Jan Nyssen, and Jean Poesen, all experts who have longstanding geomorphological experience in Ethiopia (Grove and Goudie 1971; Grove et al. 1975; Virgo and Munro 1978; Nyssen et al. 2004a, 2007a) and elsewhere. Repeat photographs were randomly presented as digital images without indicating locations; the relative position of the new and old photographs was also randomized. Only photograph pairs taken at exactly the same place, in the same season, and under the same angle were considered, and demonstrated by indicating identical objects in each pair. The immediate foreground is dependent on the exact position of the photographer; to avoid bias, it was masked for the analysis unless it had clear reference points.

Landscape Analysis

The photo-monitoring technique involved comparing on-the-ground conditions of 2006 to photographs depicting the 1975 conditions, whereby scores were assigned by the experts to various landscape features for both situations. For every couple of time-lapsed photographs, the expert panel interpreted various indicators (table 14.3) in such a way that they could select only those indicators they thought would be relevant. For these relevant indicators, they compared both photographs. The evaluation was then converted into the following numerical scores: −2 indicated the situation had strongly deteriorated; −1 indicated the situation had deteriorated; 0 was used for unchanged situations; 1, the situation had improved; 2, the situation had strongly improved. Given that the scoring method used ordinal variables, the median score per indicator was calculated for every pair of photographs, provided that at least four of the six experts thought the indicator relevant for that pair. Averages of the median scores were then calculated for each indicator for the whole set of time-lapsed photographs, and the deviation of the median from zero (no change) was tested with the *t*-test (Diem 1963). Munro et al. (2008) report a detailed analysis of homogeneous land units within the same set of photographs, whereby these land units were compared for vegetation cover and soil erosion control measures.

Table 14.3. Interpretation of landscape change in the study areas of Ethiopia ($n = 51$)

Visible Soil Erosion Indicators	n	Score	Change
Gully erosion	16	−0.1[ns]	Slightly deteriorated
Overall assessment of erosion	50	0.4[***]	Slightly improved
Land cover and protective measures			
Vegetation cover on nonarable land	48	0.7[***]	Improved
Grass, herbs, and shrubs on nonarable land	40	0.7[***]	Improved
Grass and shrubs between cultivated farm plots	20	0.9[***]	Improved
Cultivation of the steepest slopes	10	0.7[**]	Improved
Stone/soil bunds in farmland	11	1.0[***]	Improved
Overall assessment			
Vegetation cover	51	0.8[***]	Improved
Land management	50	0.7[***]	Improved

n = number of landscape sites where the phenomenon was observed by at least four of the six experts; score = average of the median scores given to all the interpreted landscapes, ranging from −2 (strong deterioration) to +2 (strong improvement), with level of significance for the deviation from a test value zero (no change) ([***]significant at 0.01 level; [**]significant at 0.05 level; [ns]not significant); change = comparison of the situation in 2006 with that in 1975.

Landscape Changes, 1975–2006

Average Landscape Changes

When evaluating environmental changes using repeat landscape photographs taken in the dry season of 1975 and 2006, our panel of six experts interpreted various indicators (table 14.3) and found unanimously that the situation has improved with respect to visible erosion phenomena, vegetation cover on nonarable land, as well as grass and shrubs between cultivated farm plots. Whereas the population of Ethiopia increased from 33 to 75 million between 1975 and 2005 (ESA 2008), land management (plate 14.5) as well as overall vegetation cover (plates 14.2 and 14.6) have improved in the study area. These changes are not climate-driven (table 14.1), but instead are the result of human intervention as is indicated in plate 14.5. Though only visible in part of the time-lapsed photographs, the status with respect to cultivation of steep slopes and building of stone bunds in farmland has also improved (plates 14.5, 14.7, and 14.8).

In the years following 1975, large environmental programs were undertaken in Ethiopia by the then Derg government (1975–1990), unfortunately in a rather top-down manner (Keeley and Scoones 2000). In most of Tigray, collective terracing activities started a decade later, in the period of civil war against the Derg regime, and still continue on a large scale today. Many people are now accustomed to seeing land covered with soil-conservation structures in northern Ethiopia, and our study shows the contrast with 30 years ago. In nearby areas where such activities were not organized (such as the Simien Mountains; Nievergelt 1998), no similar improvement can be observed.

In Tigray, environmental rehabilitation is at the top of the agenda of the regional government. The inspiration was spawned by the Tigrai Rural Development Study (TRDS), a mid-1970s baseline study during which the 1975 photographs were taken (Virgo and Munro 1978). The 1992 Symposium on Environmental Degradation held at Mekelle established the base for a new land rehabilitation approach (Aseffa et al. 1992), which stresses the necessary partnership between technicians "with the necessary knowledge, as equal partners and not as patrons" and peasant farmers "through a natural extension of the present experience with participation." Farmer-based approaches, developed in the liberated areas during the struggle against the Derg regime in the 1980s, include the integration of local and external technologies without losing the best of local practices and traditions (Berhane and Mitiku 2001, Nyssen et al. 2004b). Maintaining and enhancing farmers' participation is obviously a continuous challenge (Mitiku et al. 2006).

The median scores (table 14.3) reflect variable and sometimes extreme situations. At most sites, soil erosion has slightly decreased, but in a few cases, it has decreased strongly or even increased. With respect to overall vegetation cover, some exceptionally extreme cases of degradation exist (fig. 14.1). It appears that the traditional management of remnant forests through bylaws (Zenebe 1999) has not been sufficient to prevent them from regressing. Landscapes with deteriorated vegetation cover in 2006 invariably correspond to places where there is, or was, still a good tree cover (rift escarpment, some remnant forests around churches), and where the need for conservation seems to be less felt by population and authorities. In all other cases, the vegetation cover has improved slightly to strongly (fig. 14.1A). Similarly, overall land management has improved slightly to strongly in 85 percent of the analyzed landscapes (fig. 14.1B).

Land Management at Catchment Scale

Observations on decreased soil-erosion rates, despite strongly increased population density, become sound when considering the large-scale implementation of soil and water conservation (SWC) practices that has taken place over the last two decades. Physical structures, such as stone bunds, have been

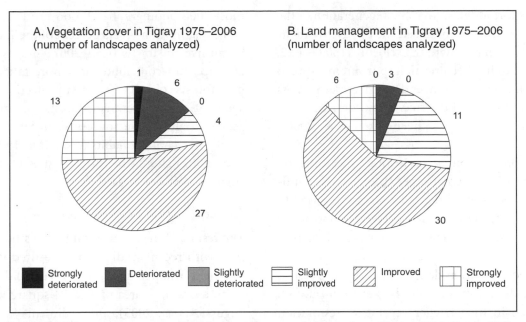

Figure 14.1. Changes of (A) vegetation cover and (B) land management in Tigray from 1975 to 2006. Figures indicate the number of monitoring sites analyzed by the expert panel.

built in all land-use types (plate 14.5), whereas trenches were established in pasture and shrublands and check dams in gullies.

Physical Conservation

Based on measurements on 202 field parcels (Nyssen et al. 2007a), 3- to 21-year-old stone bunds reduce soil loss by 68 percent. On these field parcels with stone bunds, there is an average increase in grain yield of 53 percent in the lower part of the plot, as compared to the central and upper parts (Desta et al. 2005). Taking into account the space occupied by the bunds, stone bunds led in 2002 to a mean crop yield increase from 0.58 to 0.65 metric ton per hectare. The cost of stone bund building in 2002 averaged 13.6 euros per hectare per year, which is nearly the same as the value of the induced crop yield increase (13.2 euros per hectare per year). Besides positive off-site effects such as runoff and flood regulation, the enhanced water storage in deep soil horizons on both sides of the bunds indicates that the stone bund areas can be made more productive through tree

planting. From the technical, ecological, and economical point of view, the extensive use of stone bunds, involving people's participation, is a positive operation (Desta et al. 2005, Vancampenhout et al. 2006, Nyssen et al. 2007a). Bekele and Holden (1999) and Boyd and Slaymaker (2000) have demonstrated that, given the ecosystem services rendered by the farmers through stone bund building (Robertson and Swinton 2005), the current subsidies and incentives are justifiable.

Reforestation

Eighty-one percent of household energy in Tigray is produced from firewood, and 17 percent is from crop residue and dung (CSA 2007). Rural household fuel consists of 76 percent collected firewood, whereas 74 percent of the energy used by urban households to cook and heat comes from purchased firewood and charcoal. The exclosure policy is the cornerstone in the dynamics whereby both population and vegetation density have increased. "Exclosure" denotes areas set aside where cropping and grazing are forbidden so that the regeneration of the natural vegetation is enhanced. In the Tigray highlands, vegetation regrowth in exclosures (plates 14.2, 14.5, and 14.6) has become an important measure to combat land degradation and to increase biomass production; in several districts, they cover up to 15 percent of the land (Descheemaeker et al. 2006a, 2006b).

Runoff production in exclosures, measured on runoff plots (Descheemaeker et al. 2006b), is significantly reduced when a degraded area is allowed to rehabilitate after closure. Runoff was found to be negligible when the vegetation cover exceeds 65 percent (Descheemaeker et al. 2006b).

Increased vegetation density in exclosures results in increased infiltration and higher transpiration rates, which in turn trigger vegetation restoration through increased biomass production. With vegetation restoration, water use for biomass production also becomes more efficient. The parallel increase in deep percolation is attributed to the presence of source-sink systems; the source areas produce runoff, which infiltrates in the sink areas (the exclosures) (plate 14.5). Vegetation restoration is responsible for the high infiltration capacity of the exclosure areas, but as transpiration is not increased at the same rate, the surplus infiltration drains beyond the root zone and contributes to groundwater recharge. This explains the earlier reported phenomenon of improved spring discharge in lower parts of the landscape after degraded areas were converted into exclosures in Tigray (Nyssen et al. 2002).

Besides effects on enhanced infiltration, sediment trapping, and decreased downstream flooding, exclosures provide other ecosystem services, such as growth of grass and trees, increase in wildlife and biodiversity, climate regulation, drought mitigation, and carbon sequestration.

Our very diverse photoset shows an overall improvement of vegetation cover (fig. 14.1A). Firewood is still available due to larger biomass production and to the use of bark, leaves, and branches of the now widely spread *Eucalyptus* (as can be seen on most 2006 photographs). Wood production for urban consumption—22.5 percent of the Tigray population lives in towns (CSA 2007)—takes place at a distance and contributes to deforestation of remnant forests (plate 14.9). Better protection and management of these remnant forests, enhancement of access to alternative sources of urban energy, and changes in cooking habits (Asmerom 1991, Bereket et al. 2002) should be top priorities to sustain the current positive trends.

Sediment Yield at Catchment Scale

The results of the photo-monitoring study are in accord with recent studies that measured volumes of sediment trapped in 5- to 10-year-old reservoirs (with a drainage area of 1 to 24 square kilometers; Nigussie et al. 2005), and sediment yield was related to environmental characteristics, including the catchment management activities undertaken over the last decades. Average area-specific sediment yield (SSY) is 1054 ($\pm$ 487) metric tons per square kilometer per year ($n = 10$) (Nigussie et al. 2005), which is much less than the 1,681–3,284 metric tons per square kilometer per year measured in the 1975 studies (Virgo and Munro 1978). Despite the fact that the 1975 data concern merely two rivers (catchments of 115 and 153 square kilometers) that were monitored during only one year, the difference with current rates is large enough to conclude for a significant decrease in SSY, which corresponds to the positive trends in landscape condition revealed by repeat photography.

Discussion and Conclusions

The recent active intervention by authorities and farmers to conserve the natural resources in Tigray has led to demonstrated significant improvements in terms of soil conservation, infiltration, crop yield, biomass production, groundwater recharge, and minimization of flood hazard. Results from detailed in situ studies are corroborated by analyses of landscape changes, which show that the status of natural resources has improved (and locally strongly improved) since 1975. The land rehabilitation is due both to improved vegetation cover and to the implementation of physical soil and water conservation structures. Detailed observations and model application on homogeneous land units within the same photographs (Munro et al. 2008) indicate that at present, average soil loss by sheet and rill erosion is at around 68 percent of its 1975 rate. These decreasing rates are substantiated by comparisons between sediment yields estimated in 1975 and in 2003. Exceptional degradation is still ongoing in remnant forests. As elsewhere in Tigray, conservation of vegetation cover should be strongly implemented here. A system for sustainable forest exploitation must be established.

This study invalidates hypotheses on irreversibility of land degradation in Tigray and a fortiori in less marginal semiarid areas, and on the futility of SWC programs. The study furthermore demonstrates that (1) land management has become an inherent part of the farming system in Tigray; (2) it is possible to reverse environmental degradation in semiarid areas through an active, farmer-centered SWC policy (Stocking 2003); (3) small-scale farmers may be able to stay on their land if provided with adequate levels of subsidies (Robertson and Swinton 2005, Pimbert et al. 2006), yielding an effective way to sustain the agricultural system of semiarid areas in the long term and to provide ecosystem services to the society; and (4) the "more people less erosion" paradigm (Tiffen et al. 1994) may also be valid in other, semiarid areas. In a highly degraded environment, with high pressure on the land, no alternatives remain but to improve land husbandry.

The challenges to be met in the future include (1) in situ SWC of farmland (Tewodros et al. 2009), (2) shifting from stubble grazing to stall feeding of livestock and ecologically sound grazing management of the rangelands, (3) involving local communities in decision making about resource management (Robertson and Swinton 2005, Bekele et al.

2005, Segers et al. 2008), and (4) active development of a policy for sustainable urban energy consumption.

Acknowledgments

This research was carried out in the framework of a collaborative project between Mekelle University, Relief Society of Tigray (Ethiopia), and K. U. Leuven and Africamuseum (Belgium), funded by the Belgian authorities through VLIRUOS. R. N. Munro thanks his colleagues on TRDS for discussions. Many individuals and institutions contributed to this research. The authors particularly acknowledge the Tigrayan farmers whose hard work in an adverse environment allows environmental recovery. Their endeavor is an inspiring source for scientists interested in improving land conditions.

Literature Cited

Amacher, G., Lire Ersado, D. Grebner, and W. Hyde. 2004. Disease, microdams and natural resources in Tigray, Ethiopia: Impacts on productivity and labour supplies. *Journal of Development Studies* 40:122–145.

Aseffa Abraha, Berhe Woldearegay, Tesfay Belai, and Berhane Gebregziabher. 1992. *Report of the symposium on environmental degradation*. Relief Society of Tigray, Mekelle, Ethiopia.

Asmeret Asefaw. 2001. *The impact of erosion on the terrestrial carbon reservoir: Ecological and socioeconomic implications of soil erosion in Eritrea*. Unpub. paper presented at the international conference commemorating the 10th anniversary of the independence of Eritrea: Lessons and prospects. Asmara, Eritrea.

Asmerom Kidane. 1991. Demand for energy in rural and urban centres of Ethiopia: An econometric analysis. *Energy Economics* 13:130–134.

Bekele Shiferaw, H. Freeman, and S. Swinton. 2005. *Natural resource management in agriculture: Methods for assessing economic and environmental impacts*. Wallingford: CABI Publishing.

Bekele Shiferaw, and S. Holden. 1999. Soil erosion and

smallholders' conservation decisions in the highlands of Ethiopia. *World Development* 27:739–752.

Bereket Kebede, Almaz Bekele, and Elias Kedir. 2002. Can the urban poor afford modern energy? The case of Ethiopia. *Energy Policy* 30:1029–1045.

Berhane Hailu, and Mitiku Haile. 2001. Liberating local creativity: Building on the "best farming practices" extension approach from Tigray's struggle for liberation. In *Farmer innovation in Africa: A source of inspiration for agricultural development*, ed. C. Reij and A. Waters-Bayer, 310–324. London: Earthscan.

Boerma, P. 2001. Politica forestale e cambiamenti ecologici nell'Altopiano Centrale dell'Eritea durante il colonialismo Italiano 1890–1941. *Storia Urbana* 95:15–44.

Boerma, P. 2006. Assessing forest cover change in Eritrea—a historical perspective. *Mountain Research and Development* 26:41–47.

Bonnefille, R., and A. Hamilton. 1986. Quaternary and Late Tertiary history of Ethiopian vegetation. *Symbolae Botanicae Upsalienses* 26:48–63.

Boyd, C., and T. Slaymaker. 2000. *Re-examining the "more people less erosion" hypothesis: Special case or wider trend?* Natural Resource Perspectives 63. London: Overseas Development Institute.

Brown, L. H. 1973. *Conservation for survival: Ethiopia's choice.* Addis Ababa: Haile Selassie I University Press.

Central Statistical Agency (CSA). 2007. *Basic welfare indicators.* Addis Ababa, Ethiopia: Central Statistical Agency. www.csa.gov.et (accessed 6 November 2007).

Descheemaeker, K., J. Nyssen, J. Poesen, D. Raes, Mitiku Haile, B. Muys, and J. Deckers. 2006a. Runoff on slopes with restoring vegetation: A case study from the Tigray highlands, Ethiopia. *Journal of Hydrology* 331:219–241.

Descheemaeker, K., J. Nyssen, J. Rossi, J. Poesen, Mitiku Haile, J. Moeyersons, and J. Deckers. 2006b. Sediment deposition and pedogenesis in exclosures in the Tigray highlands, Ethiopia. *Geoderma* 132:291–314.

Desta Gebremichael, J. Nyssen, J. Poesen, J. Deckers, Mitiku Haile, G. Govers, and J. Moeyersons. 2005. Effectiveness of stone bunds in controlling soil erosion on cropland in the Tigray highlands, northern Ethiopia. *Soil Use and Management* 21:287–297.

Diem, K. 1963. *Documenta Geigy: Tables scientifiques.* Bâle, Switzerland: JR Geigy SA, Département Pharmaceutique.

ESA. 2008. *World population prospects: The 2008 revision.* Population Division of the Department of Economic and Social Affairs of the United Nations Secretariat. esa.un.org/unpp (accessed 18 May 2010).

Fitsum Hagos, J. Pender, and Nega Gebresilassie. 1999. *Land degradation in the highlands of Tigray, and strategies for sustainable land management.* Socio-economic and policy research working paper 25. International Livestock Research Institute, Addis Ababa, Ethiopia. www.ilri.cgiar.org/InfoServ/Webpub/Fulldocs/WorkP 25/Acknowledgements.htm (accessed 27 July 2006).

Fournier, F. 1962. *Map of erosion danger in Africa south of the Sahara. Explanatory note.* Paris: EEC, Commission for Technical Cooperation in Africa.

Grove, A. T., and A. Goudie. 1971. Late Quaternary lake levels in the Rift Valley of southern Ethiopia and elsewhere in tropical Africa. *Nature* 234:403–405.

Grove, A. T., and O. Rackham. 2001. *The nature of Mediterranean Europe: An ecological history.* London: Yale University Press.

Grove, A. T., F. Street, and A. Goudie. 1975. Former lake levels and climatic change in the Rift Valley of southern Ethiopia. *Geographisches Jahrbuch* 141:177–202.

Hall, F. 2001. *Ground-based photographic monitoring.* USDA Forest Service General Technical Report PNW-GTR-503. Pacific Northwest Research Station, Portland, OR.

Hendrie, B. 1999. *Now the people are like a lord—local effects of revolutionary reform in a Tigray village, northern Ethiopia.* PhD dissertation, University College, London.

Hengsdijk, H., G. Meijerink, and M. Mosugu. 2005. Modeling the effect of three soil and water conservation practices in Tigray, Ethiopia. *Agriculture Ecosystems and Environment* 105:29–40.

Hurni, H. 1985. Erosion–productivity–conservation systems in Ethiopia. In *Pla Sentis, I. Soil conservation and productivity, Proceedings 4th International Conference on Soil Conservation, Maracay, Venezuela.* Caracas: Soil Conservation Society of Venezuela, 654–674.

Johnson, K. 1984. Sagebrush over time: A photographic study of rangeland change. In *Proceedings: Symposium on the biology of* Artemesia *and* Chrysothamnus, ed. E. D. McArthur and B. L. Welch, 223–252. US Department of Agriculture, Forest Service, Intermountain Research Station, Ogden, UT.

Kebrom Tekle, and L. Hedlund. 2000. Land cover changes between 1958 and 1986 in Kalu District, Southern

Wello, Ethiopia. *Mountain Research and Development* 20:42–51.

Keeley, J., and I. Scoones. 2000. Knowledge, power and politics: The environmental policy-making process in Ethiopia. *Journal of Modern African Studies* 38:89–120.

Lätt, L. 2004. *Eritrea re-photographed: Landscape changes in the Eritrean highlands 1890–2004.* MSc thesis, University of Berne, Switzerland.

Logan, W. 1946. *An introduction to the forests of central and southern Ethiopia.* Institute Paper 24. Oxford: Imperial Forestry Institute, University of Oxford.

McCann, J. 1998. *A tale of two forests: Narratives of deforestation in Ethiopia, 1840–1996.* African Studies Center Working Paper 209. Boston, MA: Boston University.

Mitiku Haile, K. Herweg, and B. Stillhardt. 2006. *Sustainable land management—a new approach to soil and water conservation in Ethiopia.* Mekelle: Mekelle University and University of Berne, Switzerland.

Moeyersons, J., J. Nyssen, J. Poesen, J. Deckers, and Mitiku Haile. 2006. Age and backfill/overfill stratigraphy of two tufa dams, Tigray highlands, Ethiopia: Evidence for late Pleistocene and Holocene wet conditions. *Palaeogeography, Palaeoclimatology, Palaeoecology* 230: 162–178.

Munro, R. N., J. Deckers, Mitiku Haile, A. T. Grove, J. Poesen, and J. Nyssen. 2008. Soil landscapes, land cover change and erosion features of the Central Plateau region of Tigrai, Ethiopia: Photo-monitoring with an interval of 30 years. *Catena* 75:55–64.

Nader, G., M. De Lasaux, and R. Delms. 1995. "How to" Monitor Rangelands. University of California Cooperative Extension Service, Handbook, Level-1. Alturas: University of California Press.

Nievergelt, B. 1998. Long-term changes in the landscape and ecosystems of the Simen Mountains National Park. *WALIA* Special issue:8–23.

Nigussie Haregeweyn, J. Poesen, J. Nyssen, G. Verstraeten, J. de Vente, G. Govers, J. Deckers, and J. Moeyersons. 2005. Specific sediment yield in Tigray—northern Ethiopia: Assessment and semi-quantitative modeling. *Geomorphology* 69:315–331.

Nyssen, J., Mitiku Haile, J. Moeyersons, J. Poesen, and J. Deckers. 2004b. Environmental policy in Ethiopia: A rejoinder to Keeley and Scoones. *Journal of Modern African Studies* 42:137–147.

Nyssen, J., Mitiku Haile, J. Naudts, R. N. Munro, J. Poe-

sen, J. Moeyersons, A. Frankl, J. Deckers, R. Pankhurst. 2009. Desertification? Northern Ethiopia rephotographed after 140 years. *Science of the Total Environment* 407:2749–2755.

Nyssen, J., J. Moeyersons, J. Poesen, J. Deckers, and Mitiku Haile. 2002. The environmental significance of the re-mobilisation of ancient mass movements in the Atbara-Tekeze headwaters, northern Ethiopia. *Geomorphology* 49:303–322.

Nyssen, J., R. N. Munro, Mitiku Haile, J. Poesen, K. Descheemaeker, Nigussie Haregeweyn, J. Moeyersons, G. Govers, and J. Deckers. 2007b. *Understanding the environmental changes in Tigray: A photographic record over 30 years.* Tigray Livelihood Papers 3. Mekelle: VLIR–Mekelle University IUC Programme and Zala-Daget Project.

Nyssen, J., J. Poesen, Desta Gebremichael, K. Vancampenhout, M. D'aes, G. Yihdego, G. Govers, H. Leirs, J. Moeyersons, J. Naudts, Nigussie Haregeweyn, Mitiku Haile, and J. Deckers. 2007a. Interdisciplinary on-site evaluation of stone bunds to control soil erosion on cropland in northern Ethiopia. *Soil and Tillage Research* 94:151–163.

Nyssen, J., J. Poesen, J. Moeyersons, J. Deckers, Mitiku Haile, and A. Lang. 2004a. Human impact on the environment in the Ethiopian and Eritrean highlands—a state of the art. *Earth Science Reviews* 64:273–320.

Nyssen, J., H. Vandenreyken, J. Poesen, J. Moeyersons, J. Deckers, Mitiku Haile, C. Salles, and G. Govers. 2005. Rainfall erosivity and variability in the northern Ethiopian highlands. *Journal of Hydrology* 311:172–187.

Pankhurst, R. 1995. The history of deforestation and afforestation in Ethiopia prior to World War I. *Northeast African Studies* 2:119–133.

Pimbert, M., K. Tran-Thanh, E. Deléage, M. Reinert, C. Trehet, and E. Bennett, eds. 2006. *Farmers' views on the future of food and small-scale producers.* London: International Institute for Environment and Development.

Progulske, D., and R. Sowell. 1974. *Yellow ore, yellow hair, yellow pine: A photographic study of a century of forest ecology.* South Dakota Agricultural Experiment Station, Bulletin 616. Brookings: South Dakota State University.

Ritler, A. 2003. *Forests, land use and landscape in the central and northern Ethiopian highlands, 1865 to 1930.*

Geographica Bernensia, African Studies Series A19. Bern, Switzerland: University of Bern.

Robertson, G., and S. Swinton. 2005. Reconciling agricultural productivity and environmental integrity: A grand challenge for agriculture. *Frontiers in Ecology and the Environment* 3:38–46.

Rohde, R., and T. Hilhorst. 2001. *A profile of environmental change in the Lake Manyara Basin, Tanzania.* International Institute for Environment and Development (IIED), Drylands Programme, Issue Paper 109, London.

Segers, K., J. Dessein, J. Nyssen, Mitiku Haile, and J. Deckers. 2008. Developers and farmers intertwining interventions: The case of rainwater harvesting and food-for-work in Degua Temben, Tigray, Ethiopia. *International Journal of Agricultural Sustainability* 6: 173–182.

Skovlin, J., and J. Thomas. 1995. *Interpreting long-term trends in Blue Mountain ecosystems from repeat photography.* USDA Forest Service Research Paper PNW-GTR-315. Pacific Northwest Research Station, Portland, OR.

Stocking, M. 2003. Tropical soils and food security: The next 50 years. *Science* 302:1356–1359.

Tewodros Gebreegziabher, J. Nyssen, B. Govaerts, Fekadu Getnet, Mintesinot Behailu, Mitiku Haile, and J. Deckers. 2009. Contour furrows for in-situ soil and water conservation, Tigray, northern Ethiopia. *Soil and Tillage Research* 103:257–264.

Tiffen, M., M. Mortimore, and F. Gichuki. 1994. *More people, less erosion: Environmental recovery in Kenya.* Chichester: Wiley.

Turner, R. M., R. H. Webb, J. E. Bowers, and J. R. Hastings. 2003. *The changing mile revisited: An ecological study of vegetation change with time in the lower mile of an arid and semiarid region.* Tucson: University of Arizona Press.

Vancampenhout, K., J. Nyssen, Desta Gebremichael, J. Deckers, J. Poesen, Mitiku Haile, and J. Moeyersons. 2006. Stone bunds for soil conservation in the northern Ethiopian highlands: Impacts on soil fertility and crop yield. *Soil and Tillage Research* 90:1–15.

Virgo, K. J., and R. N. Munro. 1978. Soil and erosion features of the Central Plateau region of Tigrai, Ethiopia. *Geoderma* 20:131–157.

Von Breitenbach, F. 1961. Forests and woodlands of Ethiopia, a geobotanical contribution to the knowledge of the principal plant communities of Ethiopia, with special regard to forestry. *Ethiopian Forest Review* 1:5–16.

Wøien, H. 1995. Deforestation, information and citations: A comment on environmental degradation in highland Ethiopia. *Geojournal* 37:501–512.

Zenebe Gebreegziabher. 1999. *Dessa'a protected area: An assessment of human impact, evolutionary pattern and options for sustainable management.* UNESCO MAB (Man and the Biosphere) Programme. www.unesco.org/mab/bursaries/mysrept/97/Zenebe/report.pdf (accessed 6 November 2007).

Chapter 15

Cattle, Repeat Photography, and Changing Vegetation in the Victoria River District, Northern Territory, Australia

Darrell Lewis

In Australia, repeat photography has been used in a number of studies, primarily to document environmental change. For example, Fensham and Holman (1998) used repeat photography as one line of evidence in their study of vegetation change on the Darling Downs in Queensland, and Start and Handasyde (2002) used the method to document changes in riverside vegetation below the Ord River dam since the dam was built. The method is also increasingly used by pastoralists and government agencies to monitor the impact of grazing and other changes in pasturelands.

Regional environmental histories have been written for various areas of southern Australia (Hallam 1975, Rolls 1984, Griffiths 2001). The baselines for these studies drew upon explorers' journals, early settlers' diaries and reminiscences, newspapers, government records, scientific studies, and early paintings. Because they dealt with areas where European settlement began long before photography was invented or commonly used, photographs play virtually no role in establishing an environmental baseline for these studies, and they incorporate relatively few repeat photographs, or none at all.

In northern Australia the situation is different. There, European occupation coincided with or postdated the availability of photography, so the opportunity exists to examine photographs to establish what the original (European contact) environment was like, and to use repeat photography to document environmental changes that have occurred afterward. As far as I can determine, my study was the first in Australia to use repeat photography to document environmental change across a large region and over almost the entire period of European occupation. It was also the first and remains the only regional environmental history for northern Australia.

The Origins of the Project

From the time I first worked in the Victoria River District in 1971, I was fascinated with the history of the region, but at that time very little had been written. In an attempt to learn more, in the early 1980s I began my own research, obtaining copies of historical documents and photographs from various archives. After 1990, I began to locate people who had worked in the district (or their descendants) and copied photographs and documents in their personal collections. Since that time I have accumulated a large amount of documentation and a collection of well over 5,000 photographs ranging in date from the 1890s to the 1970s.

Within this collection were photographs of landscapes that seemed to show differences from the same places today, so in 1994 I began to relocate the original vantage points and rephotograph the scenes. Comparison of the resulting photograph pairs often showed dramatic changes, in some cases so dramatic that I decided to enlist the expertise of a botanist with a view toward joint publication. I began by contacting Jeremy Russell-Smith, a botanist working with the Northern Territory Bushfires Council and a member of an organization funded by the Australian government, the Cooperative Research Centre (CRC) for the Sustainable Development of Tropical Savannas, which was about to begin a major scientific study of the savannas in the Victoria River District.

When I showed Dr. Russell-Smith my repeat photograph pairs, he was impressed by the changes they documented, and he invited me to join the CRC project with a view to extending my comparative photographic work. The brief I was given was to carry out a repeat photography project to provide information to scientists involved in the CRC project on changes in vegetation since European settlement. Because I already had sufficient background material, I volunteered to write an environmental history of the region (Lewis 2002).

The Study Region

The Victoria River District, located in the far northwest of the Northern Territory (fig. 15.1), comprises roughly 129,500 square kilometers of flat-topped sandstone and limestone ranges, rounded-basalt hills, and red and black soil plains. It extends from the high-rainfall coastal fringe inland to the arid sand plains on the edge of the Tanami Desert. Consequently it contains a wide variety of floristic communities: woodlands and tall grasslands in the coastal areas, open savannas and grass plains in the intermediate rainfall area, and grass plains and shrublands in the arid south (Perry 1970). In the northern areas, there are rainforest patches around permanent water sources, and further south rocky hills and ridges are often covered with *Triodia* sp. (spinifex), a spiny, arid-adapted grass.

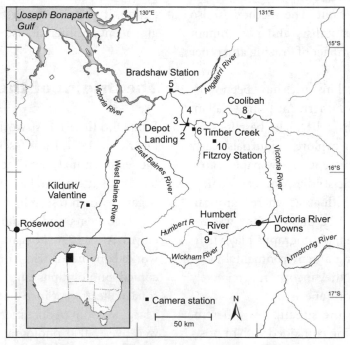

Figure 15.1. Map of the Victoria River District, Australia.

The climate is monsoonal—hot and dry for roughly half the year (April–September) and extremely hot and humid during the remainder. In July, the coldest month, temperatures range from a minimum of about 10 degrees Celsius to a maximum of 27 degrees Celsius (Plumb 1973: map 4). During the summer months (October–March) continuous high humidity is accompanied by daily temperatures of up to 45 degrees Celsius and average overnight lows of about 30 degrees Celsius (Plumb 1973). By the end of the winter dry season, desiccation may be extreme, and all but the major water sources have dried up, ground cover is sparse, and the soil is dusty and cracked. During October, November, and December, massive thunderheads appear, bringing violent electrical displays and erratic, turbulent winds that often produce short, fierce dust storms followed by heavy downpours. The storms become more frequent until, finally, the great monsoon rains arrive in January and February.

Average rainfall ranges from 700 millimeters in the north to 330 millimeters in the south (Plumb 1973). These figures mask the true situation as rainfall can vary significantly from year to year. During one summer only one monsoonal event may occur, while another summer may experience several monsoonal events, so that one wet season may be comparatively dry and another wet season may experience major flooding. If overall rainfall is poor, environmental conditions toward the end of the following dry season can be harsh. On rare occasions, the wet season can fail completely; when this happens conditions during the following dry season are disastrous for cattle and for the environment in general (Lewis 2002). By European Australian standards, in a normal year seasonal conditions for humans are extremely harsh. In fact, Lee's (1969) study of human adaptation in tropical environments indicates that in this area there is an average of 225 days per year when heat and humidity combine to exceed human comfort levels, a longer period than anywhere else in Australia.

The region was one of the last parts of Australia to be occupied by Europeans, with the first settlers ar-

riving in 1883 (Makin 1970). From their viewpoint, it was a cattleman's paradise of vast, thinly wooded grasslands, abundantly watered with large waterholes, numerous springs, and generally reliable monsoon rains. The banks of the rivers and creeks were steep and densely covered with reeds, vines, and other growth highly palatable to cattle (Lewis 2002).

While undoubtedly it was wonderful cattle country, the settlers failed to understand that this "cattle paradise" was the result of millennia of Aboriginal land management, particularly their use of fire to keep the country open by suppressing the growth of scrub and trees or to burn firebreaks to protect valued fire-sensitive plants. The net result of Aboriginal burning was to produce a mosaic of patches of country burnt at different times over the previous few years (Jones 1975, Pyne 1991, Vigilante 2001). As happened elsewhere in Australia, the pastoralists wanted to preserve grass for their cattle, so when they found Aborigines burning the country they drove them away, often killing those who resisted (Blair 1889, Rose 1991). They then instituted their own burning practices, which were quite different from those of the Aborigines (Lewis 2002).

The region soon became famous for its huge and now legendary stations (ranches) such as the great Victoria River Downs (also known as VRD), which in its heyday covered more than 31,100 square kilometers (Makin 1970). While there have been some significant changes in land tenure over the past 30 years, the Victoria River District remains predominantly cattle country today.

From the beginning the settlers ran their cattle on what is now known as the "open range" system. This entailed merely turning the cattle loose on the most favorable areas and providing very little infrastructure, usually only a few huts and yards, and a bullock and horse paddock, but no boundary or internal fencing. With nothing to stop the cattle wandering at will, as their numbers increased they spread into new areas of the station and quickly developed an annual pattern of movement, regulated by the alternating wet and dry seasons. When the rains came, they spread out across the land, and later as the smaller waters dried up they gradually retreated to larger waters until, by the end of the dry season, most had retreated to the river frontages. If the previous wet season had been good and the first rains of the next wet season came early, the number of cattle on the frontage was somewhat reduced, but if the last wet season was poor and the next rains came late, virtually the entire herd congregated on the river. On Victoria River Downs, this meant that at the end of each dry season anything from 50,000 to 140,000 cattle could be concentrated on the river, with obvious implications for damage to riverine vegetation and erosion of riverbanks.

With no boundary fences or internal fencing for the control of the herd, cattle became wild and very difficult to handle (Watson 1895, Wise 1929). Breeding could not be controlled and wild bulls soon became a major concern, along with large numbers of feral donkeys and horses. To give some indication of the scale of the problem, from the 1930s to the 1960s there were an estimated 20,000 cleanskin (unbranded) bulls on one of the bigger stations (*Hoofs and Horns* 1947, Lewis 1960). On another station, 27,000 wild horses and donkeys were shot in 1980–1983 (Rosewood station 1889–1998). In short, for most of its European history, the region was grossly overstocked with vast numbers of uncontrolled wild cattle, horses, and donkeys. Environmental damage in the form of erosion quickly became obvious (Blair 1889, Kyle-Little 1928, Maze 1945), but other changes and impacts took place "slower than the eye can see," and thus went unnoticed, or were ignored (Lewis 2002).

At the beginning of the twenty-first century the banks of the rivers and creeks are no longer steep, nor are they covered with reeds and vines, but much of the Victoria River District could still be described as thinly wooded grassland. From written descriptions alone it can be difficult to tell if there has been any major change in tree numbers since the settlers and cattle arrived, but repeat photography can discern this environmental change.

The Photographs

To begin this project, I first sought out additional photographs in state and scientific archives, private collections, old magazines, books, and journals. Combined with photographs from my private collection, I was able to draw together a collection of over 150 images I believed could be useful in a repeat photo project. In addition, I expanded my collection of historical records and also collected scientific papers relevant to the area.

In the majority of instances, very little was known about who took the original photographs, what type of camera or film was used, and what time of day, which day, or which month the original photo was taken. Exceptions were confined mainly to photographs from more recently established collections, such as that from the Australian Government Bureau of Mineral Resources, where 35 millimeter Minolta and Pentax cameras were used, and the name of the photographer, the location, and the date of the photo were recorded. Sometimes the precise year the photo was taken was unknown but it was bracketed between two known dates. For example, an undated photo may have come from the collection of a station manager who was in the district for a known period.

Photograph Sources

The earliest images were two high-quality photographs exposed on glass-plate negatives in 1891, only seven years after the first settlers arrived in the region. The backdrop in these particular images is a steep hillslope across the river from the Depot Landing, the place on the Victoria River where goods for the inland stations were boated in from Darwin (fig. 15.2A; Makin 1970). This was land that in 1891 had not yet been taken up or stocked with cattle, and the photographs therefore may have been taken to document, in essence, a pre-European landscape. In the photographs it can be seen that part of the hillslope had been burnt, quite possibly by Aboriginals. Because of its location opposite the Depot Landing, this particular hillslope formed the background to a series of photographs taken over the following 60 years, providing a unique record of changes through time (e.g., figs. 15.2B, 15.3A,B).

The cameras used ranged from large box cameras producing glass negatives to "Box Brownies," medium-format (57 millimeter) cameras, and modern 35-mm single-lens reflex (SLR) cameras. As a result there is considerable variation in the quality of

A

B

Figure 15.2. Depot Landing, Victoria River, Australia.
A. (1891). This view of the Depot Landing on the Victoria River shows bare stony areas on the largely treeless hillslope across the river that are the result of fire, possibly from traditional Aboriginal burning. (P. Foelsche, PH 275/49, Northern Territory Library).
B. (24 September 1998). Erosion of the riverbank is evident, and the hillslope supports a very large number of trees not present in 1891. (D. Lewis).

A

B

Figure 15.3. Depot Landing, Victoria River, Australia.
A. (ca. 1926). This view of the Depot Landing jetty on the Victoria River was taken more than three decades after figure 15.2A. The hillslope across the river is still largely treeless and there is a mangrove species (dark vegetation) along the water's edge. (Courtesy R. Durack).
B. (24 September 1998). The riverbanks have eroded significantly, the mangrove species has disappeared, and trees have covered the hillslope. Comparison of figures 15.2 and 15.3 indicates that these changes occurred in the twentieth century. (D. Lewis).

images and very few are of such quality that grass species and annual shrubs can be identified with certainty. Here, I focused the study on changes in tree cover or in the tree-species mix evident between the originals and the repeat photographs.

Apart from variations in quality and differences resulting from the different cameras and films used, short- or long-term seasonal factors undoubtedly influenced the apparent differences between the original photographs and my repeat photographs. For example, a run of years with higher than usual wet-season rainfall could lead to a thickening of tree cover and increased canopy size, while a run of relatively dry wet seasons or even the complete failure of the "wet" will lead to the deaths of trees and a reduction in tree canopy size. While such factors could influence the difference between the original and the repeat photographs, by compiling a series of repeat shots in similar topographic and ecological areas such differences can be averaged out.

Methods

After gathering suitable photographs, I began the task of relocating the vantage point of each one. In most cases this was not too difficult because the photo was either of a distinctive feature in the landscape (sometimes named on the photo), or there was sufficient written information with it to enable relocation of the scene. In addition, when I began the project I had over 24 years of experience working and traveling in most parts of the region and thus had an in-depth knowledge of the district. (For additional information on repeat photography technique, see Boyer et al., chapter 2; Hanks et al., chapter 3; Klett, chapter 4).

When I reached the general location where the original photo was taken, I walked around the area until the overlapping hills in the background or other features matched as closely as possible with those in the photograph. In most instances, this required only a short walk from my vehicle, although in some cases it required climbing several hundred meters up steep hillsides. Sometimes it was impossi-

ble to find the original location, while in others it was possible to determine the exact spot from which the original photograph was taken. An example of the latter is figure 15.4A, which was taken from a large, flat rock projecting from a hillside. At the time the original photo was taken, this rock provided a convenient and unimpeded view, but when I was there in 1997 the view was obscured by trees growing in front of the rock. The repeat photograph had to be taken slightly higher up the slope; even then several tree branches had to be removed before a clear view could be obtained (fig. 15.4B).

In almost every case, clues in the original photograph enabled the correct viewing angle to be determined and the distance from the subject to be established within a matter of meters, but in some instances it was impossible to be sure how far away from the subject the original vantage point was. Considering that tree numbers and size of tree canopy were the focus of my research interest, a majority of the original vantage points were located within an acceptable margin of error.

Once the original vantage point was fixed as closely as possible, I rephotographed the scenes with a 35-millimeter Minolta SLR camera, using a macro 50-millimeter lens and Ilford Delta 100 black-and-white film. I used a compass to determine the viewing angle and a Global Positioning System (GPS) to pinpoint the location on the relevant 1:100,000 topographic maps. Along with a set of the resulting photograph pairs, these data have since been archived and will be available for future monitoring of the vegetation at the various repeat photograph locations or camera stations.

Results

In broad terms, the repeat photographs show a significant increase in the size and density of trees along the rivers and creeks and on the riverine flats and plains out to the foot of the nearest ranges (figs. 15.4–15.9). Tree numbers may have begun to increase soon after European settlement, but repeat

A

B

Figure 15.4. Depot Landing, Victoria River, Australia.

A. (ca. 1914). The Australian government boat *Leichhardt* at the Depot Landing, Victoria River. The dark vegetation along the water's edge is a mangrove species, and the riverine flats are relatively open with sporadic woody vegetation. (Courtesy of R. Johns).

B. (1 June 1995). The mangroves have disappeared, replaced by *Parkinsonia*, a nonnative weed. In addition, tree growth across the riverine flats has increased significantly except for an area close to the base of the range that has been cleared for an airstrip. (D. Lewis).

A

B

Figure 15.5. Bradshaw Station, Victoria River, Australia.

A. (ca. 1914). The *Leichhardt* and other boats are moored at the old Bradshaw station landing. The riverbanks are relatively open and allow a view a considerable distance toward the base of the range, and the old Bradshaw homestead is clearly visible at the right. (P. Foelsche, R7/91, National Library of Australia).

B. (8 June 1995). An increase of trees along the riverbank has obstructed the view toward the base of the range and has nearly hidden the old Bradshaw homestead from view. (D. Lewis).

photography, backed up by oral testimony, suggests that any such increase was slow during the first 60 or more years and has accelerated over the last 30 to 40 years (Lewis 2002).

In the hill country, the results were ambiguous and probably depended upon a variety of factors, including fire history, climatic events, and geology. Most of the repeat pairs from the hill country did not show any major change, but there were some notable exceptions. For example, the sandstone hillslope opposite the old Victoria River Depot Landing has considerably more timber today than in photographs taken in the 1890s and 1920s (figs. 15.2 and 15.3). This hillslope is very steep and rocky, especially toward the top, where it is capped by a low cliff, and it is unlikely to have suffered more than minimal

A

B

Figure 15.6. Timber Creek, Victoria River District, Australia.
A. (1950). This view northwest from the hillside above the two old Timber Creek police stations shows Timber Creek flowing across the middle of the scene. The treeless area at the top right has been cleared for the original airstrip, and the Victoria River is at the base of the distant range. (Courtesy B. Mettam).
B. (4 June 1995). The trees along Timber Creek have grown taller and thicker while the old airstrip has been invaded by trees and bushes. Tree cover has increased across the entire riverine plain. (D. Lewis).

A

B

Figure 15.7. The confluence of Kildurk–Valentine creeks, Victoria River District, Australia.
A. (1947). This view shows the junction of Kildurk and Valentine creeks on Rosewood station. (Courtesy C. Stone).
B. (8 September 1999). The tree cover along Valentine Creek has increased considerably, and there has been a general increase in tree density across the riverine flats. (D. Lewis).

A

B

Figure 15.8. Coolibah Homestead, Victoria River District, Australia.
A. (1959–1960). Ernie Rayner and horses outside Coolibah homestead, with Wondoan Hill visible to the west across the open grass plain. Besides the animals, the lack of trees in the view is noteworthy. (Courtesy E. Rayner).
B. (12 November 1996). A significant increase of trees is evident on the formerly open grass plain. (D. Lewis).

A

B

Figure 15.9. Humbert River, Victoria River District, Australia.
A. (1957). This view, from a hillside across Humbert River, shows the country around Humbert River homestead. (E. Kettle, PH 0127/0128, Ellen Kettle Collection, Northern Territory Library).
B. (28 August 2000). A significant increase in tree cover along the river and across the riverine plain is evident. (D. Lewis).

impact from European livestock. In addition, its location across the river from the old Depot store renders it unlikely to have been impacted by early European land uses, such as timber harvesting for building purposes. Similarly, the trees on the steep sandstone hill adjacent to the old Timber Creek police station appeared to have increased in number and to have larger canopies today than was the case in 1910. Lastly, there had been an increase in the number and size of trees on a limestone range on Mistake Creek station since 1971 (see Lewis 2002).

In two instances there has been a loss of tree cover on hills. Photographs taken in 1953 on the plateau immediately above the narrow part of Jasper Gorge, at the northern end of Victoria River Downs, show a greater number of trees than are there today. The main road from the Victoria Highway to VRD passes through this gorge, and the loss of trees may be due to a relatively high frequency of "hot" fires, deliberately or accidentally lit by travelers or campers. The other instance where there appears to have been a reduction in vegetation is on a small limestone hill near Tower Hill on Fitzroy station. In 1968, this hill bore a scattering of bushes or small trees, but a repeat photograph taken in 1999 shows that the number of trees and bushes had severely declined. The hill is now nearly bald though the surrounding country shows no discernible change (Lewis 2002, fig. 15.10).

Stimulated by my preliminary findings, Fensham and Fairfax (2003) used aerial photographs taken decades apart to study the vegetation at selected sites throughout the Victoria River District. This was effectively a repeat photography exercise using aerial photographs rather than oblique angle photographs (see McClaran et al., chapter 12). They found that in 18 percent of sites, cover had declined substantially; in 38 percent, cover was relatively stable; and at the remaining 44 percent of sites, there had been an increase in cover. Overall, the trend was for increasing tree cover, with by far the greatest increase (38 percent) in riverine areas, confirming the results of my study.

Discussion

While repeat photography and oral history sources leave no doubt that there has been a significant thickening of tree cover in riverine areas within the past 30 to 40 years, the reason for this is unclear. The most compelling reason would appear to be related to climate change spurred by the global increase in greenhouse gases, including carbon dioxide, which suggests that annual rainfall in northern Australia will increase (Polley 1997, Polley et al. 1997, O'Donnell 2001).

Fossil fuel burning over the past 200 years has caused an increase of about 30 percent in levels of global atmospheric carbon dioxide, and studies have shown that, among other changes, this stimulates

A

B

Figure 15.10. Fitzroy station, Victoria River District, Australia.
A. (1968). This view shows a small hill in a valley on Fitzroy station. (Photographer unknown, neg. M796, frame 28, © Commonwealth of Australia (Geoscience Australia).
B. (28 July 1999). A significant reduction in the number of bushes and small trees on the hill since 1968 is evident, though the reason for this decline is unclear. In general, tree and shrub cover in the view has not changed significantly in the intervening 31 years between photographs. (D. Lewis).

vegetation growth (Polley 1997, Polley et al. 1997, O'Donnell 2001). The fact that tree numbers in the Victoria River District do not appear to have increased significantly between 1883 and World War II may be because the recruitment of seedlings was limited by factors such as regular burning, cattle and feral animal impacts, and the annual dry season drought; fewer trees means that during droughts there is less competition for scarce groundwater, which enables the existing trees to survive (Cook and Liedloff 2000). Levels of carbon dioxide have accelerated in the postwar period and continue to rise, and this may explain the apparent significant increase in tree numbers in the past 30 or 40 years.

A predicted outcome of climate change is that average rainfall across northern Australia will increase, and there is evidence that this increase has already begun. Hennessy et al. (1999) looked at rainfall records from 300 sites around Australia compiled between 1910 and 1995. Their study found that in the Victoria River region, average annual rainfall has increased by 15 percent since about 1970 and also that there has been an average increase of 10 percent in the number of rainy days per year. This increased rainfall could have enabled more trees to survive the annual drought in riverine areas of the district.

Apart from climate change, other possible causal factors can be suggested, including changes in fire

regimes, changes in stocking rates, a change in the type of cattle on the stations, reduction in the numbers of feral animals, local extinctions of native animals, and intensification of land use. Climatic records in the Victoria River District extend over little more than a century, far too short a time period for any understanding of long-term cycles. It may be that the region is approaching a peak in a rainfall cycle of unknown length and unrelated to atmospheric carbon dioxide, with a corresponding increase in tree cover, and that eventually there will be a swing the other way, with a reduction in tree cover.

From their study using aerial photographs to study vegetation in the Victoria River District, Fensham and Fairfax (2003) concluded that the increase in riverine tree numbers they observed was primarily the result of tree recovery since a major drought, with regrowth fueled by the generally above average rainfall since about 1970. The most severe and widespread drought in north Australian history occurred in 1951–1952 when the wet season failed completely (Lewis 2002), and this fact lends weight to Fensham and Fairfax's conclusion. However, they did not rule out the possibility that land-management changes can have an important effect on woody vegetation cover.

Fire is a crucial element in the floristic structure of Australia, and it is possible that the change from traditional Aboriginal burning regimes to European pastoralists' burning regimes has led to or contributed to the changes documented here. However, the question arises why these changes did not occur, or occurred only slowly, during the first 60 to 80 years of European settlement, when Aborigines were no longer engaged in traditional burning, and accelerated in the postwar period. My subjective impression is that, since the Second World War, the use of fire by Victoria River pastoralists has declined, and the use of fire-ploughs or graders to create firebreaks has increased. This needs confirmation, but if it is correct the reduction in burning may have contributed to the increase in tree cover.

Changes in stocking rates may also have affected the recruitment or suppression of tree seedlings. Ernie Rayner (pers. comm., 1996), who worked as stockman on Victoria River Downs in the early 1960s, believes that stocking rates were then much heavier than they are today, and local Aborigines have made similar observations. Their claims are almost impossible to confirm but are nevertheless likely to be correct. In the early 1960s, cattle on VRD (and other stations in the district) were still being run on the open-range system, and the herd was completely out of control. At the time, estimates of cattle numbers varied from 90,000 to 140,000 or more, but in the late 1990s, when internal fencing had brought the herd under control, its size could be reliably determined to be 80,000 (station manager Jim Coulthart, pers. comm., 2001).

Along with the huge numbers of uncontrolled cattle, for most of the period of European settlement there were huge numbers of feral donkeys and brumbies (wild horses) in the district. These animals certainly added enormously to the generally heavy stocking rates on the stations and undoubtedly intensified the effects of the normal and extreme droughts. Like the cattle, they would have contributed to erosion and may have grazed emerging shrubs and trees, suppressing their establishment.

Intensification of land use involving extensive internal fencing and supply of subartesian waters began on VRD and elsewhere in the district in the 1960s, and a second phase occurred during a campaign to eradicate the cattle diseases brucellosis and tuberculosis in the late 1970s and early 1980s. This intensification led to more uniform grazing across the stations so that areas previously lightly grazed or grazed for a limited period each year came to be grazed more heavily throughout the year, and grazing pressure along river frontages and around other permanent water sources was reduced. It might be expected that increased grazing pressure would suppress recruitment of seedlings, and, conversely, reduced grazing pressure would allow more seedlings to survive, but the reality is not so clear-cut. Repeat photography has shown that there has been an increase in tree numbers along the river frontages but also in the areas where grazing pressure now continues throughout the year. It is possible to suggest reasons why this has happened.

From the early 1970s, Victoria River District stations began to replace the original shorthorn cattle (*Bos taurus*) with Brahman cattle (*Bos indicus*). Local cattlemen say that Brahman cattle graze differently from shorthorns, foraging further from water and eating a wider variety of vegetation. Brahman cattle browse "top feed" (shrubs and trees) more than shorthorn cattle, and this might be expected to suppress edible shrubs and trees, but when Ernie Rayner visited the station after an absence of more than 30 years, he noticed a dramatic increase in tree cover, including *Terminalia volucris* (rosewood), a valued top feed.

This increase in *Terminalia* and tree numbers in general could be because of the grazing habits of the Brahman cattle; seeds may have dropped onto their bodies as they browsed and been carried into open areas or been deposited there in their droppings. It may also be related to a phenomenon in north Queensland, a study of which demonstrated that the change to Brahman cattle from European breeds and the adoption of feed supplements led to better survival and growth rates, and thus to an increase in grazing pressure. As a result, some areas were overgrazed, with consequent loss of herbaceous cover, increased soil erosion, and shrub invasion (Gardener et al. 1990). A similar finding was made in a study in the United States where Brown and Archer (1990) concluded that in the prairie lands, heavy grazing by cattle could lead to invasion by woody plants. However, in the Victoria River District the thickening of tree cover occurred in riverine areas, which had experienced decades of repeated overgrazing, so the question remains as to why the increase in trees has been most significant during the postwar period.

Experiments on the Kidman Springs Research Station have demonstrated that if European livestock breeds are removed from an area, there is a substantial increase in the density of trees and bushes (Foran et al. 1985, Bastin and Andison 1990, Bastin et al. 2000). A similar increase occurred on sections of Victoria River frontage where livestock was excluded by fencing in the 1990s (Lewis 2002). Unfortunately, no experiments have been conducted on the effects of fire on the vegetation in the ungrazed areas, so whether this increase was due to the absence of grazing, the absence of fire, or a combination of both remains unclear.

Elderly Aborigines know of a number of native animals and birds once found in the district that are either extinct or present only in small numbers. These include flock pigeons (*Phaps histrionica*), which were formerly present in tens of thousands, and also a range of small marsupials, including bandicoots (*Perameles* spp.), possums (*Trichosurus* spp.), native cats (*Dasyurus* spp.), various marsupial "mice" and "rats," and hare wallabies (*Lagostrophus* spp.). Whether any of these animals played a role in controlling the increase of tree species is unclear, but it is known that in southeast Australia, rat-kangaroos (*Potorous* spp.) played an important role in keeping acacias and native pine under control by eating the seeds and seedlings. When the rat-kangaroos disappeared, these plants rapidly took over large areas of grazing land (Pyne 1991). Similar conversions have been demonstrated in southeastern Arizona involving kangaroo rats (*Dipodomys* spp.) and perennial and annual grasses (Brown and Heske 1990).

Conclusions

In my study, I think that repeat photography convincingly demonstrates that since European settlers and their cattle, donkeys, and horses arrived in the Victoria River District, and particularly or perhaps only during the postwar period, there has been a general increase in tree cover in riverine areas. This change would have been virtually impossible to detect if only written records had been consulted, and this result clearly demonstrates the value of repeat photography as a tool in studies of environmental change.

The reason for the increase in tree numbers remains unknown. Various factors may have a bearing on this question, and it will only be answered when sufficient research has been carried out by scientists from a number of different disciplines. It seems likely that humanly induced global climate change may be the primary cause, but it is possible that long-term seasonal cycles, changed fire regimes, extreme droughts, the numbers and types of animals grazing the country, local extinctions of certain small marsupials, and other factors may have played a role. At least one thing is clear: if both European livestock and fire are excluded from a given area, vegetation of all types increases very quickly. By and large, ecological change in the region has occurred "slower than the eye can see," and continual monitoring via repeat photography is required to ensure such changes do not go unnoticed in the future.

Acknowledgments

I gratefully acknowledge the Cooperative Research Centre for the Sustainable Development of Tropical Savannas, Darwin, Australia, which funded the study upon which this chapter is based. My thanks also go to Ernie Rayner and Jim Coulthart for their assistance with personal recollections and current knowledge of the Victoria River region, and to Reg Durack (deceased), Robert Johns, Bert Mettam,

Ernie Rayner, and Clive Stone for permission to use photographs from their personal collections.

Literature Cited

Bastin, G. N., and R. Andison. 1990. Kidman Springs country—10 years on. *Range Management Newsletter* 90:15–19.

Bastin, G. N., J. Ludwig, R. Eager, and A. Liedloff. 2000. Vegetation recovery: Kidman Springs exclosure photos over 25 years. *Range Management Newsletter* 00(2):1–5.

Blair, B. to Goldsbrough Mort & Co. October 24, 1889. Goldsbrough Mort & Co: Sundry papers re CB Fisher and the Northern Australia Territory Co, 1886–1892. 2/876/7, Noel Butlin Archives, Canberra, ACT.

Brown, J., and S. Archer. 1990. Water relations of a perennial grass and seedling vs. adult woody plants in a subtropical savanna, Texas. *Oikos* 57:366–374.

Brown, J. H., and E. J. Heske. 1990. Control of a desert–grassland transition by a keystone rodent guild. *Science* 250:1705–1707.

Cook, G., and A. Liedloff. 2000. Simulating the effects of the last hundred years of fire management and rainfall variability in North Australia. In *Australian Rangeland Society centenary symposium*, ed. S. Nicolson and J. Noble, 55–59. Broken Hills, New South Wales: Australian Rangeland Society.

Fensham, R. J., and R. Fairfax. 2003. Assessing woody vegetation cover change in north-west Australian savanna using aerial photography. *International Journal of Wildland Fire* 12:359–367.

Fensham, R. J., and J. Holman. 1998. The use of the land survey record to assess changes in vegetation structure: A case study from the Darling Downs, Queensland, Australia. *Rangelands Journal* 20:132–142.

Foran, B., G. N. Bastin, and B. Hill. 1985. The pasture dynamics and management of two rangeland communities in the Victoria River District of the Northern Territory. *Australian Rangeland Management* 7:107–113.

Gardener, C., J. McIvor, and J. Williams. 1990. Dry tropical rangelands: Solving one problem and creating another. *Proceedings of the Ecological Society of Australia* 16:279–286.

Griffiths, T. 2001. *Forests of ash: An environmental history.*

Sydney, New South Wales: University of Cambridge Press.

Hallam, S. 1975. *Fire and hearth.* Canberra, ACT: Australian Institute of Aboriginal Studies.

Hennessy, K., R. Suppiah, and C. Page. 1999. Australian rainfall changes, 1910–1995. *Australian Meteorological Magazine* 48:1–13.

Hoofs and Horns. December 1947:14.

Jones, R. 1975. The Neolithic, Palaeolithic and the hunting gardeners: Man and land in the Antipodes. In *Quaternary studies*, ed. R. Suggate and M. Cresswell, 21–34. Wellington: The Royal Society of New Zealand.

Kyle-Little, S. (Culkah). 1928. In north Australia, pt 1. *Pastoral Review* Sept. 15:884–885.

Lee, D. H. K. 1969. Variability in human response to arid environments. In *Arid lands in perspective*, ed. W. G. McGinnies and B. J. Goodman, 234. Tucson: University of Arizona Press.

Lewis, D. 2002. *Slower than the eye can see: Environmental change in northern Australia's cattle lands.* Darwin, Northern Territory: Cooperative Research Centre for the Sustainable Development of Tropical Savannas.

Lewis, G. 1960. VRD station report for Hooker Pastoral Company Pty. Ltd., station reports 1959–68. 119/15, Noel Butlin Archives, Canberra, ACT.

Makin, J. 1970. *The big run: The story of Victoria River Downs station.* Sydney, New South Wales: Rigby.

Maze, W. H. 1945. Settlement in the East Kimberleys, Western Australia. *Australian Geographer* 5:1–16.

O'Donnell, K. 2001. Rising CO_2: What's in store for the savannas? *Savanna Links* 17:10.

Perry, R. 1970. Vegetation of the Ord-Victoria area. *CSIRO Land Research Series* 28:104–119.

Plumb, T., ed. 1973. *Atlas of Australian resources.* Canberra, ACT: Department of Minerals and Energy.

Polley, H. 1997. Implications of rising atmospheric carbon dioxide concentration for rangelands. *Journal of Range Management* 50:562–577.

Polley, H., H. Mayeux, H. Johnson, and C. R. Tischler. 1997. Viewpoint—atmospheric CO_2, soil water, and shrub/grass ratios on rangelands. *Journal of Range Management* 50:278–284.

Pyne, S. 1991. *Burning bush.* Seattle: University of Washington Press.

Rolls, E. 1984. *A million wild acres.* Sydney, New South Wales: Penguin Books.

Rose, D. 1991. *Hidden histories: Black stories from Victoria River Downs, Humbert River and Wave Hill stations.* Canberra, ACT: Aboriginal Studies Press.

Rosewood station. 1889–1998. Yearly rainfall and station journal. Hooker Pastoral Company, records relating to Rosewood station, 1908–1975, NTRS 9. Northern Territory Archives Service, Darwin, Northern Territory.

Start, A., and T. Handasyde. 2002. Using photographs to document environmental change: The effects of dams on the riparian environment of the lower Ord River. *Australian Journal of Botany* 50:465–480.

Vigilante, T. 2001. Analysis of explorers' records of aboriginal landscape burning in the Kimberley region of Western Australia. *Australian Geographical Studies* 39:135–155.

Watson, J. to Goldsbrough Mort. & Co. December 5, 1895. Goldsbrough Mort & Co: Board papers, 1893–1927. 2/124/1659, Noel Butlin Archives, Canberra, ACT.

Wise, F. J. S., agricultural adviser to Sir Charles Nathan, Perth, 15-8-1929. A494/1, Item 902/1/82, Australian Archives, ACT.

Chapter 16

People, Elephants, and Habitat: Detecting a Century of Change Using Repeat Photography

David Western

The East African savannas support the richest variety and greatest abundance of large mammals anywhere on Earth. Amboseli Game Reserve in the 1960s was referred to as the jewel in the crown of Kenya's protected areas because of its spectacular concentration of wildlife and stunning *Acacia xanthophloea* (fever tree acacias) set below the snows of Kilimanjaro. Then, to the consternation of tourists and conservationists, the woodlands began dying, threatening many wildlife species (Simon 1962). Blame was leveled at the Maasai pastoralists, leading to pressure on the Kenya government to declare Amboseli a national park. A detailed experimental study vindicated the Maasai and showed the complex interplay of elephants and people governing the ecology of Amboseli (Western and Maitumo 2004). The study underscored the importance of long-term ecological monitoring and illustrates the role of repeat photography in visually documenting complex ecosystem processes.

The Amboseli ecosystem, which includes Amboseli National Park, lies due north of Kilimanjaro on the Kenya–Tanzania border (fig. 16.1). The vegetation is shaped by climate and modified by pastoralism and large wild ungulates. In recent centuries, Maasai pastoralists have been the dominant human influence (Western 1973). Like other savanna ecosystems (Cole 1986, Deshmukh 1986), the ecology of Amboseli is complex and dynamic.

Two keystone actors dominate Amboseli's ecology: the African elephant (*Loxodonta africana*) and livestock, including cattle, sheep, and goats (Western and Maitumo 2004). Elephants tend to reduce woody vegetation and encourage grasslands (Laws 1970, Cumming et al. 1997). Livestock tend to reduce grass cover and induce woody growth (Pratt and Gwynne 1977). This interaction of elephants and livestock creates a shifting patchwork of woodlands and grasslands where herds are still migratory. The shifting mosaic contributes to biotic diversity (Western and Maitumo 2004).

Amboseli highlights the problems of conserving biota in African savannas. In colonial times, the prevailing view held that pristine savannas could be conserved by creating national parks and banishing humans (Simon 1962). In East Africa, approaching independence and customary land claims limited the area of national parks. The land constraints and lack of ecological research resulted in national parks that focused on the most conspicuous wildlife herds. Most parks consequently protect dry season concentration areas, leaving pastoralists the wet season range.

The division of savanna ecosystems into wildlife

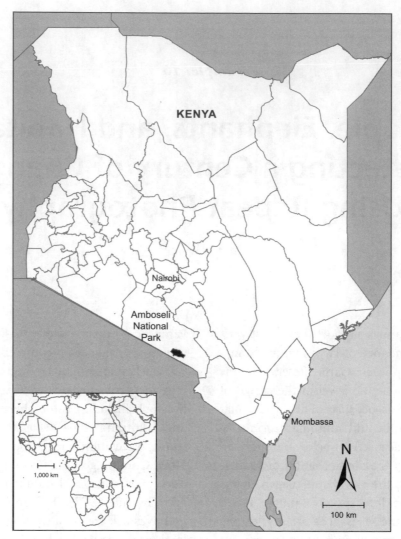

Figure 16.1. Location map of Amboseli National Park, Kenya.

and livestock ranges severed the intertwined migrations. The ecological consequences are only now becoming apparent. Western and Gichohi (1993) conclude that dividing wildlife and livestock into smaller seasonal domains causes ecological dislocation, lowers biotic diversity, and decreases drought resilience either side of park boundaries. The ecological dislocation was not apparent at first because little research took place before the creation of parks. Even after research got under way, most studies treated parks as "natural" enclaves and ignored the role of pastoralism (Homewood and Rogers 1991). The savannas were generally regarded as biologically rich because of their long-term ecological stability

(Botkin 1990). Later vegetation studies showed flux and change to be typical of savanna ecosystems, leading to lively debate over whether the changes are natural or caused by human activity (Botkin 1990). The answer bears on whether national parks should be managed in the interest of wildlife, habitat, and biotic diversity (Dublin et al. 1990, Cumming et al. 1997).

One exception to the exclusion of traditional human activity in savanna ecosystems research is the long-term Amboseli Research and Conservation Program. Begun in 1967, the study looked at the structure, dynamics, and changes resulting from the interactions of wildlife and subsistence pastoralists

(Western 1973) a full decade before the creation of Amboseli National Park and has run continuously since then. The 40-year study spans the transition from free-ranging interactions of livestock and wildlife to the creation of Amboseli National Park and, more recently, land subdivision and sedentarization (Western and Nightingale 2004). The vegetation changes have been monitored continuously, using permanent sampling plots and GIS (geographic information system) habitat mapping (Western 2007).

The Amboseli study therefore offers an opportunity to compare opportunistic repeat photography with vegetation changes systematically mapped over several decades. How well does repeat photography capture the measured changes? Can repeat photography give insights into the causes of change? To what extent can repeat photography be used to infer longer-term habitat changes decades before ecological research began? In this chapter, I address these questions using ground and low-level aerial photographs.

Study Area

"Amboseli" refers to a shallow trough formed by a Pleistocene lake that formed to the north of Mt. Kilimanjaro on the Kenya–Tanzania border (Williams 1967). The trough, known as the Amboseli Basin, lies at an altitude of 1,200 meters. Rainfall averages between 250 and 300 millimeters and falls in two seasons, the short rains in November and December and the long rains in March–May (Altmann et al. 2002). The vegetation is classified as semiarid rangeland (Ecological Zone V) by Pratt et al. (1966). Aquifers carrying rainwater from the forests of Kilimanjaro discharge into the Amboseli Basin, creating a series of swamps and a high water table that supports shallow-rooted hydrophilic woodlands dominated by *Acacia xanthophloea*. The ecological setting is described in detail by Western and van Praet (1973), Western (1973, 1975, 2007), and Lindsay (1994).

The basin is a dry season concentration area for large wild ungulates that disperse over an area of

8,500 square kilometers in the rains. The Maasai pastoralists subsisted on cattle, sheep, and goat herds until the last two decades. Maasai migrations track those of wildlife ungulates through the seasons (Western 1975). An area of 388 square kilometers, covering much of the dry season range, was set aside as a national park in 1974. The remainder of the ecosystem is divided up among six group ranches. Small-scale farms have been established on the slopes of Kilimanjaro and in swamps north of the mountain in the last two to three decades (Western and Nightingale 2004).

Methods

A map of 28 vegetation zones in the 600-square-kilometer study area was drawn up in 1967 using Poore's (1966) "successive approximation" technique. A description of the vegetation composition of each zone is given in Western (1973, 2007). Several studies have looked at the nature and causes of habitat change (Western and van Praet 1973, Western and Maitumo 2004, Western 2007).

In comparing repeat photography with long-term monitoring data, I restrict the analysis to Amboseli National Park because it is so well documented photographically. I have selected four of my own photographs taken over the last four decades to explore how well repeat photography captures the quantified habitat changes.

Results

A Half Century of Habitat Change

Plate 16.1 presents a series of four of the nine sequential GIS maps given in Western (2007). Of the six habitats in the national park, dense woodland shows the largest loss over the past half century, followed by the open woodland. Open woodland expanded and replaced the dense woodland for a decade or more, before contracting sharply in turn. The changes in percentage habitat cover are shown

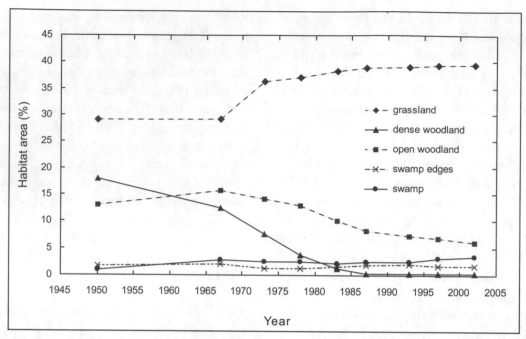

Figure 16.2. The relative changes in habitat cover in the Amboseli Basin between 1950 and 2002, derived from GIS maps (plate 16.1).

in figure 16.2. The loss in both open and dense woodlands between 1967 and the early 1980s was confined to *Acacia xanthophloea*, which occurred in monotypic stands throughout the central basin (Western and van Praet 1973). By the mid-1980s, the *A. xanthophloea* woodlands were restricted to three isolated patches in the central and eastern park. An exception to the loss was a small plot of regenerating woodland in the center of the park where an herbivore exclosure was constructed in 1981 (Western and Maitumo 2004). By 2002, the *A. xanthophloea* woodlands were extinct within the park, except for the rapidly disappearing Ol Tukai grove. Here the *A. xanthophloea* are partially protected by thickets of *Phoenix reclinata* (date palm). Another small area of regenerating *A. xanthophloea* is found on the southern edge of the park, close to Maasai settlements. The only other woodland remaining in the park consists of *Acacia tortilis*, a species that spreads in a long narrow band along the shoreline of the ancient Pleistocene lakebed. *A. tortilis* woodland showed little shrinkage between 1950 and the early

1980s but has since begun contracting rapidly (Western 2007).

The habitat showing the largest expansion is *Suaeda-Salvadora* scrub. Absent in 1950, a small island of *Suaeda-Salvadora* scrub had colonized the central basin by 1967, replacing a former patch of mixed dense and open woodland. By 1983, *Suaeda-Salvadora* scrub had spread extensively, replacing the former dense and open woodlands. By 2002, the *Suaeda-Salvadora* scrub had spread in a few areas but contracted in its northern range abutting the grasslands. Grasslands have also expanded over the last half century. The expansion traced the receding woodlands in the central portion of Amboseli and, by 2002, invaded the northerly patches of *Suaeda-Salvadora* scrub.

The two large permanent swamps—Longinye and Enkongo Narok—emerge from springs in the southern basin. Both have expanded greatly over the last half century. The tall dense stands of *Papyrus* and *Cyperus immensus* that dominated the swamp in 1967 had been largely trampled and eaten down by elephants by 2003 (Sarkar 2006). A number of smaller isolated swamps appeared in the southern basin as the woodlands thinned and died out (Sarkar 2006, Western 2007).

A small contraction of the swamp-edge habitat has occurred since 1950, largely due to the expansion of the permanent swamp. In the early 1970s, the swamp edge comprised shrub thickets dominated by *Solanum incanum* (sodom apple), *Pluchea dioscordis*, *Triplocephalum* spp., *Withania somnifera*, and *Abutilon* spp., interspersed with mats of *Cynodon dactylon*, a stoloniferous grass. During the drought of the mid-1970s, heavy browsing reduced the shrub thickets to isolated pockets and encouraged the expansion of a uniform sward of *Cynodon dactylon* grassland.

Habitat Change Detection Using Repeat Photography

The habitat changes documented in the Amboseli Basin over the last 50 years are best captured by low-level oblique photographs taken from Observation Hill, a small volcanic intrusion in the center of the national park. Figure 16.3 shows a paired compari-son taken from Observation Hill in 1967 and 2007, looking southwest toward the Ilmarishari Hills in the background. The view covers a cross section of the open woodland in the foreground and dense woodland below Ilmarishari Hills. The 40-year sequence captures the documented disappearance of *Acacia xanthophloea* woodlands (Western 2007). The remaining woodland is confined to the thinning belt of *Acacia tortilis* at the base of the Ilmarishari Hills.

In 1967, open stands of *A. xanthophloea* dominated most of the southern basin, and dense stands dominated the area around the Ilmarishari Hills. Several points can be deduced from the photograph. First, the level of mortality is related to woodland density, progressing from the open woodlands in the foreground to denser woodlands in the far distance. A large number of dead standing and fallen trees can be seen in the nearest grove. Fewer dead trees are visible in the distant groves. Second, the age structure of the trees—smaller in the foreground, larger in the background—suggests an age gradation from older trees at the edge of the basin to younger in the center. Third, the absence of young trees points to a lack of recruitment over many years. The high mortality rate

A

B

Figure 16.3. Looking southwest from Observation Hill, Amboseli National Park, Kenya.
A. (1967). This view, looking southwest from Observation Hill, shows open stands of *Acacia xanthophloea* (fever tree) dominating most of the southern basin, with dense stands below the Ilmarishari Hills. The trees in the middle distance are in a more advanced state of desiccation and collapse than those in the far distance.
B. (2007). Note the replacement of *Acacia xanthophloea* woodlands and expansion of *Suaeda-Salvadora* scrub. The line of *Acacia tortilis* trees in the far distance is thinning.

and absence of recruitment suggest a rapid collapse of the woodlands. Fourth, the top-down collapse of branches suggests that water transport to the distal branches is impaired. Finally, the density of understory scrub vegetation—*Suaeda-Salvadora*—is highest where the woodlands have died back and lowest among denser groves.

The woodland collapse and replacement by *Suaeda-Salvadora* shown by repeat photographs capture the habitat changes described by Western (1973) and Western and van Praet (1973). The 2007 photograph shows complete replacement of *A. xanthophloea* woodlands by *Suaeda-Salvadora* scrub, as predicted by Western and van Praet (1973) and documented in the GIS map series (plate 16.1).

A second sequence of photographs (fig. 16.4), taken on the same dates as the first pair, looks east from Observation Hill across the Enkongo Narok Swamp to the Ol Tukai Orok grove of trees. The 1967 photograph shows the swamp in the foreground, dominated by the sedge *Cyperus immensus*. *Cyperus* is heavily grazed and trampled in the foreground, but less so at the far edge of the swamp. A narrow band of dying *A. xanthophloea* with *Suaeda-Salvadora* understory borders the swamp. A narrow

finger of alkaline grassland separates the swamp and dense stands of *A. xanthophloea* in Tukai Orok grove. Although the name Ol Tukai refers to the Maasai name for *Phoenix reclinata*, the palms are hidden by the dense outer thicket of *A. xanthophloea*. The woodland understory is relatively open with no *Suaeda* or *Salvadora* present.

By 2007, the swamp sedges have all but disappeared and been replaced by open water. The fringing belt of *A. xanthophloea* has also disappeared, along with the *Suaeda-Salvadora* scrub. *A. xanthophloea* in the Ol Tukai grove has shrunk to isolated trees scattered among the denser and squatter *Phoenix* palms, now clearly visible. Scattered clumps of *Suaeda-Salvadora* scrub have invaded the outer fringes of the woodland. A narrow band of *Acacia tortilis* can be seen in the far distance. A large, open grassy plain has enveloped Ol Tukai woodland and displaced the *Suaeda-Salvadora* scrub along the edge of the swamp.

This set of repeat photographs captures the habitat changes in the central park documented by the 50-year GIS mapping (plate 16.1). The changes include the expansion of open water in the swamp, the contraction of *A. xanthophloea* woodlands to a small

Figure 16.4. Looking east from Observation Hill, Amboseli National Park, Kenya.
A. (1967). This photograph looks east from Observation Hill across the Enkongo Narok Swamp to the Ol Tukai Orok grove. The swamp is dominated by the sedge *Cyperus immensus*.
B. (2007). The dense *Acacia xanthophloea* thickets of Ol Tukai Orok grove on the far side of the swamp largely disappeared over the 40-year interval, and *Phoenix reclinata* replaced it as the dominant woody plant. The swamps showed signs of heavy elephant disturbance in 1967 and gave way to open water by 2007.

remnant grove in Ol Tukai, and the expansion of open plains and contraction of *Suaeda-Salvadora* scrub at its northern margins.

Another set of repeat photographs (fig. 16.5), taken at my first research camp in 1967 and repeated in 2007, captures the vegetation changes in the central Amboseli Basin. The 1967 photograph shows the *A. xanthophloea* trees dying from the top branches downward and a lack of regeneration. The trees in the foreground are heavily gouged and debarked by elephants. The understory growth beyond my camp is dominated by *Solanum incanum* scrub. The edge of the grove in the foreground shows clumps of *Salvadora* bordering the grasslands to the right. By 2007, the woodlands have died out and been replaced by swamp. The tall swamp sedge, *Cyperus immensus*, is heavily grazed down. Most clumps are reduced to stubble and overgrown by matted *Cynodon dactylon* grass and smaller sedge species. In the far distance, to the right in the photo, a dense thicket of regenerating *A. xanthophloea* marks the site of an experimental restoration plot enclosed by a solar electric fence (Western and Maitumo 2004).

The photograph matches capture the collapse of the woodlands detailed by Western and van Praet (1973), and the expansion of the swamps shown in the GIS maps (plate 16.1). The close-up photographs capture the *A. xanthophloea* forests at the point of collapse. The photographs show *Suaeda-Salvadora* scrub replacing the *A. xanthophloea* woodlands and alkaline plains expanding into the northern margins of *Suaeda-Salvadora* scrub. The photographs capture the expansion of swamp and depict heavy browsing and grazing pressure in the park.

Cause of Habitat Change

The three sets of repeat photographs show the scale of woodland collapse and habitat change but do not give an unambiguous cause. By using multiple replicates of a time series of photographs over several decades, it would be possible to establish a strong correlation between tree mortality and elephant damage, as was done using spatial sampling of *A. xanthophloea* across the basin in the 1970s (Western 1973, Western and van Praet 1973). However, even such a quantitative time series of opportunistic photographs would do no more than suggest plausible hypotheses. Given the strong correlations between

Figure 16.5. The author's campsite in Amboseli National Park, Kenya.
A. (1967). The author's campsite was located in a grove of *Acacia xanthophloea* already showing signs of decay. All trees show extensive debarking by elephants and collapsing apical branches. Thickets of *Solanum incanum* (sodom apple) constitute the understory below the camp, while clumps of *Salvadora* grow at the edge of the grove (foreground).
B. (2007). The woodlands have died out and Longinye Swamp has expanded northward and inundated the campsite. The *Cyperus immensus* is heavily grazed down and smaller prostrate sedges cover the swamp. The dense stand of trees in the right background is the experimental woodland restoration plot established in 1981 (see fig. 16.6).

woodland growth, rainfall patterns, water table, salinity levels, and herbivory (Young and Lindsay 1988, Western and Maitumo 2004), opportunistic repeat photographs cannot readily distinguish cause and effect.

Selective repeat photography can, however, capture the results of field experiments designed to test cause and effect in large-scale habitat changes. Figure 16.6 shows repeat low-level aerial photographs of the 20-year woodland restoration experiment in the center of Amboseli National Park (Western and Maitumo 2004). In the 1981 photo, a ditch excluding elephants is visible behind my research house; the tourist lodge is in the background. The same scene in 2003 (fig. 16.6B) shows the regeneration of trees confined to the area within an electric fence around the research house and lodge. The impact of the electric fence in restoring the woodlands eliminates both climate and soils as explanations of woodland regeneration. An eye-level electric wire selectively excluding elephants shows the role of elephants in preventing woodland regeneration and, by inference, its collapse. A similar experiment near the Ilmarishari Hills shows the recovery of tall sedges within two years of selective elephant exclusion (Sarkar 2006). In both cases, selective repeat photog-

raphy captures the results of naturalistic experiments in untangling the cause and effect of habitat change.

Inferring Longer Changes through Repeat Photos

Having shown how well repeat photography matches habitat changes monitored quantitatively over four decades, I now extend its reach to detect changes long before systematic vegetation studies began. To do so, I located several collections of photographs taken in the early 1900s. The best collection is Martin and Osa Johnson's, shot in 1921 (Johnson 1935). The Johnson collection has over 60 photos, including panoramic views from the top of Observation Hill. I used his 1921 shots from Observation Hill and repeated them in 2003 to detect centennial vegetation changes.

The 1921 photograph (fig. 16.7) shows dense *A. xanthophloea* woodland flanking Enkongo Narok Swamp, a small alkaline plain in the middle-ground, and a pocket of *Salvadora* in the foreground. The most striking feature of the 1921 woodlands is the number of young *A. xanthophloea*. Comparing the trees in the Johnson photograph to the 20-year

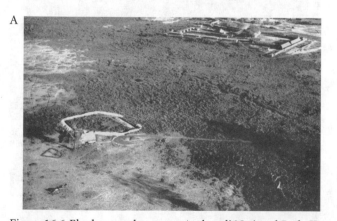

Figure 16.6. Elephant exclosures at Amboseli National Park, Kenya.
A. (1981). This pair, taken from a low-level aircraft, shows the recovery of *Acacia xanthophloea* woodlands when protected from herbivores. There is an elephant-excluding ditch (white pentagon) at the back of the author's research house.
B. (2003). The original elephant ditch was expanded into a woodland restoration program using a solar electric fence, aimed at reestablishing woodlands around the tourist lodges in central Amboseli National Park.

A

B

Figure 16.7. Looking south of Observation Hill, Amboseli National Park, Kenya.
A. (1921). Martin Johnson captured this view looking south of Observation Hill. Note the relative youth of the *Acacia xanthophloea* by comparing with the 20-year-old trees in figure 16.6. (M. Johnson, Martin and Osa Johnson Library).
B. (2003). By 2003, the woodlands had disappeared, except for small fringing groves outside the park boundary on the far side of the swamp close to permanent Maasai settlements. Enkongo Narok Swamp, beyond the avenue of trees in the foreground, was far smaller in 1921 and covered by dense scrub rather than sedge. (R. M. Turner).

woodland regeneration experiment (fig. 16.6), few trees in the 1921 photographs can be older than 25 to 30 years. Trees along the swamp fringe are larger and older than those in the central basin. Note too the lack of sedges, absence of surface water, and dominance of shrubs in the swamp, indicating that Enkongo Narok was far smaller and surrounded by a swamp-edge thicket in the 1920s.

A review of Johnson's 60 Amboseli photographs confirms the prevalence of young trees across the basin in the 1920s. The largest trees were clustered around the springs of the southern basin, suggesting a founder population that spread northward around 1900. There are no visible dead trees, either standing or fallen. Given that dead *A. xanthophloea* last up to 30 years in the dry alkaline soils of Amboseli, I infer that the *A. xanthophloea* woodlands were confined to the swamp's edges in the late 1800s and several decades beforehand. In contrast, there is no indication that *A. tortilis* stands were young in the early 1900s. The *A. tortilis* woodlands suggest a wide age structure and show no signs of elephant damage.

What then accounts for the spread of *A. xanthophloea* a century ago? With the exception of a single tree in the left foreground, elephant damage is absent. Few of Johnson's entire photographic collection show signs of elephants or elephant damage. Strikingly, there is little elephant dung either, in stark contrast to the ubiquity of dung in photographs taken in the latter half of the twentieth century. The lack of elephants is confirmed by Johnson's narrative account (Johnson 1935). He mentions only two elephants seen in many days of filming. They were so rare, in fact, that he had to wait for hours to film two bulls as they emerged from the swamps. The dearth of elephants is further confirmed by Johnson's photographs of the Amboseli swamps. All show tall, undisturbed beds of *Cyperus immensus* and no signs of elephant grazing or trampling.

The hint of woodland expansion in the 1920s shown by Johnson's photographs is borne out by other photographers at the turn of the twentieth century. German photographer Behrens's 1910 images, which I inspected at the Royal Geographical Society in London, show denser, younger stands of *A. xanthophloea* regeneration than Johnson's photographs and no sign of elephant damage. Joseph Thomson, the first European explorer to pass through Amboseli, describes the basin in 1883 as a barren plain covered by a juniper-like bush, presumably *Suaeda*, with scattered swamps covered by tall reeds (Thomson 1885). Though an avid hunter, he made no mention of elephants in Amboseli.

Summary of Photographic Evidence

Pulling together the pictorial evidence, we can deduce that the *Acacia xanthophloea* woodlands in Amboseli were confined to the spring margins in the late 1800s and began spreading across the basin about that time due to the scarcity of elephants. By the 1950s, the woodlands had spread to the central basin. Shortly after the establishment of Amboseli Game Reserve in the late 1940s, a rising elephant population began visibly depleting the woodlands. The woodland destruction and loss began in the thinner, younger groves in the central basin and progressed southward to the basin edge (Western and van Praet 1973). The more elephant-resilient *Suaeda-Salvadora* scrub proliferated as the woodland receded. Swamp expansion coincided with the woodland recession. A compressed elephant population in the 1970s (Western and Maitumo 2004) grazed and trampled down the tall sedge, creating large, open expanses of water. Rainfall records over the last 35 years show no significant increase that would explain the expanding swamps (Altmann et al. 2002). It seems more likely that the swamp expansion resulted from a rising water table as the combined transpiration from woodlands and tall sedges fell with increased herbivory.

These broad conclusions accord well with the historical evidence of a sharp drop in elephant populations in eastern and central Africa by the mid-1880s. Håkansson (2004) shows that elephants were rare from the coast to Kilimanjaro by the mid- to late 1800s, due to the impact of the slave and ivory trade (Parker and Amin 1983). Like Thomson, other early explorers passing through Tsavo, Amboseli, and central Kenya in the late 1800s rarely if ever mention elephants. The plight of elephants was of such concern at that time that the eight colonial powers of Africa convened a wildlife convention at Lancaster House in London to protect the elephant and other species (Simon 1962). The evidence of a rising elephant population in Amboseli during the first half of the twentieth century fits a pattern documented in southern Africa (Cumming et al. 1997). The increased elephant populations in the park in the second half of the twentieth century have been documented by regular censuses (Western 1975, 1994; Western and Lindsay 1984).

It is possible that elephant numbers in Amboseli during the sixteenth and seventeenth centuries were sufficiently high to contain the spread of *Acacia xanthophloea* across the basin. Based on the size and age-spread of the long-lived *Acacia tortilis* groves in 1921, we can deduce that the elephant numbers were not as concentrated year-round as the populations from the 1970s onward. It is likely that Maasai settlements in the southern basin protected *A. xanthophloea* pockets around the permanent swamps prior to the 1880s.

The separation of the formerly synchronous migrations of elephants and livestock into two separate domains either side of the park boundary in recent decades has uncoupled the interactions that maintained the patchwork dynamics of Amboseli vegetation (Western and Gichohi 1993, Western and Maitumo 2004). The decline in habitat diversity in Amboseli National Park has caused a significant loss of large ungulate diversity (Western 2007), most notably a decrease in browsing species and increase in grazers (Western and Nightingale 2004). The large-scale ecological change bears on the conservation of Amboseli's biological diversity. Similar declines in woody vegetation and loss of biotic diversity have been widely reported for African parks wherever elephant populations have been compressed (Laws 1970, Cumming et al. 1997). In the case of Amboseli, selective electric fencing can protect woodland refugia and conserve threatened plant and animal species (Western and Maitumo 2004).

The increased density of livestock in the semiarid savannas over the last century has also caused large-scale habitat changes across Kenya rangelands, typi-fied by increased woody vegetation (Turner et al. 1998). The divergent trends toward grasslands inside parks and bushlands outside have large-scale regional repercussions, given the loss of habitat heterogeneity and biotic diversity in both cases (Western 1989). These large-scale trends, apparent in the repeat photography of Turner et al. (1998), will be documented in a countrywide photo-repeat study of Kenya undertaken by David Western and Raymond Turner (in preparation).

Discussion and Conclusions

Amboseli National Park offers a unique opportunity to test the reliability of repeat photography as a method of detecting long-term habitat changes. The results show that even a few opportunistically repeated photographs can capture the main habitat changes of the last half century. The photographs show sufficient detail of herbivore–plant interactions to deduce plausible hypotheses about the causes of change. Selective photographs of naturalistic experiments offer a powerful tool for detecting habitat change. By calibrating repeat photography against documented changes over several decades, it is possible to infer centennial-scale changes from the earliest existing photos.

The compelling picture of habitat change captured in this repeat photography study could be greatly improved by matching many thousands of photographs taken by Amboseli visitors since the 1940s. Such a large time-series would make it possible to plot the changing relative frequencies of plant and animal species over several decades. The utility of such repeat photography could be improved by setting up photo-repeat stations linked directly to the ecological sampling design and long-term monitoring plots. Such efforts will be greatly eased by the use of digital photography and GPS technology, although preservation of the digital images remains a concern (Boyer et al., chapter 2).

Acknowledgments

This study was supported by the National Geographic Society and the Liz Claiborne–Art Ortenberg Foundation. I am grateful for the assistance of many Amboseli wardens over the years, and for the help of research assistant David Maitumo and Victor Moses in preparing the manuscript. I would also like to thank the Martin and Osa Johnson Library in Chanute, Kansas, for locating the Johnson photographs and making them available to me; and Ray and Jeanne Turner and Samantha Russell for their part in repeating the Johnson photographs. All photographs were taken by the author unless otherwise noted.

Literature Cited

Altmann, J., S. C. Alberts, S. A. Altmann, and S. B. Roy. 2002. Dramatic change in local climate patterns in the Amboseli Basin, Kenya. *African Journal of Ecology* 40:248–251.

Botkin, D. B. 1990. *Discordant harmonies: A new ecology for the twenty-first century*. New York: Oxford University Press.

Cole, M. M. 1986. *The savannas*. London: Royal Holloway and Bedford College, University of London.

Cumming, D. H. M., M. B. Fenton, I. L. Rautenback, R. D. Taylor, G. S. Cumming, M. S. Cumming, J. P. Dunlop, A. G. Ford, M. D. Hovorka, D. S. Jonston, M. Kalcounis, Z. Mahlangu, and C. V. R. Portfos. 1997. Elephants, woodlands and biodiversity in southern Africa. *South African Journal of Science* 93:231–236.

Deshmukh, I. 1986. *Ecology and tropical biology*. Oxford: Blackwell Scientific Publications.

Dublin, H. T., A. R. E. Sinclair, and J. McGlade. 1990. Elephants and fires as sources of multiple stable states in the Serengeti-Mara woodlands. *Journal of Animal Ecology* 59:1147–1164.

Håkansson, N. T. 2004. The human ecology of world systems: The impact of the ivory trade. *Human Ecology* 32:561–591.

Homewood, K. M., and W. A. Rogers. 1991. Maasailand ecology: Pastoralist development and wildlife conservation in Ngorongoro, Tanzania. Cambridge: Cambridge University Press.

Johnson, M. 1935. *Over African jungles*. New York: Harcourt, Brace and Company.

Laws, T. M. 1970. Elephants as agents of landscape change in East Africa. *Oikos* 21:1–15.

Lindsay, K. W. 1994. *Feeding ecology and population demography of African elephants in Amboseli, Kenya*. PhD dissertation, University of Cambridge.

Parker, I., and M. Amin. 1983. *Ivory crisis*. London: Chatto and Windus.

Poore, M. E. D. 1966. The method of successive approximation in descriptive ecology. In *Advances in ecological research*, Vol. 1, ed. J. B. Cragg, 35–68. London: Academic Press.

Pratt, D. J., P. J. Greenway, and M. D. Gwynne. 1966. A classification of East African rangeland, with an appendix on terminology. *Journal of Applied Ecology* 3:369–382.

Pratt, D. J., and M. D. Gwynne, eds. 1977. *Rangeland management and ecology in East Africa*. London: Hodder and Stoughton.

Sarkar, S. 2006. *Long- and short-term dynamics of the wetlands in the Amboseli savanna ecosystem, Kenya*. PhD dissertation, Ontario, Canada: University of Waterloo.

Simon, N. 1962. *Between the sunlight and the thunder*. London: Collins.

Thomson, J. 1885. *Through Masailand*. London: Case.

Turner, R. M., H. A. Ochung', and J. B. Turner. 1998. *Kenya's changing landscape*. Tucson: University of Arizona Press.

Western, D. 1973. *The structure, dynamics and changes of the Amboseli ecosystem*. PhD dissertation, University of Nairobi, Kenya.

Western, D. 1975. Water availability and its influence on the structure and dynamics of a large mammal community. *East African Wildlife Journal* 13:265–286.

Western, D. 1989. The ecological role of elephants in Africa. *Pachyderm* 12:42–45.

Western, D. 1994. Ecosystem conservation and rural development: The case of Amboseli. In *Natural connections: Perspectives in community-based conservation*, ed. D. Western, M. Wright, and S. C. Strum, 15–52. Washington, DC: Island Press.

Western, D. 2007. A half century of habitat change in Amboseli National Park, Kenya. *African Journal of Ecology* 45:302–310.

Western, D., and H. Gichohi. 1993. Segregation effects and the impoverishment of savanna parks: The case for an ecosystem viability analysis. *African Journal of Ecology* 31:269–281.

Western, D., and K. Lindsay. 1984. Seasonality, migration and herd dynamics in a savanna elephant population. *African Journal of Ecology* 22:229–244.

Western, D., and D. Maitumo. 2004. Woodland loss and restoration in a savanna park: A 20-year experiment. *African Journal of Ecology* 42:111–121.

Western, D., and D. Nightingale. 2004. Environmental change and the vulnerability of pastoralists to drought: The Maasai in Amboseli, Kenya. In *Africa environment outlook, case studies: Human vulnerability to environmental change*, 31–50. London: Earthprint on behalf of the United Nations Environment Programme.

Western, D., and C. van Praet. 1973. Cyclical changes in the habitat and climate of an East African ecosystem. *Nature* 241:104–106.

Williams, J. G. 1967. The Kilimanjaro volcanic rocks of the Amboseli area, Kenya. PhD dissertation, University of Nairobi, Kenya.

Young, T. P., and K. W. Lindsay. 1988. The role of even-aged population structure in the disappearance of *Acacia xanthophloea* woodlands. *African Journal of Ecology* 26:69–72.

Chapter 17

Repeat Photography and Low-Elevation Fire Responses in the Southwestern United States

Raymond M. Turner, Robert H. Webb, Todd C. Esque, and Garry F. Rogers

In the western United States, wildland fire is a major land-management issue, but for different reasons in different ecosystems. Fires are an important part of ecosystem processes in *Pinus ponderosa* (ponderosa pine) forests and the higher-elevation desert grassland (Bahre 1985; McPherson 1995; Turner et al. 2003; McClaran et al., chapter 12); the long-term anthropogenic use of fire to manage landscapes is well known in many parts of the world (Veblen, chapter 13), particularly in desert and savanna landscapes (Lewis, chapter 15). Fire suppression, in particular, has long been considered a primary reason for conversion of perennial desert grasslands to savannas, allowing woody vegetation to encroach where mostly herbaceous species once grew. Land managers commonly use fire to reduce woody vegetation in the Southwest either by setting controlled fires or by allowing wildfires to burn.

At lower elevations in the Great Basin, Mojave, and Sonoran deserts, areas that once may have had too little fuel to sustain significant fires are now burning. Much of the reason for the increased fire occurrence is considered to be the fine fuel from nonnative annual and perennial species, primarily grasses (Young and Evans 1973, Rogers 1982, Esque and Schwalbe 2002, Brooks et al. 2004), but also from various forbs, such as *Brassica tournefortii* (Sa-

haran mustard) (T. C. Esque, pers. comm., 2008). However, the role of native annual species in fueling lower-elevation desert fires is not well-enough known to be incorporated into management planning (Zouhar et al. 2008). In this chapter, we review and add to the documentation of fire occurrence in selected parts of the Sonoran, Mojave, and Great Basin deserts.

The increase in desert fires has spawned widespread concern that type conversions, typically from shrublands to annual grasslands, may be occurring in the Mojave Desert (Brooks et al. 2004), the Sonoran Desert (Schmid and Rogers 1988), and the Great Basin Desert (Rogers 1982). Many of the dominant species in these deserts are not well adapted to fire, and widespread mortality with little or no resprouting is typical (Rogers and Steele 1980), although in a study of burn response among both dominant and minor species in the Sonoran Desert, 88 percent of this larger group sprouted one year after a wildfire (Wilson et al. 1995). As a result, recovery following fires is an extremely long process that has been compared with recovery from other disturbances that involve severe soil compaction or disruption (Webb et al. 2009). Typical recovery times range from around a century for total perennial cover to many centuries for species composition in the Mojave and

Transition Great Basin deserts; similar estimates have not been made in the Sonoran Desert.

Repeat photography is one tool well suited for documenting the short-term effects of fire on desert ecosystems as well as the long-term prospects for recovery. Photography graphically documents fire effects, and in some instances can add information that cannot easily be obtained using other techniques, such as permanent vegetation plots or satellite imagery. In this chapter, we report several case studies of the use of repeat photography, combined with other techniques, to document fire effects and recovery in parts of the Great Basin, Mojave, and Sonoran deserts (fig. 17.1). The case studies are too few in number to test general propositions for the desert regions we discuss, but they do provide additional answers to some important questions.

Among these are the questions of whether desert vegetation is being irreversibly altered by fires and whether increases in nonnative annual species create ecosystem-type conversions. The case studies also provide an opportunity to test our assertion that repeat photography is a useful tool for documenting fire effects in deserts.

Fire in the Sonoran Desert

The Sonoran Desert occurs in the American states of California and Arizona and extends south into Baja California and Sonora, Mexico (Turner and Brown 1994). Annual rainfall in this desert ranges from about 25 millimeters in the driest parts to over 300 millimeters near Tucson. The plants that dominate the typical Sonoran Desert landscape include *Larrea tridentata* (creosotebush); *Carnegiea gigantea* (saguaro), a massive columnar cactus; *Ferocactus* spp. (barrel cactus); *Opuntia* spp. (pricklypear); *Cylindropuntia* spp. (cholla); and leguminous trees, including *Prosopis velutina* and *P. glandulosa* (mesquite), *Olneya tesota* (ironwood), and *Parkinsonia microphylla* (foothill paloverde) and *P. florida* (blue paloverde). This array of species is not well adapted to fire (Rogers and Steele 1980). Our case studies illustrate what happens when areas that once supported these species are burned.

Fire frequency in southeastern Arizona during the last years of the nineteenth century was studied by thorough analyses of the region's newspaper accounts (Bahre 1985). These early accounts recorded frequent fires in coniferous forests and oak–juniper woodlands of the region, but fire in Sonoran Desert vegetation was rarely documented. Studies examining a later time period (1955 through 1983) in Arizona's Tonto National Forest have shown that the area burned in the Sonoran Desert segment of that national forest was actually about the same as in nondesert habitats (Rogers 1986). Fires from all causes increased during the period, but the most significant increase was in the number of human-caused desert fires.

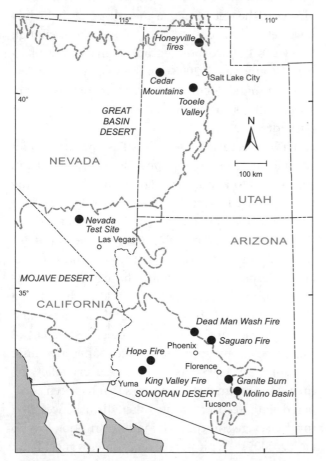

Figure 17.1. Map of the southwestern United States showing the location of historical fires.

As human occupation of the Sonoran Desert has increased, the potential for increased fire frequency resulting from accidental human ignitions has increased. Fire's temporal and spatial variability appears to be highly sensitive to changes in human activities, as well as to shifts in climate (Alford et al. 2005). Added to this increased fire potential are incursions of nonnative species, particularly *Pennisetum ciliare* (buffelgrass), *Pennisetum setaceum* (fountain grass), and *Enneapogon cenchroides* (soft feather pappusgrass), all originally from Africa. *P. ciliare*, in particular, is considered to be an extreme threat to the Sonoran Desert (Búrquez-Montijo et al. 2002, Esque et al. 2006). In addition to these perennial species, an array of annual nonnative species contributes to increased fire potential, including *Bromus madritensis* ssp. *rubens* (red brome) and *Schismus* spp. (Mediterranean grasses) (Esque and Schwalbe 2002).

Fire at the Upper Elevation Limit of *Carnegiea*

Across the northern part of the range of *Carnegiea*, true desert gives way to semidesert grassland at about 915 meters on level to gently sloping terrain (Turner and Brown 1994). Sonoran Desert iconic species such as *Carnegiea* may continue to survive upslope for another 550 meters on steep, south-facing slopes where the lethal effects of freezing temperatures at this higher elevation are reduced because of the favorable exposure and where the beneficial effects of thermal inversion ameliorate otherwise harsh conditions. However, at these higher elevations, various grass species are favored, and the Sonoran Desert plants are often surrounded by dense grass, a fire-prone element not common in *Carnegiea*'s home at lower elevations. Prior to the initiation of inadvertent and purposeful fire suppression activities that arose at the time of heavy human occupation of the area, these semidesert grassland areas were engulfed by fires at regular intervals (Bahre 1985, Davis et al. 2002). This raises the question of whether the presence of *Carnegiea* in the lower fringe

of the semidesert grassland is artificial and related to fire suppression.

Carnegiea's response to fires is unlike most of the woody species with which it grows. Instead of being consumed by flames and perhaps burning to the ground, badly scorched *Carnegiea* individuals remain standing, albeit without spines and with the surface tissue in various stages of decline (Esque et al. 2004). The severely burned plants may survive for four or more years before finally succumbing (Rogers 1985, Narog and Wilson 2002). Once lost from an area, reestablishment of *Carnegiea* is a slow process because this species does not produce long-lived seeds that are embedded in the soil seed bank and, in addition, any new plants appearing in the postfire habitat would not be sexually mature for 30–35 years (Steenbergh and Lowe 1977). With a fire return interval of once every decade or two, *Carnegiea* and other similar fire-sensitive desert plants would not survive in these habitats (Rogers 1985). This reasoning suggests that *Carnegiea* stands occurring at elevations well above the upper limit of the Sonoran Desert owe their presence in these areas to human-induced fire-frequency decline.

In the early 1960s, J. Rodney Hastings and Raymond M. Turner became intrigued by a small group of *Carnegiea* growing on the south-facing slope overlooking the Molino Basin campground in the Santa Catalina Mountains near Tucson (fig. 17.1). These *Carnegiea*, at an elevation of 1,500 meters, were the last ones to be seen as one drives up the Mount Lemmon Highway toward the summit. At first, Hastings and Turner focused their attention on *Carnegiea*'s response to freezing temperatures; the subject of fire effects arose much later. Set upon understanding the temperature conditions of this upper-limit habitat, they installed a recording thermograph next to the uppermost *Carnegiea*. In March 1961, a photograph was taken to record this installation (fig. 17.2A).

The next winter, on 11 January 1962, minimum temperatures at the Tucson weather station reached −6° Celsius, and subfreezing temperatures persisted for 19 hours (Bowers 1980–1981). The *Carnegiea*

Figure 17.2. Molino Basin, Santa Catalina Mountains, Arizona, USA.
A. (6 March 1961). A recording thermograph was installed next to a *Carnegiea gigantea* (saguaro) at its upper-elevation limit (1,500 meters) in order to better understand the local temperature conditions. (R. A. Dodge).
B. (31 January 1962). This image was captured two weeks after a catastrophic freeze. The apex of the *Carnegiea*, its frost-sensitive part, is shrunken and covered with necrotic tissue. (J. R. Hastings).
C. (31 October 1970). Although still standing, the *Carnegiea* has not grown. The constriction partway up the *Carnegiea* probably dates to an earlier freeze, possibly in 1937 or 1949, when low temperatures equaled or exceeded those of 1962. The thermograph was removed several years before this photograph was taken. (J. R. Hastings).
D. (12 January 1976). In 1975, the *Carnegiea* was burned and killed. By 2006, no *Carnegiea* are present on this slope, which is now densely covered with nonnative *Enneapogon cenchroides* (soft feather pappusgrass). (R. M. Turner, Stake 770).

shown in figure 17.2A and others nearby were damaged; a photograph taken on 31 January 1962 records the damage to the subject plant (fig. 17.2B); on this date, its apex is shrunken and is covered by necrotic tissue. Because of its unfavorable surface to volume ratio, the apex of these cacti is the most sensitive to freezing temperatures, and often a plant's apex will show the only frost injury the plant experiences. The original photograph was repeated several times over the next years; these records showed that this *Carnegiea* did not grow again after the 1961

freeze. In October 1970, for example, after almost nine postfreeze years, the plant was still standing, had not grown, and showed a few additional necrotic areas (fig. 17.2C). The constriction halfway up the *Carnegiea*'s stem is also probably the result of some previous freeze, most likely either in 1937 or 1949, when low temperatures equaled or exceeded those of 1962 (Steenbergh and Lowe 1977).

Tough times for this population of *Carnegiea* were not over in 1962. In 1975, fire burned across the *Carnegiea*'s habitat, and the plant was finally killed as shown in a photograph taken on 12 January 1976 (fig. 17.2D). This former *Carnegiea* habitat overlooking Molino Basin was almost completely *Carnegiea*-free following the 1962 freeze; the fire of 1975 completed the plant's demise and the last *Carnegiea* on this slope were eradicated. There are no *Carnegiea* present today on the south-facing slope overlooking Molino Basin. A visit to this site in February 2006 revealed an interesting addition not seen before: the slope, now heavily covered by grasses, supports an abundant population of *Enneapogon cenchroides*, a nonnative species from East Africa and the Indian subcontinent. This grass was first found in Molino Basin in 1976 and was found only here and nearby for the next 20 years. Then it spread rapidly, both east and west, and today is found about 160 kilometers away in both directions.

A more dramatic effect of fires was seen downslope from the Molino Basin overlook when a fire burned across the slope above the Mount Lemmon Highway at an elevation of about 1,128 meters (at about highway kilometer 4.5). This fire burned in 1987 across a south-facing exposure on which many *Carnegiea* had become established. A series of panning photographs was taken in December 1987 to capture the fire's effect. In one of the captured views are seen about a dozen fire-scarred plants that have all but succumbed (fig. 17.3A). Three years later, in March 1991, those plants had collapsed (fig. 17.3B) and only a single living *Carnegiea* could be seen at the right, extending above the distant ridge top. This scene was recaptured in August 2006, revealing the single ridge top *Carnegiea* as the lone survivor (fig. 17.3C). Several small *Carnegiea* have become established since the fire. Although no *Enneapogon* was found at this site, *Pennisetum ciliare*, another nonnative African grass, was abundant and can be seen surrounding the *Celtis pallida* (desert hackberry) in the foreground. The abundant fuel is a hazard to the new crop of *Carnegiea* making a comeback at the site.

During the years when upper-elevation *Carnegiea* were being photographed, the question slowly emerged concerning the plant's role in this habitat far above the upper edge of the Sonoran Desert proper, *Carnegiea*'s true home. To address this question, there arose the need to roughly estimate the age of each *Carnegiea* in the original photograph based on each plant's height. Height and *Carnegiea* age are closely related (Hastings and Alcorn 1961, Steenbergh and Lowe 1983, Pierson and Turner 1998), and we used height versus age data from a long-term *Carnegiea* study plot at the base of the Rincon Mountains in nearby Saguaro National Park (Turner 1992) to estimate ages for the *Carnegiea* along the Santa Catalina Highway.

Ranging poles were placed close to the original position of two clusters of *Carnegiea* in order to later measure, in the darkroom, the heights of those plants (fig. 17.3C). Using the photographic enlarger, the new photograph was adjusted to exactly the same scale as the original after which it was possible to measure the heights of seven *Carnegiea* in the older photo using the superimposed image that included the ranging pole. The results showed that those *Carnegiea* varied in height from 2.9 meters to 5.2 meters, indicating that the plants varied in age from 40 to 89 years. Two multibranched, older plants in the view were not measured, but their ages probably did not exceed 175 years.

Thus, in this small part of the larger population of *Carnegiea*, most plants had become established in the 1890s or later, with a few having been present by the early 1800s. The large younger segment of the population would have become established after the area had been opened to livestock grazing in the late nineteenth century. The few older plants that were present prior to that time probably occupied open rocky terrain lacking fine fuels that carry fires. Measurements were also made of the few small, recent arrivals in the area and found that their ages ranged from 11 to 17 years, showing that all became established after the last fire in 1987. Seeds for these new arrivals would probably have come from the many nearby scorched survivors of the 1987 fire.

The Granite Burn Near Florence, Arizona

In June 1979, the Granite Burn scorched 11,330 hectares south of Florence, Arizona (McLaughlin

Figure 17.3. Santa Catalina Highway burn, Arizona, USA.

A. (27 December 1987). This image was taken downslope from the Molino Basin overlook a few months after a fire burned across a south-facing slope. The *Carnegiea gigantea* are showing the effects of being scorched but are still standing. (R. M. Turner).

B. (23 March 1991). Just over three years later, the *Carnegiea* have collapsed. A single living *Carnegiea* is visible on the ridge top on the right side of the image. (R. M. Turner).

C. (12 August 2006). Two ranging poles were placed near the positions of two sets of *Carnegiea* in order to determine their heights, and hence approximate ages. Most were established in the 1890s, after livestock grazing and the removal of fine fuels had begun in the area. Several small *Carnegiea* have become established since the fire; one is to the left of the left-hand ranging pole, another is hidden behind the *Parkinsonia florida* (blue paloverde) at the center. *Pennisetum ciliare* (buffelgrass), an invasive nonnative grass, is abundant at this site and surrounds the *Celtis pallida* (desert hackberry) in the foreground. (R. M. Turner, Stake 1314f).

and Bowers 1982; fig. 17.1). This fire is thought to have been promoted by two previous winters of exceptionally high rainfall, stimulating a heavy growth of native annual plants that provided a peak of dry fuel in late spring of 1979. A lightning strike on 29 June is believed to have set the fire. McLaughlin and Bowers (1982) followed the response of the desert plants 7 months and 19 months after the fire; after 19 months, large *Carnegiea* (greater than 2 meters in height) had a mortality rate of 11 percent and *Parkinsonia microphylla* had a mortality rate of 63 percent. In January 1984, remeasurement of the Granite Burn (Rogers 1985) showed that *Carnegiea* postfire mortality continued to rise beyond the 19-month levels.

A photographic record was initiated seven months after the Granite Burn when Turner established a camera station with eight panning views around one point. These photographs have been repeated three times since then, and postfire changes over a quarter of a century can be identified by examining these scenes. Figure 17.4 shows how Sonoran Desert vegetation responds 7 months, 15 months, 6.5 years, and 26.5 years after the fire. In figure 17.4A, a badly scorched Sonoran Desert assemblage comprising *Carnegiea*, *Ferocactus* sp., *Parkinsonia microphylla*, *Larrea tridentata*, and *Cylindropuntia versicolor* (staghorn cholla) is apparent.

The slow response of many desert plants to burning is revealed eight months later (fig. 17.4B), when the fire's toll becomes more evident. At this time, many of the *P. microphylla* have fallen and the *Carnegiea* are in advanced stages of decline. After 6.5 years (fig. 17.4C), an open plant community of widely scattered low shrubs is all that remains of the former arborescent community. The only *Carnegiea*

Figure 17.4. Granite Burn near Florence, Arizona, USA.
A. (1 February 1980). In late June 1979, the Granite Burn swept through the desert near Florence. This photograph, taken seven months after the fire, shows charred but still standing *Carnegiea gigantea*, *Ferocactus* sp. (barrel cactus), *Parkinsonia microphylla* (foothill paloverde), *Larrea tridentata* (creosotebush), and *Cylindropuntia versicolor* (staghorn cholla). (R. M. Turner).
B. (24 September 1980). Evidence of the fire is more obvious in this image taken later in the year, when scorched plants have collapsed. (R. M. Turner).
C. (11 February 1987). An open community of low shrubs has replaced the former arborescent community in the 6.5 years since the fire. The only *Carnegiea* visible are across the road, which served as a firebreak. (R. M. Turner).
D. (4 January 2006). More than 25 years later, some *P. microphylla* are present, but few cacti. In bottomlands, *Carnegiea* may disappear from the landscape as frequent human access increases the likelihood of fire. (R. M. Turner, Stake 989h).

to be seen are across the distant road, which acted as a firebreak. After 26.5 years (fig. 17.4D), *P. microphylla* that had been killed back to ground level have resprouted and provide a meager reminder of the former arborescent desert. Decades will go by before many of the other former inhabitants will be visible across the landscape. There is a real threat of local extirpation of *Carnegiea* in areas of smooth topography, such as the Granite Burn, where frequent human access increases the likelihood of fire and where

patches of unburned plants are not assured of protection from future fires (Rogers 1985, Alford et al. 2005).

The Saguaro Fire East of Phoenix, Arizona

Following a wetter than normal season during the winter of 1973–1974, fine fuel production by annual plants, mostly the nonnative *Erodium cicutarium* (filaree) and *Bromus madritensis* ssp. *rubens*, was very high. As a result, during late spring and throughout the summer of 1974, fires were numerous in the Arizona Uplands of the Sonoran Desert. The Saguaro Fire (fig. 17.1) was ignited on 19 June 1974 and was assumed to be human caused because of its proximity to a busy highway. Before the fire, the most abundant plant was *Ambrosia deltoidea* (triangle-leaf bursage), a small globular perennial shrub about 0.5 meters tall that formed the grainy matrix connecting larger shrubs and small trees. Shortly after the fire, permanent observation plots and transects were established in burned and adjacent unburned vegetation (Rogers and Steele 1980), and numerous photographs were taken. Observations were repeated during the winters of 1977 and 1982 and in May 1986. Figure 17.5A, taken on 18 February 1978, almost four years after the fire, shows very little progress toward reestablishing the original vegetation.

During the summer of 1986, a second fire burned the foreground and middle of the scene and burned patches of vegetation around *Parkinsonia microphylla* on the steeper and rockier slope below the ridgeline that crosses the center of the photograph. Figure 17.5B, taken almost 22 years after the second fire, shows that most of the area remains barren with almost no *Ambrosia* recovery. Vegetation on the rocky slope is recovering but is still less continuous

than in 1978. The three ridgeline *Carnegiea* died after the second fire. *Larrea* and *Acacia constricta* (whitethorn acacia) occur on low mounds originally occupied by *P. microphylla* along a small wash crossing the foreground of the scene. The two have become the most prominent plants in the scene.

The removal of *Ambrosia* and reduction of larger woody species changed the texture of the landscape, making the areas between small trees and large shrubs much smoother. In the foreground of figure 17.5B, the pale stems of *Sphaeralcea* sp. (globemallow) and *Acacia* spp. are protruding from dense clusters of annuals. Clusters of annuals also occupy some of the low mounds that were occupied by shrubs before the fires. The typical annual cluster, composed chiefly of the native annual *Amsinckia* (fiddleneck), is more than 0.6 meter tall. Similar clusters were frequently composed entirely of introduced mustards with a base strata of nonnative *Erodium cicutarium*. A developing carpet of annual plants in open areas between shrubs and mounds is dominated by nonnatives, including *Erodium*, *Euphorbia* sp., and leguminous species. The annual plant cover is not continuous but is probably sufficient to carry fire; in 1977, a wet El Niño year (Bowers 2005), *B. madritensis* ssp. *rubens* produced a very dense stand across the scene, but fire did not occur.

A

B

Figure 17.5. Saguaro Fire, Arizona, USA.
A. (18 February 1978). In 1974, a fire swept through this part of the Sonoran Desert, killing *Carnegiea*, *Parkinsonia*, and various *Cylindropuntia*. *Ambrosia dumosa* (white bursage), a small hemispheric shrub, was eliminated from most of the scene. (G. F. Rogers, 43a).
B. (29 February 2008). Recovery following a second fire in 1986 has been very slow. At the time of this photograph, a developing carpet of annuals gave a soft appearance to the ground. (G. F. Rogers, 43b).

The added risk of human-caused ignitions has increased as the regional population has grown.

The Dead Man Wash Fire North of Phoenix, Arizona

Ambrosia deltoidea and the similar-sized *Encelia farinosa* (brittlebush) can establish seedlings and approach their prefire density within a few years (Rogers and Steele 1980). At the Dead Man Wash Fire, about 40 kilometers north of Phoenix (fig. 17.1), permanent quadrats and point-centered quarter transects were established in burned and adjacent unburned vegetation following a fire that burned on 8 May 1974. As at the Saguaro Fire site, annual plant biomass (especially that of introduced *Erodium cicutarium* and *Bromus madritensis* ssp. *rubens*) was very high before the fire. Various features of the plants in the quadrats and on the transects were measured during May through December, 1974, and again in March 1979. Figure 17.6A shows an unburned control quadrat on a rocky south-facing slope. The site burned again on 4 July 1979. This second fire encompassed the entire site, including the originally unburned quadrats and transects. Both fires were believed to be human caused.

Figure 17.6. Dead Man Wash Fire, Arizona, USA.
A. (18 February 1978). This south-facing slope is a control area established in an unburned patch in 1974. *Ambrosia* and *Cylindropuntia acanthocarpa* are the most abundant plants, but *Ferocactus* (immediate right foreground), *Carnegiea*, and *Parkinsonia microphylla* are prominent. This plot burned in 1979. (G. F. Rogers, 42a).
B. (18 January 1982). The 1979 fire killed some of the *Parkinsonia* and *Carnegiea* and most other species. Regrowth of *Ambrosia* is apparent throughout the scene. (G. F. Rogers, 42b).
C. (29 February 2008). *Ambrosia*, *Parkinsonia*, and *Carnegiea* have recovered to near their prefire density, although other species, including *C. acanthocarpa*, have not regrown. (G. F. Rogers, 42c).

The second fire killed and consumed most of the plants in the south-slope unburned "control" plot shown in figure 17.6A. When the quadrat was photographed and remeasured on 18 January 1982, 2.5 years after the second fire, small *Ambrosia* were abundant throughout the scene (fig. 17.6B). *Ambrosia* recovery, as shown in figure 17.6C, was probably facilitated by the rocky character of the rubble-clad volcanic slope, which is similar to the one in the middle of figure 17.5 at the Saguaro Fire site. The *Carnegiea* and *P. microphylla* populations on this slope also fared well, but in extensive level areas skirting the slopes burned during the Dead Man Wash Fire, both were almost eliminated. Other species common before the fire have sharply declined in all habitats: *Cylindropuntia acanthocarpa* (buckhorn cholla), originally abundant in figure 17.6A, is no longer present, and *Ferocactus cylindraceus* (barrel cactus), also originally common, is now rare. *Cylindropuntia bigelovii* (teddy bear cholla), originally dense on some ridge tops and lower slopes, has almost disappeared at both the sites of the Saguaro and Dead Man Wash fires. The abundance and species of annuals present at the Dead Man Wash Fire site in 1994 and 2008 were similar to those at the Saguaro Fire site, and sufficient quantities of annuals to carry fire were present in 2008.

The King Valley Fire Northeast of Yuma, Arizona

In late September to early October 2005, a fire swept north from the Yuma Proving Grounds of southwestern Arizona into the Kofa National Wildlife Refuge (Kofa NWR), traveling mainly along desert wash systems (Webb et al. 2007; fig. 17.1). The ignition point was related to military weapons testing at Yuma Proving Grounds. This fire burned through xeroriparian systems in washes (fig. 17.7) as well as low-elevation desert ecosystems in King Valley, a major area of designated wilderness in the southern part of the Kofa NWR. Using satellite imagery, Webb et al. (2007) determined that 9,255 hectares of Sonoran Desert habitat had burned.

The burned plants were mostly low-cover *Larrea*

scrub with scattered *Parkinsonia microphylla*; the more heavily impacted wash environments (fig. 17.7) had significant tree cover, including *Olneya, P. florida, P. microphylla, Chilopsis linearis* (desert willow), and/or *Psorothamnus spinosus* (smoke tree). The fine-fuel loading for the fire was mostly *Plantago insularis*, a native forb, and lesser amounts of *Schismus* sp., a nonnative annual grass, both of which had grown at high densities in response to the extremely wet winter of 2004–2005. Mean annual rainfall in this area ranges from 95 to 161 millimeters.

That winter, which was influenced by El Niño conditions in the Pacific Ocean (NOAA Earth System Research Laboratory Physical Sciences Division), had the highest rainfall totals in a half century of measurements in western Arizona (R. H. Webb, unpublished data, 2008). At Kofa Mine (57-year record), rainfall from October 2004 to April 2005 was 362 millimeters, the highest winter total in the record, 31 percent higher than the second-highest winter total (1993), and 3.9 times the average winter rainfall of 93 millimeters.

T. C. Esque (unpublished data, 2008) examined mortality rates of various species after the fire in much the same way as McLaughlin and Bowers (1982). He defined scorching as surficial damage—especially to extremities—causing browning of foliage. Charring was defined as deeper tissue damage to stems, trunk, and roots as manifested by blackening due to incineration. Within two years of the fire, *Olneya* had 50 percent mortality with only 30 percent scorching, and 70 percent mortality with only 10 percent charring. *Parkinsonia* sp. had a much higher variability in response, with about 50 percent mortality with 50 percent scorching but 90 percent mortality with only 10 percent charring. Although *Parkinsonia* spp. appear to be fire intolerant in King Valley, McLaughlin and Bowers (1982) found that 25 percent of *P. microphylla* resprouted following the Granite Burn. Rainfall in 2006 and 2007 following the King Valley Fire was well below average. Under these drought conditions, insufficient time has elapsed to evaluate whether *Parkinsonia* spp. will respond similarly to the Granite Burn or whether droughts inhibit root-crown resprouting in this species.

Figure 17.7. King Valley Fire, Kofa National Wildlife Refuge, Arizona, USA.

A. (28 June 2006). This northeast view shows unburned site US1 in King Valley. The plants in the view are *Larrea tridentata* (creosotebush), *Ambrosia dumosa* (white bursage), and *Encelia farinosa* (brittlebush, all in the foreground) and *Parkinsonia microphylla* (foreground). (D. E. Boyer).

B. (15 May 2008). In the two years between matches, very little has changed in this view. (D. E. Boyer, Stake 4904b).

C. (28 June 2006). On about 1 October 2005, a fire started from military ordnance on the Yuma Proving Grounds and swept northward into King Valley. This view, taken nine months after the fire, shows that xeroriparian trees, particularly *Olneya tesota* (ironwood), *Parkinsonia microphylla*, and *Parkinsonia florida*, provided fuel to enable the fire to bypass low-biomass upland vegetation, which included *Larrea* and other xerophytic species. This northerly view shows scorched *P. microphylla* trees with live crowns (midground), an *Olneya* (behind *Parkinsonia* at left center), and two *Larrea* in the left foreground. (D. E. Boyer).

D. (1 February 2007). Seven months later, the *Parkinsonia* are fully dead, and the *Olneya*, although badly scorched, has regrown leaves. One of the two *Larrea* in the left foreground is also still alive. (D. E. Boyer).

E. (15 May 2008). Fifteen months later, the *Olneya* at the left background has fully leafed out, and other trees in the distance have also grown. The badly scorched trees in the foreground continue to decompose. (D. E. Boyer, Stake 4905a).

Repeat photography (fig. 17.7) shows that evaluation of mortality of xeroriparian trees in the Sonoran Desert cannot be evaluated immediately after a fire. *Parkinsonia* spp. that were deeply charred basally, and were growing where a thick growth of annual plants had increased fine fuel over the intershrub areas, appeared to have live crowns nine months after the fire (fig. 17.7C). However, 16 months after the fire, the same *Parkinsonia* spp. were clearly dead above ground level (fig. 17.7D). The "delayed mortality" in these trees is similar to the fire effects at the Granite Burn, where mortality also increased with time following fire.

The Hope Fire Southeast of Quartzsite, Arizona

The Hope Fire was a small (243 hectares), human-caused burn on 18 February 2006 on the Ranegras Plain in the northeastern corner of the Kofa NWR (fig. 17.1), a broad expanse of low-relief, fine-grained sediments that erode into shallow gullies (fig. 17.8). As for the King Valley Fire, the fuel for this fire was the dried remains of annuals, mostly *Plantago insularis*, a native forb, and lesser amounts of *Schismus* spp. Because it occurred during the winter, when temperatures are cool and the humidity is rel-

Figure 17.8. Hope Fire, Kofa National Wildlife Refuge, Arizona, USA.
A. (25 April 2006). In February 2006, the Hope Fire burned in the northern end of the Kofa NWR. This view shows the edge of the fire and the unburned area on the left, which has abundant fine fuels consisting of the native herb *Plantago insularis* and sparse perennial shrubs. (T. C. Esque).
B. (27 June 2006). This monitoring photograph, taken a few meters away from fig. 17.8A, documents both regrowth of vegetation, particularly *Larrea*, and changes of the gully at the center of the image. (D. E. Boyer).
C. (10 July 2008). This match of fig. 17.8B shows *Larrea* putting on new growth. The headcut on the arroyo has extended over a meter to the south (the right of the photograph) and has deepened. (D. E. Boyer, Stake 4903).

atively high, the Hope Fire is considered to be a "cold fire" that mostly consumed dried plant material and subshrubs. *Larrea* in the path of the fire survived (fig. 17.8A).

Randomly placed vegetation monitoring plots were established on upland sites to detect changes in the plant community after the Hope Fire. There were three transects each on paired burned/unburned sites. Belt transects for measuring plant density were 2 by 100 meters with a line-intercept transect established on one side to measure cover (table 17.1). Species richness (number of species per 200 square meters) was measured within the belts (table 17.1). Perennial plant species found in the area included *Larrea*, *Lycium andersonii* (wolfberry), *Ambrosia deltoidea*, *Cylindropuntia ramosissima* (pencil cholla), and *Argythamnia claryana* (desert silverbush).

Cover in the Hope Fire area is low in general (mean cover 7.3 percent) and was decreased by 50 percent immediately after the fire (table 17.1). Unlike the other fires discussed in this chapter, all of the shrub mortality occurred within 2 years of the fire. Plant density was also low with only 11.3 live perennial plants/100 square meters on unburned transects; burning reduced plant density by 60 percent. Species richness in this area is also quite low (1–2, table 17.1) and dominated by *Larrea*. Resprouting and regrowth of perennial plants, particularly *Larrea* in 2007, indicate a small amount of recovery (fig. 17.8C). Of more long-term concern are activation and movement of the gully, which appeared to be inactive in 2006 but which enlarged by about 1.5 meters by 2008 (fig. 17.8C).

Fire in *Coleogyne* Assemblages on the Nevada Test Site, Mojave Desert

The Mojave Desert occurs mostly in southeastern California but extends into southern Nevada, southwestern Utah, and northwestern Arizona (fig. 17.1). Annual rainfall in this desert ranges from 34 to 310 millimeters per year with high interannual variability (Hereford et al. 2006). The Mojave Desert is primarily a shrubland (Turner 1994) with several important dominants that include *Larrea* at lower and middle elevations and *Coleogyne ramosissima* (blackbrush) at higher elevations transitional to the Great Basin Desert (Beatley 1975). As with the Sonoran Desert, annual nonnative species, including *Bromus madritensis* ssp. *rubens* and *Schismus* spp., contribute to increased fire potential (Esque and Schwalbe 2002). The Mojave Desert is noted for its extensive displays of native annual wildflowers, which are typically at their maximum expression during high rainfall winters associated with El Niño conditions in the Pacific Ocean (Bowers 2005).

Fire frequency in the Mojave Desert has increased with time, resulting in the burning of 12,200 hectares per year from 1980 to 2004 (Brooks and Matchett 2006). Most of these fires occurred in "middle-elevation shrublands" (Brooks and Matchett 2006) that are dominated by *Coleogyne*, with human-caused fires outnumbering lightning-caused fires by 2:1. Between 1980 and 2004, 240,000 hectares of "middle-elevation shrubland" burned in the Mojave Desert, making this vegetation type the

Table 17.1. Cover on burned and nearby unburned line-intercept transects (100 meters long) at the Hope Fire in 2006 and 2007

Treatment	Year	Cover (percent)	SE (±1)	Density (plants/100 square meters)	SE (±1)	Species Richness (species/200 square meters)	SE (±1)
Unburned	2006	7.3	1.2	11.3	3.0	1.6	0.7
Burned	2006	2.4	1.9	2.7	0.9	1.0	0.0
Unburned	2007	6.5	1.2	13.0	3.0	2.0	1.0
Burned	2007	0	0	4.0	0.6	1.6	0.3

SE, standard error.

most susceptible to fire in terms of area burned in this region. *Coleogyne* typically occurs at high densities with considerable herbaceous cover, including contributions by the nonnative *Bromus madritensis* ssp. *rubens*. *Coleogyne* assemblages are considered to be one of the most flammable types in the Mojave Desert because the high density of plants, a high fuel bulk density, and the prevalence of annuals make for high fine-fuel loading (Hansen and Ostler 2004). Fire typically kills *Coleogyne*, and the recovering ecosystem tends to have greater biodiversity with more perennial species as well as nonnative annuals than before (Brooks and Matchett 2003).

In the 1950s, a series of fires occurred in *Coleogyne* assemblages in Mid Valley on the Nevada Test Site (fig. 17.1). Janice C. Beatley established permanent vegetation plots on the Nevada Test Site in 1963, including three paired unburned–burned plots in *Coleogyne* assemblages and two sets in Mid Valley (plots 18–19, 39–40, and 41–42; Webb et al. 2003). In the southern part of Mid Valley, a lightning-caused fire in the summer of 1959 burned a large area of *Coleogyne* assemblages, and Beatley established plots 39 (unburned) and 40 (burned) in 1963 and photographed plot 40 in 1964 (fig. 17.9A). The prefire vegetation assemblage was dominated by

Figure 17.9. Coleogyne assemblage, Nevada Test Site, USA.

A. (1964). This monitoring plot for a 1959 lightning-caused fire in Mid Valley was established in 1963 and photographed in 1964. The fire swept through an assemblage dominated by *Coleogyne ramosissima* (blackbrush), which is visible in the midground. The dark plants in the midground are *Yucca brevifolia* (Joshua tree). (J. C. Beatley, 40-4).

B. (10 May 2000). After 41 years, perennial cover had reestablished but with little *Coleogyne*. The plants in the previously barren area include *Achnotherum speciosum* (needlegrass), *Thamnosma montana* (turpentine bush), and *Ephedra nevadensis* (Mormon tea). The *Yucca* in the midground have grown substantially with many new individuals. (R. H. Webb).

C. (7 September 2005). In August 2005, a lightning-ignited fire swept through this area and into the previously unburned area. (R. H. Webb, Stake 4026b).

Coleogyne, with about 33 percent total cover and 17 percent cover of *Coleogyne* in 1963 (Webb et al. 2003).

From 1963 through 2005, cover of perennial vegetation in the burned area was measured with line intercepts, and total cover increased from about 10 percent to 25 percent, with most of the increase coming from short-lived shrub species and native perennial grasses (Webb et al. 2003; Webb, unpublished data, 2005). A repeat photograph in 2000 shows this cover increase (fig. 17.9B). Webb et al. (2009) report that, in the absence of continued disturbance, burned areas will recover total perennial cover on average in about a century in the Mojave Desert. Comparison of figures 17.10A and 17.10B also indicates that *Yucca brevifolia* (Joshua tree) have greatly increased in density and stature in the midground, an outcome that could only be determined from repeat photography in this valley.

In August 2005, following above-average winter precipitation, a lightning-caused fire again swept through the southern part of Mid Valley, burning both plots. Approximately 400,000 hectares of the Mojave Desert burned that summer, primarily in Nevada, Utah, and Arizona (Brooks and Matchett 2006). From October 2004 through April 2005, precipitation at the Mid Valley climate station (1964–2005) was 406 millimeters, or 2.3 times the normal winter precipitation and the highest total in the 41-year record. The fire was sustained by a high density of nonnative annual grass *B. madritensis* ssp. *rubens*, the native annual forb *Amsinckia tesselata*, and the biennial native forb *Sphaeralcea ambigua* (L. A. DeFalco, unpublished data, 2007). Postfire cover in the recovering plot 40 was reset to less than 1 percent, and the *Yucca* were mostly killed (fig. 17.9C). About two-thirds of plot 39 was burned, including some prominent *Yucca* (fig. 17.10B).

Both fires affecting the area near plots 39 and 40—in 1959 and 2005—occurred after above-average precipitation related to El Niño conditions caused greatly increased growth of annual vegetation, including both native and nonnative species. Both fires were lightning sparked, indicating that human land-use practices had minimal influence. However, both fires may well have been affected by

A

B

Figure 17.10. Coleogyne assemblage, Nevada Test Site, USA.
A. (10 May 2000). This permanent plot, established in 1963, served as a control for a nearby fire-recovery plot (see fig. 17.9). The shrubs are mostly *Coleogyne ramosissima*, which dominates large areas of the Mojave Desert; one individual of *Ephedra nevadensis* appears in the left foreground, and *Yucca brevifolia* appears throughout the midground. (R. H. Webb).
B. (7 September 2005). In August 2005, a lightning-caused fire swept through this valley, burning approximately two-thirds of the 30- by 30-meter plot. The *Yuccas* were scorched or fully burned and appear dead, although some basal resprouting occurred in nearby, older burned areas. (R. H. Webb, Stake 4029).

the additional fine fuel provided by nonnative annual grass species, suggesting that the relation between above-average precipitation and nonnative vegetation may strongly affect the potential for future fires in this valley.

Fire in the Great Basin Desert

Occurring entirely within the western United States, the Great Basin Desert occupies portions of nine states, including most of Nevada and Utah (fig. 17.1). It is typified by extensive arid shrublands and woodlands with intermingled grasslands spread across broad valleys that gradually slope upward from valley floors through lower valleys, upper valleys, and foothills to the summits of widely spaced mountain ranges. Because of salt accumulation in undrained depressions in valley floors, a 1-meter rise in elevation can produce striking vegetation changes. Low rainfall and relatively small areas of irrigated farming have limited land use for most of the region to cattle and sheep grazing.

Upper-valley and foothill vegetation dominated by *Artemisia tridentata* (sagebrush), perennial grass

(e.g., *Agropyron spicatum* [bluebunch wheatgrass]), *Juniperus osteosperma* (Utah juniper), and *Pinus* spp. (pinyon) is more productive than lower-valley sites and can burn without additional fuel provided by ephemeral herbs. The higher forage value of perennial grass has long inspired management efforts to eradicate shrubs and trees by using fire or mechanical and chemical techniques. Burning probably began not long after settlement in the mid-1800s (Pickford 1932), and continues today.

Valley floors often contain bare salt flats surrounded by sparse vegetation consisting of salt-tolerant perennial species such as *Allenrolfea occidentalis* (iodine bush) and *Distichlis spicata* (saltgrass). Plant cover is discontinuous, and even in wet years when annual plants increase, fire is uncommon. *Sarcobatus vermiculatus* (greasewood) dominated vegetation, transitional between valley floors and the less saline lower-valley areas, is susceptible to fire. Across the broad alluvial plain surrounding Delta, Utah, extensive areas of *Sarcobatus* have been replaced by the annual *Descurainia* sp. (mustard) following fires.

Repeat photography (Rogers 1982) indicates that fire frequency might be increasing in these lower valleys. Figure 17.11, from the west slopes of the Cedar

A

B

Figure 17.11. Eastern edge of the Great Salt Lake Desert of western Utah, USA.
A. (1901). This view of the foot of the Cedar Mountains shows abundant *Atriplex confertifolia* (shadscale) in the foreground. (G. K. Gilbert, 1841, USGS Photographic Library).
B. (1 September 1976). The light foreground area has *Bromus tectorum* with a few darker *Halogeton glomeratus*, both nonnatives. It is possible that *A. confertifolia* in the foreground was removed by livestock, but its replacement by annuals increases the risk of fire. In 1986 or 1987, a large fire swept across the foreground, but its effects have not been assessed. The trees on the mountain slopes have increased, which is common in the Great Basin (Rogers 1982). (G. F. Rogers, 201).

Mountains on the eastern edge of the Great Salt Lake Desert (fig. 17.1), illustrates the replacement of *Atriplex confertifolia* (shadscale) by nonnative annuals dominated by *Bromus tectorum* (cheatgrass), with some *Halogeton glomeratus* (halogeton), *Ranunculus testiculatus* (bur buttercup), and *Descurainia* sp. The foreground of figure 17.12, a site at the south end of Tooele Valley, Utah (fig. 17.1), shows the replacement of *Kochia americana* (greenmolly), a common upper-valley dominant, by *B. tectorum*. During the decade following the second photographs in figures 17.11 and 17.12, fires burned thousands of hectares across the lower-valley areas shown in the photographs. The photographs have not been repeated, but cursory inspection of the sites indicates that annual plants might have replaced the original shrubland vegetation.

The twentieth-century spread of nonnative annuals added a new and unintended result from burning. Young et al. (1971), Young and Evans (1973), and others reported that burned native vegetation was sometimes replaced by nonnative annuals, representing a type conversion from shrubland to annual grassland. On the ancient beach face across the upper section of figure 17.12, *Artemisia* has been completely replaced by *B. tectorum*. The scenes in figure 17.13, at a site near Honeyville in north-central Utah (fig. 17.1), show that *Artemisia* recovered following a fire that occurred before 1901, only to be burned again. The small herbs carpeting the ground in 1901 could not be identified. In 1978, *B. tectorum* formed an almost complete cover at this site. This example suggests the possibility that, like the Mojave Desert, fire in the Great Basin Desert may promote type conversions with fire-enhancement feedback mechanisms that suppress woody shrubs.

Discussion and Conclusions

Repeat photography provides a unique perspective for viewing the effects of fire in the Great Basin, Sonoran, and Mojave deserts of the southwestern United States. Our 11 examples show that fire—and

Figure 17.12. South Mountain, Tooele Valley, Utah, USA.
A. (3 June 1912). This view of South Mountain shows extensive stands of bottomland vegetation in the foreground, interrupted by two prominent anthills. The shrub *Kochia americana* (greenmolly) extends from the foreground to the background slopes covered with *Artemisia*. (H. L. Shantz, d-1-12, University of Arizona Herbarium).
B. (12 June 1978). Fires have swept through this valley repeatedly in the twentieth century. Annual vegetation dominated by nonnative *Bromus tectorum* (cheatgrass) has replaced the native shrubland. This set and others in bottomland vegetation (e.g., Rogers 1982, plate 4) confirm that repeated fires can convert bottomland shrublands into annual grasslands. (G. F. Rogers, 113).

Figure 17.13. Honeyville, Utah, USA.

A. (1901). This view shows the effects of a recent fire on the eastern edge of the Bear River Valley and the Wellsville Mountains. A few *Artemisia tridentata* remain, but most of the scene appears to be covered by annual plants. (G. K. Gilbert, 3483, USGS Photographic Library).

B. (20 July 1968). *Artemisia* now covers most of the scene. This indicates that this type of vegetation may be able to recover from fire in less than a century without additional disturbance. (R. M. Turner, Stake 500).

C. (3 September 1978). Another fire has removed the cover of *Artemisia*. The scene is dominated by *B. tectorum* and small amounts of other herbaceous species. The livestock trailing on the hillslopes in the background is more prominent as a result of the conversion of perennial to annual vegetation. (G. F. Rogers 282).

fire suppression in the case of Molino Basin—have long-term effects that are related to the interrelation among native and nonnative species and climate. In three of the four examples where precipitation amounts are known, fire, in what appeared to be never- or infrequently burned ecosystems, was carried on fine fuels enhanced by above-average winter precipitation. In only one of the examples (Mid Valley on the Nevada Test Site), nonnative species provided the most significant fine fuels, although our examples also show that nonnatives clearly play a significant role in repeated burning of the Sonoran and Great Basin deserts. Although we do not discount the role of nonnatives—particularly annual grasses—in increasing the fire potential in previously unburned landscapes of the Mojave and Sonoran deserts (Brooks and Esque 2002), total annual plant production appears to be the key to at least some fires.

The ignition source of fires is also extremely important, and at least three of the nine examples we discuss were caused by lightning. Most ecologists believe that fire burning large areas of the low-elevation parts of the Sonoran Desert of western Arizona was unprecedented in the nineteenth and early twentieth centuries, and its potential is now increased because of human land-use practices (Brooks and Esque 2002). The King Valley Fire example shows the interrelation between high seasonal precipitation, the accumulation of large amounts of fine fuel on the landscape, and ignition by military weapons testing. Without the change in ignition source, fire likely would never have occurred in this part of the Sonoran Desert under the present climate region, and the plants are not adapted for recovery from burning. The amount of time required for significant recovery of this ecosystem cannot be accurately estimated.

At higher elevations or on rocky slopes in the Sonoran Desert, limited recovery has occurred following fires, as two of our examples show. In all cases, low-slope alluvial surfaces show little recovery, particularly of *Carnegiea*. In five of six examples from this desert, mortality continued long after the fire had ceased, and we believe this illustrates the lack of fire adaptation for most of the species in this desert. Thus fire is seen to play an increasingly important role in shaping the distribution pattern of plants lacking fire resistance, such as *Carnegiea*, both within the Sonoran Desert proper and at elevations above that desert, where the frequency of severe freezes becomes important. Conservation of desert vegetation depends on learning what drives this potential alteration so that proper management response can be made. Repeat photography can amplify the findings of traditional permanent plot analysis in this search.

Our experience with paired burned–unburned Beatley plots on the Nevada Test Site suggests that repeated fires may now occur where dense *Coleogyne* assemblages were once prevalent. Of the three sets of paired plots established to monitor recovery from fires in the 1950s, two sets (plots 39–40 and 41–42) were partially or totally burned in 2005, and the third set (plots 18–19) narrowly missed being burned. Because fire reduces perennial vegetation cover, the cover of annuals and biennials increases, providing a greatly increased fine-fuel loading in these ecosystems. The high establishment and growth rates of *Yucca brevifolia*, as well as its ultimate demise in the 2005 fire, would not have been captured in the permanent vegetation plots in Mid Valley; repeat photography was required to document this long-term landscape change and its relation to fires.

In the Great Basin Desert, vast areas formerly occupied by a dynamic mosaic of shrubs, grasses, and trees are now occupied by *Bromus tectorum* and other nonnative annuals. The cycle of type conversion caused by the spread of nonnative annuals and the increase in fire frequency is well documented in the upper-valley shrublands and woodlands and might be occurring in the lower-valley *Atriplex* shrublands as well. Despite the harsher environment and smaller amounts of vegetation, the photographs show that large areas of *Atriplex* have been replaced by nonnative annuals. The inescapable conclusion is that introduced annuals adapted to frequent disturbance will continue to replace native vegetation, probably at an accelerating rate. For upper-valley *Artemisia* and *Juniperus* vegetation in the Great Basin, the next challenge is to find ways to replace *B. tectorum* and other annuals with native perennials (Cox and Anderson 2004). For lower-valley areas more research on the necessary conditions and extent of shrub replacement by annuals is needed.

Fire, like any other powerful event, affects individual species, emergent community values, and environmental conditions. Evidence from all the North American deserts indicates that fire causes long-term changes in species composition, diversity, and productivity. At the Saguaro Fire, Rogers and Steele (1975) found that when the entire fire area was considered, the reduction of total plant density and cover was not associated with reduced species evenness; the total number of species was only reduced by about 10 percent. The most striking change over the twice-burned Saguaro and Dead Man Wash fires is that the spatial scale of the community has changed. If we sampled either site today without knowing the fire history, our results would not reveal the degree of species impoverishment that had occurred. We would not realize how much total biomass had declined, and, if we neglected to study microtopography and soils, we would not realize that fertile mounds built over centuries by the litter and dust accumulated under large plants were now exposed to trampling, insolation, and wind and water erosion.

In all of our examples, the potential for recovery is clearly low, given that even without future disturbances, hundreds to thousands of years may be required to reestablish perennial vegetation to its unburned state. Exceptions apparently may occur on rocky slopes, where more microsites could preserve a higher diversity of propagules to enable recovery. The establishment of nonnative vegetation, and the increased fire hazard that this vegetation creates, strongly suggests that, once burned, these desert ecosystems likely will not recover, but instead will shift to a different vegetation type controlled by more frequent fires.

Acknowledgments

Numerous individuals have contributed to the work described here. J. Rodney Hastings, who died in 1974, initiated the work on *Carnegiea* using repeat photography. Jeanne Turner assisted with photography at the Sonoran Desert sites as did Mary Schmid at the Mount Lemmon Highway site. Diane Boyer, Peter Griffiths, and Lindsay Smythe contributed to work related to the King Valley Fire. Lesley DeFalco, Dustin Haines, Philip Medica, Mimi Murov, Helen Raichle, and Sara Scoles all contributed to measurements made in Mid Valley on the Nevada Test Site. Walter P. Cottam and Denise Rogers assisted with the Great Basin fieldwork. All Rogers photographs are courtesy of the photographer; all other photographs are courtesy of the USGS Desert Laboratory Repeat Photography Collection, unless otherwise indicated.

Literature Cited

Alford, E., J. H. Brock, and G. J. Gottfried. 2005. Effects of fire on Sonoran Desert plant communities. In *Connecting mountain islands and desert seas: Biodiversity and management of the Madrean Archipelago II*, comp. G. J. Gottfried, B. S. Gebow, L. G. Eskew, and C. B. Edminster, 451–454. USDA Forest Service Proceedings RMRS-P-36. Rocky Mountain Research Station, Fort Collins, CO.

Bahre, C. J. 1985. Wildfire in southeastern Arizona between 1859 and 1890. *Desert Plants* 7:190–194.

Beatley, J. C. 1975. Climates and vegetation pattern across the Mojave/Great Basin Desert transition of southern Nevada. *American Midland Naturalist* 93:53–70.

Bowers, J. E. 1980–1981. Catastrophic freezes in the Sonoran Desert. *Desert Plants* 2:232–236.

Bowers, J. E. 2005. El Niño and displays of spring-flowering annuals in the Mojave and Sonoran deserts. *Journal of the Torrey Botanical Society* 132:38–49.

Brooks, M. L., C. M. D'Antonio, D. M. Richardson, J. B. Grace, J. E. Keeley, J. M. DiTomaso, R. J. Hobbs, M. Pellant, and D. Pyke. 2004. Effects of invasive alien plants on fire regimes. *BioScience* 54:677–688.

Brooks, M. L., and T. C. Esque. 2002. Alien annual plants and wildfire in desert tortoise habitat: Status, ecological effects, and management. *Chelonian Conservation and Biology* 4:330–340.

Brooks, M. L., and J. R. Matchett. 2003. Plant community patterns in unburned and burned blackbrush (*Coleogyne ramosissima* Torr.) shrublands in the Mojave Desert. *Western North American Naturalist* 63:283–298.

Brooks, M. L., and J. R. Matchett. 2006. Spatial and temporal patterns of wildfires in the Mojave Desert, 1980–2004. *Journal of Arid Environments* 67:148–164.

Búrquez-Montijo, A., M. E. Miller, and A. Martinez-Yrizar. 2002. Mexican grasslands, thornscrub, and the transformation of the Sonoran Desert by invasive exotic buffelgrass (*Pennisetum ciliare*). In *Invasive exotic species in the Sonoran region*, ed. B. Tellman, 126–146. Tucson: University of Arizona Press.

Cox, R. D., and V. J. Anderson. 2004. Increasing native diversity of cheatgrass-dominated rangeland through assisted succession. *Journal of Range Management* 57:203–210.

Davis, O. K., T. Minckley, T. Moutoux, T. Jull, and B. Kalin. 2002. The transformation of Sonoran Desert wetlands following the historic decrease of burning. *Journal of Arid Environments* 50:393–412.

Esque, T. C., and C. R. Schwalbe. 2002. Alien annual grasses and their relationships to fire and biotic change in Sonoran Desertscrub. In *Invasive exotic species in the Sonoran region*, ed. B. Tellman, 165–194. Tucson: University of Arizona Press.

Esque, T. C., C. R. Schwalbe, D. F. Haines, and W. L. Halvorson. 2004. Saguaros under siege: Invasive species and fire. *Desert Plants* 20:49–55.

Esque, T. C., C. R. Schwalbe, J. A. Lissow, D. F. Haines, D. Foster, and M. C. Garnett. 2006. Buffelgrass fuel loads in Saguaro National Park, Arizona, increase fire danger and threaten native species. *Park Science* 24:33–37.

Hansen, D. J., and W. K. Ostler. 2004. *A survey of vegetation and wildland fire hazards on the Nevada Test Site*. US Department of Energy Report DOE/NV/11718-981, Las Vegas, NV.

Hastings, J. R., and S. M. Alcorn. 1961. Physical determinations of growth and age in the giant cactus. *Journal of the Arizona Academy of Science* 2:32–39.

Hereford, R., R. H. Webb, and C. Longpré. 2006. Precipitation history and ecosystem response to multidecadal precipitation variability in the Mojave Desert and vicinity, 1893–2001. *Journal of Arid Environments* 67:13–34.

McLaughlin, S. P., and J. E. Bowers. 1982. Effects of wildfire on a Sonoran Desert plant community. *Ecology* 63:246–248.

McPherson, G. R. 1995. The role of fire in the desert grasslands. In *The desert grassland*, ed. M. P. McClaran and T. R. van Devender, 130–151. Tucson: University of Arizona Press.

Narog, M. G., and R. C. Wilson. 2002. *Delayed mortality: Saguaro cacti are still dying 10 years after wildfire!* Fifth Symposium on Fire and Forest Meteorology, November 16–20, 2003. Orlando, FL. Boston: American Meteorological Society.

NOAA Earth System Research Laboratory Physical Sciences Division. Multivariate ENSO Index. www.cdc.noaa.gov/people/klaus.wolter/MEI/.

Pickford, G. D. 1932. The influence of continued heavy grazing and promiscuous burning on spring–fall ranges in Utah. *Ecology* 13:159–171.

Pierson, E. A., and R. M. Turner. 1998. An 85-year study of saguaro (*Carnegiea gigantea*) demography. *Ecology* 79:2676–2693.

Rogers, G. F. 1982. *Then and now: A photographic history of vegetation change in the central Great Basin Desert*. Salt Lake City: University of Utah Press.

Rogers, G. F. 1985. Mortality of burned *Cereus giganteus*. *Ecology* 66:630–632.

Rogers, G. F., 1986. Comparison of fire occurrence in desert and non-desert vegetation in Tonto National Forest, Arizona. *Madroño* 33:278–283.

Rogers, G. F. and J. Steele. 1975. Investigations of the effect of fire on species diversity in the Sonoran Desert. *Journal of the Arizona Academy of Science* 10: Proceedings Supplement.

Rogers, G. F., and J. Steele. 1980. *Sonoran Desert fire ecology*. USDA Forest Service, Rocky Mountain Forest and Range Experiment Station, General Technical Report RM-81. Fort Collins, CO.

Schmid, M. K., and G. F. Rogers. 1988. Trends in fire occurrence in the Arizona Upland Subdivision of the Sonoran Desert, 1955 to 1983. *Southwestern Naturalist* 33:437–444.

Steenbergh, W. F., and C. H. Lowe. 1977. *Ecology of the saguaro, part 2: Reproduction, germination, establishment, growth, and survival of the young plant*. US National Park Service Scientific Monograph Series 8. Tucson, AZ.

Steenbergh, W. F., and C. H. Lowe. 1983. *Ecology of the saguaro, part 3: Growth and demography*. US National Park Service Scientific Monograph Series 17. Tucson, AZ.

Turner, R. M. 1992. Long-term saguaro population studies at Saguaro National Monument. In *Proceedings of the symposium on research in Saguaro National Monument, 23–24 January 1991*, ed. C. P. Stone and E. S. Bellantoni, 3–11. Tucson, AZ: National Park Service,

Rincon Institute, Southwest Parks and Monuments Association.

Turner, R. M. 1994. Mojave Desertscrub. In *Biotic communities, southwestern United States and northwestern Mexico*, ed. D. E. Brown, 157–168. Salt Lake City: University of Utah Press.

Turner, R. M., and D. E. Brown. 1994. Sonoran Desertscrub. In *Biotic communities, southwestern United States and northwestern Mexico*, ed. D. E. Brown, 181–221. Salt Lake City: University of Utah Press.

Turner, R. M., R. H. Webb, J. E. Bowers, and J. R. Hastings. 2003. *The changing mile revisited: An ecological study of vegetation change with time in the lower mile of an arid and semiarid region*. Tucson: University of Arizona Press.

Webb, R. H., J. Belnap, and K. A. Thomas. 2009. Natural recovery from severe disturbance in the Mojave Desert. In *The Mojave Desert: Ecosystem processes and sustainability*, ed. R. H. Webb, L. F. Fenstermaker, J. S. Heaton, D. L. Hughson, E. V. McDonald, and D. M. Miller, 343–377. Reno: University of Nevada Press.

Webb, R. H., P. G. Griffiths, C. S. A. Wallace, and D. E. Boyer. 2007. *Channel response to low-elevation desert fire: The King Valley Fire of 2005*. US Geological Survey Data Report DS 275.

Webb, R. H., M. B. Murov, T. C. Esque, D. E. Boyer, L. A. DeFalco, D. F. Haines, D. Oldershaw, S. J. Scoles, K. A.

Thomas, J. B. Blainey, and P. A. Medica. 2003. *Perennial vegetation data from permanent plots on the Nevada Test Site, Nye County, Nevada*. US Geological Survey Open-File Report 03-336.

Wilson, R. C., M. G. Narog, A. L. Koonce, and B. M. Corcoran. 1995. Postfire regeneration in Arizona's giant saguaro shrub community. In *Biodiversity and management of the Madrean Archipelago: Southwestern United States and northwestern Mexico*, tech. coord. L. H. DeBano, P. H. Ffolliott, A. Ortega-Rubio, G. J. Gottfried, R. H. Hamre, and R. H. Edminster, 424–431. USDA Forest Service General Technical Report RM-GTR-264. Fort Collins, CO.

Young, J. A., and R. A. Evans. 1973. Downy brome—intruder in the plant succession of big sagebrush communities in the Great Basin. *Journal of Range Management* 26:410–415.

Young, J. A., R. A. Evans, and J. Major. 1971. Alien plants in the Great Basin. *Journal of Range Management* 24:194–201.

Zouhar, K., J. K. Smith, S. Sutherland, and M. L. Brooks. 2008. *Wildland fire in ecosystems: Fire and nonnative invasive plants*. US Forest Service General Technical Report RMRS-GTR-42, v. 6. Ogden, UT.

PART V

Cultural Applications

Use of repeat photography for addressing questions concerning human geography and anthropology is relatively new. Although there has been a sustained effort to use this technique in an artistic form (Klett, chapter 4), until recently few have used this technique to examine changing land-use and settlement patterns from the human perspective. One example of the use of repeat photography to document changing patterns of land use coupled with the human experience is provided by Bass (chapter 20), who used parts of Honduras for his examples. Other examples come from Scotland, where Rohde (chapter 18) shows how changes in land-use practices (the decline of crofting) affect the landscapes in rural settings or near small towns, and Moore (chapter 19), who discusses how the Scottish Highlands are changing in response to urban growth, land-use practices, and climate change.

The use of photographic evidence to precisely locate the paths of expeditions is also a new application of repeat photography. Hanks et al. (chapter 3) pioneered the use of high-resolution geographic information system (GIS) techniques to locate camera stations prior to fieldwork, but their motives for this work were to discover the route of anthropological expeditions in the late 1920s in northern Arizona and southern Utah. The potential for this applica-

tion increases for expeditions conducted prior to the development of photography, and this application is restricted to those expeditions who employed artists to capture the scenery. Jonas (chapter 21) shows the problems associated with relocating the scene depicted in nineteenth-century artwork, including distortions and fanciful representations of vegetation, but he also documents some artwork that is dimensionally correct with a probable faithful depiction of the natural environment.

Relocation of the routes of historical expeditions may be somewhat esoteric, but preservation of archaeological sites is mandated by law in many countries. The ravages of time have removed many sites, particularly along watercourses, but as Howard et al. (chapter 22) document, many sites remain standing, albeit without roofs and with erosion of walls or foundations. Photographs taken at or near the time of discovery can be used to aid restoration or stabilization efforts. These photographs can also be used to document change in the surrounding landscape, as Howard et al. (chapter 22) illustrate with wildland fire impacts. All of the cultural applications presented in this part show the diversity of applications of repeat photography in the social sciences, while reinforcing the long-established use of this technique in documenting landscape change.

Chapter 18

Written on the Surface of the Soil: Northwest Highland Crofting Landscapes of Scotland during the Twentieth Century

Richard F. Rohde

The history of the economical transformation which a great portion of the Highlands and Islands has during the last century undergone does not repose on the loose and legendary tales that pass from mouth to mouth; it rests on the solid basis of contemporary records, and if these were wanting, *it is written in indelible characters on the surface of the soil.* (Parliamentary Papers 1884: 2; emphasis added)

"Crofts" are uniquely Scottish land units in the Highlands. The "crofting way of life" and its associated landscapes are iconic points of reference to the historical and contemporary identity of Scotland today (Smout 1994). Indeed, the past injustices of the Highland Clearances and the Crofters Holdings (Scotland) Act of 1886, which gave security of tenure to the peasant population of the area, are deeply embedded in the Scottish collective memory (Hunter 1976). There are many detailed accounts of this period and the subsequent history of land and agrarian reform that evolved during the last 120 years. Rather than revisit this well-worn historical narrative, this chapter will explore the socioeconomic and agrarian transformation that has come about over the last 75 years or so by tracing the evidence for these changes in the landscape itself, as seen in repeat photographs.

Today, some 17,000 crofts occupy an area of 8,000 square kilometers, mainly within the rugged landscapes of the Northwest Highlands and Islands. Typically, crofting townships consist of a scattering of simple whitewashed houses set among small arable fields, within a larger pastoral landscape of common grazing for sheep and cattle. In most instances, these townships are located within, or are surrounded by, large, privately owned sporting estates. The wet and windy climate and geography dominated by acidic rock, shallow soils, and treeless heath make this an unpromising and uncompromising place to extract a living from the land. Many of the mainland glens that were cleared to make way for southern sheep farmers and sporting estates in the nineteenth century remain empty to this day. However, the crofting legislation enacted at the end of the nineteenth century effectively "fossilized" the pattern of land occupancy and resulted in a diverse rural economy with a high population density (Brown and Slee 2004).

The repeat photographs presented here are concentrated within the Skye and Lochalsh District of Scotland, which might be characterized as typical of the core crofting areas (fig. 18.1). As such, these images are typical of northwest coast township landscapes. Apart from one photo by Erskine Beveridge taken in 1898, all the original photographs used in this study were made by Robert Moyes Adam (1885–1967; see also Moore, chapter 19), during the 1930s.

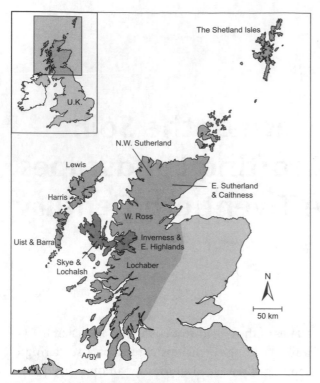

Figure 18.1. Map of crofting areas of Northwest Scotland showing Skye and Lochalsh District (after www.crofting.org; accessed 14 October 2006).

His photographic record of the rural landscapes of Scotland during the first half of the twentieth century is perhaps the most comprehensive of its kind, consisting of around 15,000 negatives now held in the library of St. Andrews University. His photographs of crofting townships depict a cultural landscape on the cusp of transition from the deep rural peasant subsistence agriculture of the late nineteenth century to a version of modern agriculture based on subsidies and economies of scale. He made at least 10 trips to the Skye and Lochalsh area between 1926 and 1943, most often in the late summer or early fall when crofting activities were clearly visible in the form of haystacks, corn stooks, and patches of unharvested potatoes. Typically during each visit to the area, he would expose more than 50 glass half-plates depicting the area's landscapes, although only a small fraction of these were of human

settlements. His images of crofting landscapes are of a quality and consistency that make it a pleasure to follow his footsteps into the past.

This chapter focuses on an analysis of change evident in five case study sites drawn from 36 photographs depicting 19 crofting townships (fig. 18.2). They show not only the environmental effects of changing agricultural practice but also hint at the underlying socioeconomic and political forces that accompanied agrarian change. The processes of de-agrarianization manifest differently across the developed and developing world (Bryceson 1996), and the crofting landscapes of the Northwest Highlands and Islands are one particular manifestation of this process. Population decline, which reached its nadir in the 1970s (Bryden and Houston 1976), has been reversed partly due to the influx of southerners over the last 30 years (Short and Stockdale 1999). Tourism is now the largest sector of the local economy, but in relation to national standards, the area remains marginal with high levels of poverty and unemployment (Chapman and Shucksmith 1996, Pacione 1996). The nine matched photos presented here illustrate three distinct processes that were

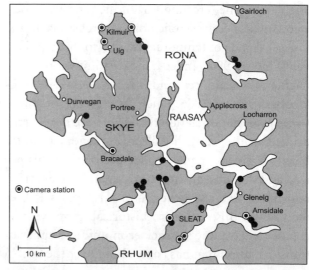

Figure 18.2. Map of the Northwest Highlands and Islands depicting the locations of all repeat photographs used in this study.

"written in indelible characters" across the landscape of crofting townships throughout the twentieth century: human population change, agrarian change, and parallel socioeconomic transformation.

Background

The Northwest Highlands have been continuously occupied since around 8,500 years BP when the first small bands of Mesolithic hunter-gatherers arrived. Arable farming has been practiced for at least 5,000 years, and the Neolithic, Bronze Age, and Iron Age cultures that preceded the Roman occupation of southern Scotland can be traced in numerous standing stones, brochs, hut circles, and souterrains scattered across the Northwest Highland landscape (Simmons 2001). During the last two millennia, a succession of cultures, including Pictish and Viking, have left their traces in place names and archaeological sites. Later medieval castles and Christian cathedrals such as Iona and Skeabost are testament to the organization of the regional polity connected by the waterways of the west coast.

Climate change too has left its mark on the landscape. The relatively benign conditions of the Holocene climatic optimum (8,000–4,500 years BP), when temperatures were up to 2.5 degrees Celsius higher than today, gave way to the cooler, wetter, peat-forming conditions around 4,000 years BP, coinciding with a regional decline in tree cover eventually resulting in a treeless landscape in the Hebrides by around 2,500 years BP (Hirons and Edwards 1990, Fossitt 1996, Seppä et al. 2003, Tipping et al. 2006). At the onset of the Iron Age (ca. 3,000 years BP), anthropogenic and natural processes had already resulted in a reduction in woodland cover in the Northwest Highlands (Armit and Ralston 2003). Untangling the combined effects of climate change and anthropogenic disturbance is complex, but their combined impact on vegetation and soils over the last two millennia has been deleterious (Tipping 2003). Climatic oscillations during this period (e.g.,

the Romano-British warm period, deterioration during the Dark Ages, the Medieval Warm Period, and the Little Ice Age) have undoubtedly resulted in phases of agrarian abandonment and recolonization of Northwest Highland landscapes. In common with all other parts of the British Isles, the Northwest Highlands have been significantly transformed by human intervention over the last several thousand years (Blundell and Barber 2005). However, the landscapes of the Northwest Highland crofting areas have attained their mythical status of "wilderness" only recently as a result of nineteenth-century Clearances, sporting estates, tourism, and conservation propaganda (MacDonald 1998, Mackenzie 2006).

The period preceding the Highland Clearances of the early nineteenth century probably left the most lasting impact on the environment. Population reached a peak in the early nineteenth century as landlords moved people out of inland glens into miserable overcrowded townships that became the defining characteristic of the crofting system until it was reformed in the 1880s. Townships were laid out, often superimposed on more ancient field systems, across old enclosures, and disregarding head dikes (Dodgshon 1994). As populations expanded, every scrap of land capable of cultivation was put to use, and mixed herds of cattle, goats, and sheep were shepherded between summer and winter grazing land. Migrant labor, seasonal herring fishing, shellfish gathering, and kelping became essential livelihood activities in the subsistence economy (Hunter 1976).

After decades of struggle, reform finally came with the passage of the Crofters Holdings Act of 1886, which gave security of tenure, rights to compensation for permanent improvements, the right to a fair rent arbitrated by an independent Crofters Commission, and the right to bequeath tenancy to a family member. Typically, new crofting townships were planned on a grid or strip pattern, depending on topography, giving each household a roughly equal portion of arable land (between 0.5 and 2.0 hectares) and rights to a limited number of grazing animals on common land as well as rights to defined areas to dig peat for fuel.

Crofting Township Case Studies

At the time of Adam's visits to the Northwest Highlands, the crofting society he photographed was on the cusp of change. It was a time when the "crofting way of life" came close to its apotheosis as an integrated and unique cultural expression. A number of folklorists, ethnographers, musicologists, photographers, and writers from the outside world were attracted to the Northwest Highlands and the Hebrides at this time, and several stayed and helped lay the foundations of the Gaelic revival. It also coincided with the economic hardships of the Depression, which forced many urban migrants back onto the resources of the family crofts (Cameron 1998).

Although the region's demographic nadir would not be reached for another 40 years, poverty and poor living conditions were widespread, and there were few economic opportunities apart from crofting. The "Highland problem," as it became known to developers in the following decades, was a result of the complex history of socioeconomic change: the demise of the clan system of communal landownership, the Clearances, the introduction of sheep, the Napier Commission, and the Crofting Acts that followed. Frank Fraser Darling, one of the most influential writers on Highland affairs during the 1940s and 1950s, defined the problem as a lack of resources and capital, as well as widespread depopulation. There was not enough arable land in the crofting areas to enable economic-sized holdings for each croft, a lack of capital meant that crofters could not improve their land, and "in order to gain capital crofters migrated to urban centers making it impossible to cultivate their land" (Fraser Darling 1955: 10).

The overpopulation, famines, destitution, squatting, rack-renting, and insecurity of tenure of the early and mid-nineteenth century had been overcome by the time of Adam's visits. But by the 1950s,

Fraser Darling found the problem to be moving "in the other direction," where there was excessive depopulation on the northwest mainland to the extent of endangering normal social life, and where the croft land and common grazing were deteriorating and used well below their agricultural potential (Fraser Darling 1955: 12). This is the cusp of change, between overpopulation and depopulation, that Adam captured in his township images.

Arnisdale and Corran

Two small villages of Arnisdale and Corran lie within 2 kilometers of each other, clustered around the bay on the north shore of Loch Hourn and on either side of the Arnisdale River and surrounded on all sides by Arnisdale Estate. The photographs I present here depict Arnisdale itself, but the historical narrative includes Corran as these two settlements are in reality one community that developed around the herring industry in the late eighteenth century. A census from 1836 puts the population of Corran and Arnisdale at 600, but even then it was in decline due to emigration (Miers 2006).

I include two repeat photos of Arnisdale in order to give an indication of the processes of social and material change preceding the time when Adam visited the area. Within 10 years of the passing of the Crofters Holdings Act in 1886, improvements of living conditions proceeded throughout the crofting areas. Erskine Beveridge's photo taken in 1898 (plate 18.1A) is a good example of how this change came about and contextualizes Adam's photo taken 41 years later (plate 18.2A). Like many other west coast crofting settlements of the nineteenth century, the land provided for small crops of potatoes and fodder for a few livestock, but in the main, livelihoods were won from the sea. During this period, Loch Hourn was renowned for its annual herring fishing between July and October when up to "a couple of thousand fishermen would come from all parts of Scotland" (Parliamentary Papers 1884: 2044).

At the time of the Napier Commission, the crofting families of Arnisdale had no more than half a hectare of arable land apiece and no access to hill grazing. Only one family kept a cow by permission of the proprietor of Arnisdale Farm.

Testimony of the poor living conditions was given to the Commissioners regarding the so-called turf and rush thatched "black houses," so named due to the fact that they had internal fires but no chimneys and often no windows (see fig. 18.3). Typically they housed both humans and animals, and their interiors were notoriously damp, dark, and dirty:

> We built them ourselves, for the proprietor gave us no assistance. . . . We spoke to the factor about a change of site for the houses and as an answer he asked us whether the sea was coming over the floors of them. We said it sometimes did; and he told us then that we should put back doors upon them, and when the sea came in that we could run away. (Parliamentary Papers 1884: 2027)

The dwellings under construction in Beveridge's 1898 photo are forerunners of the improved "white houses," subsidized by the Congested Districts Board, the Department of Agriculture, and the Crofters Commission, that became ubiquitous throughout the crofting region during the first half of the twentieth century.

In the 1890s, 45 houses were occupied in Arnisdale and Corran by a population approaching 200 people (Parliamentary Papers 1895, English 2000). By the time Adam took his photo in 1939, this number had nearly halved; many of these houses would have been occupied by elderly men and women, although the school remained open with a handful of pupils until 1954; today, only 16 houses are permanently inhabited by 32 people. These include 10 households occupied by the retired descendants of the original crofting families and four by young couples with eight school-aged children. However, this is a welcome improvement from the period between 1990 and 2000 when there were no school-aged children in the village. During the last 20 years, several

derelict houses have been bought and restored as holiday homes, and most of the year-round residents of the village have also modernized and extended their original croft houses.

At the time of Adam's photo, employment in the village would have revolved around the Arnisdale and adjacent estates, involving stalkers, stockmen, boatmen, housekeepers, two school teachers, a shopkeeper, postman, and laborers for the recently established Forestry Commission. Employment today consists of one estate worker, one contract worker in the oil industry, one construction worker and one support staff for a local wind-farm company, a part-time postmistress, a nurse, one bed-and-breakfast business during the summer months, a seasonal tearoom, a school bus driver, and one self-employed seasonal boatman ferrying hill-walkers to more remote parts of Loch Hourn.

During the period between the repeat photographs, crofts have gone fallow and unused as crofters have retired or died. By the 1970s, there were only 10 active crofters, and today none; the last of the livestock were sold in 2006. This is not untypical of marginal and remote townships where agrarian decline has been inexorable during the whole of the twentieth century. Today, a slow renewal of the local economy by diverse enterprises such as fish farming, wind farms, tourism, and service-related self-employment means that the marginal agriculture underlying the "crofting way of life" is less relevant than ever before.

Tarskavaig

The village of Tarskavaig was created as a result of forced removals from other parts of Skye during the early nineteenth century. The arable land was divided into 31 lots, and by 1883, 45 families subsisted on remittances from migrant labor and the produce from mixed agricultural smallholdings varying in size from 0.6 to 1.4 hectares (Parliamentary Papers 1884: Q5535). In 1931, the village was still agriculturally active and several new white houses indicate a degree of improvement in living conditions at that time (see plate 18.3A).

Adam's photo of Tarskavaig (plate 18.3A) found its way into various publications in the following decades (e.g., Barnett 1933). Fraser Darling used several of Adam's photos depicting the townships of Sleat in his seminal books *Crofting Agriculture* (1945) and *West Highland Survey* (1955), where he commented that, unlike many other parts of Skye, Tarskavaig had good grazing and well-drained arable land of light loam and represented "an idyllic picture of a crofting township" (Darling 1955: 39). Adam's image does indeed seem idyllic: it depicts one of those relatively rare dry, clear days of autumn with the hay, oats, and barley harvest neatly stacked to dry before being moved into larger stooks or indoors for the winter. Even the smallest fields and most marginal arable land were worked and the adjacent hill grazing kept free of bracken. Seven houses, two of which were new government-subsidized white houses, two partially modernized black houses with corrugated iron roofs, and three traditional black houses of dry-stone-rubble walls with thatched roofs, are evident in this part of the township.

> This would be good early potato ground, and seaweed would not be hard to get. The crofts look trim and well kept. The oat crop also looks a good one. The bracken is evidently creeping in, from the appearance of the foreground. Many crofters in such a district as West Sutherland would think themselves fortunate with stretches of arable as good as this. Small mechanized agriculture is possible here, and vegetables for the Skye tourist trade would do well. (Fraser Darling 1945: 74)

The most obvious contrast with 2006 is the absence of stooks and hayricks or potato fields. The precarious and labor intensive business of drying fodder and grain has been superseded with the aid of a large tractor and big baler to make silage for cattle.

The recollections of Flora MacLean, who was born in Tarskavaig in 1949, confirm that little changed during the 20 years following Adam's visit. Flora recalled that when she was a child, all the crofts were ploughed with horses and cut with scythes, the oats were made into stooks, and hay was dried just as in Adam's photo. Most crofters kept a cow and follower, and the village population worked the sheep and planted potatoes together. Change came during the 1950s and 1960s as the old people died, crofts became disused, and many of the houses stood empty. She herself went away to work in the south after leaving school, but came back in the 1970s.

> The decline in the village continued until about 15 years ago when people started coming back. Four or five houses are now occupied by people who do not croft or have a plot of land. Some of these work at the nearby Gaelic college and have family ties to the area. Young people have come back and almost all the houses are still in possession of the original crofting families. Those that are complete incomers have integrated well. (F. MacLean, pers. comm., 26 October 2006)

Today, only one farmer makes a substantial part of his income from the land. He has a tractor and round baler, and cuts grass for four or five other crofters who still keep a few cattle. Deagrarianization has followed contemporary economies of scale relying on capital-intensive big-bale silage technology by one farmer where previously there were over 30. The more marginal lands have been left fallow, and only the larger fields are cut (fig. 18.3B). About 10 of the 30 houses in the village are used intermittently during holidays by the relatives of original crofting families or are holiday houses used by outsiders. The other houses are occupied more or less all year round by locals. The village is a healthy, thriving community today—back from the brink of depopulation. The land is no longer the mainstay of the economy since people no longer accept that level of hardship and poverty. A diversity of employment opportunities, many of them related to the seasonal tourist trade in the area, sustain many similar communities in Skye and Lochalsh.

Figure 18.3. Calligarry township, Scotland, UK.
A. (21 September 1931). This image of Calligarry township depicts one of the few "black houses" described in the text and still inhabited at the time. Within 20 years, none were inhabited in the parish and only a handful in Skye itself (Barron 1985). (R. M. Adam, RMA-H2752).
B. (29 August 2006). The "black house" and thatched byre are today barely recognizable ruins. The more marginal fields have been abandoned, and natural woodland regeneration is evident as a result. The open spaces between the wooded areas and in the drier sections of the fallow fields are now overrun with *Pteridium aquilinum* while *Juncus effusus* have invaded the rougher and damper ground. Today, the croft land is farmed in the same way as described for Tarskavaig—a few crofters with tractors and big-bale silage machines crop the grass early in the summer. The bales are wrapped in plastic and stacked at the sides of fields close to where cattle will be fed in the winter months. Anyone who has tried to make hay during the wet west Highland summers will agree that this modern technique is a great advance on the hay cutters, turners, and balers of the more recent past, to say nothing of hand scything, raking, and stooking of the nineteenth and early twentieth century. Forestry Commission plantations of *Picea sitchensis* (Sitka spruce) and *Pinus contorta* (contorta pine) that have transformed the open moorlands of the Highlands during the twentieth century are visible in the distance. (R. F. Rohde, Sco 40b).

Calligarry and Ardvasar

Before 1803 there were only seven families in the Calligarry township—in that year it was divided into 18 lots—everyone that sent a son to his lordship's regiment would get one . . . five entire lots are held by as many tenants, the remaining thirteen being subdivided into as many as six divisions. There are two cottars in the township that have no land at all and cannot even keep a hen. (Parliamentary Papers 1884: 274)

From testimony to the Napier Commission in 1883 it is clear that no one in either Calligarry or Ardvasar had kept sheep since about 1800. The tacksman of Ord (whose farm extended to 2,400 hectares with 9 hectares of it arable) employed a few crofters as ploughmen. About 10 or 12 people were employed at Armadale Castle (the Clan Donald seat) each summer in the gardens, and others went south to seek employment. "They go to the Irish coast fishing—the younger ones; and a number of girls go to be house servants in the towns; and those that don't do that go to work with farmers in the Lothians and on the east coast of Scotland" (Parliamentary Papers 1884: 313).

The establishment of larger croft holdings and substantial common grazings after the Crofters Holdings Act of 1886 meant that a higher standard of living based on small-scale agriculture was possible. Fraser Darling's comments on the crofting landscape depicted in Adam's 1931 image of Calligarry (fig. 18.3A) show some insight into the agrarian change taking place at the time:

Some of the hay is in large quoils, some in small quoils, some lying in the swathe and some uncut. Tripods in the middle distance indicate that the big quoils are being made with air space inside them. They would dry even better if they were lifted clear of the ground. This township has obviously shrunk since the days when much of the ground behind the crofts was cultivated. The lazybeds evidently supported a large population, for all those acres of *feannagan* [so-called lazy-beds] in the photograph represent hard labor. (Fraser Darling 1945: 44)

The trend that Fraser Darling noted in the 1931 photograph has continued into the present (see fig. 18.3B).

Figure 18.3A depicts one of the few black houses still inhabited at the time. Within 20 years, none were inhabited in the parish and only a handful in Skye itself (Barron 1985). The relatively large and fertile crofts of the Sleat peninsula provided partial livelihoods for crofters at the time of Adam's photos, but similar to previous eras, in the 1960s a family needed another source of income apart from the croft. But given a large enough landholding, "a young man would be justified in working well and making crofting his full-time occupation, because present day Government subsidies and good proceeds for stock and wool bring in more income now

than ever before" (Barron 1985: 513). During the 1960s electrification and rural water supplies were completed, and most houses had piped water and modernized bathrooms.

The photo of Ardvasar township (fig. 18.4) shows similar trends in agricultural practice: only the larger fields are used for grazing or silage production resulting in increased woodland, invasion of *Pteridium aquilinum* (bracken) and *Juncus effusus* (rushes), as well as the development of new modern bungalows close to the main road linking Skye to the mainland ferry.

The change in cattle numbers in the Sleat Parish over the last two centuries is indicative of the change in agricultural practice from intensive mixed farming centered on cattle to extensive sheep farming (Dodgshon 1993). In 1795, there were about 2,600 cattle and almost no sheep; in 1963, there were only 749 beef cattle and 15,010 sheep (Barron 1985), and by 2005 these numbers had fallen to 607 and 10,437, respectively (SEERAD 2007).

Drynoch

In common with the townships just described, the crofters of the parish of Bracadale were confined to

Figure 18.4. Ardvasar township, Scotland, UK.
A. (21 September 1931). This view of Ardvasar township shows enclosed grazed fields in the foreground and crofts in the midground, overlooking the Sound of Sleat and Knoydart. (R. M. Adam, RMA-H2751).
B. (29 August 2006). Note the heavily grazed sward in the foreground, the increase in *Juncus effusus* and *Pteridium aquilinum* in the midground, and woodland incursions around arable fields in the distance. These are combined symptoms of extensification of land use associated with sheep farming and a decline in arable production. The monoculture forestry plantation in the left distance is nearing maturity when it will be clear-cut and exported primarily for pulp. (R. F. Rohde, Sco 41b).

pitifully small patches of land while the bulk of the area was divided into six enormous sheep farms in the mid-nineteenth century, of which Drynoch was one (Hunter 1976). At the time of the Napier Commission, only 11 families remained in Drynoch "very much decreased," all being cottars with no land of their own and not allowed to keep livestock, "apart from five families allowed to keep 1 cow and 2 or 3 sheep—all tethered" (Parliamentary Papers 1884: 336). Most of the evictions had taken place in the early nineteenth century, and the remaining hill ground was taken away from the crofters in the 1840s. Those tenants who were not removed were required to labor for the family of the lowland proprietor of the sheep farm.

Unlike the townships in Arnisdale and Sleat, land-tenure reform did not reach Drynoch until 1923 when Drynoch Farm came into the hands of the Board of Agriculture for Scotland and was put under crofting tenure (Cameron 1996). By the time of the third Statistical Report compiled during the 1950s, the crofts of Drynoch were described as being

in the main of such a size as to afford their occupants a very fair standard of living. There is no great poverty in the parish. The smallholders, who form the greater bulk of the population, are now in a comparatively prosperous condition, and many of them own motor cars, tractors and other farm implements. Ploughing by horses is becoming rarer and rarer. (Barron 1985)

Overall, the major changes suggest that this is a highly productive and valued agricultural landscape. Large crofts that were created with some consideration for economic viability in the twentieth century have survived with little change (fig. 18.5).

Kilmuir District

In 1847, the 18,600 hectare Kilmuir Estate had been sold by Lord MacDonald and turned over to several large sheep farms and the crofting tenantry confined to small coastal settlements (Barron 1985). During the land wars of the 1880s, Kilmuir was known to be

Figure 18.5. Drynoch township, Scotland, UK.
A. (24 September 1936). Drynoch township is an ancient cultural landscape with extensive preclearance dikes, which are apparent against hillslopes in the midground and distance. Sheep are grazing in the foreground near the footstones of a preclearance dwelling that are hidden within a patch of *Juncus effusus*. To the left of frame in Glen Drynoch, abandoned homesteads and precrofting field systems are widespread. (R. M. Adam, RMA-H5226).
B. (23 September 2006). Large fields on the lower flood plain, where there are now six new large agricultural sheds corresponding to six large croft holdings, are in use for big-bale silage and suckler beef production. Only two of the six crofts on the opposite slope are now cropped for silage; the others are either abandoned or used only for grazing. Other changes include the extension of arable fields into the lower floodplain, the planting of conifer shelter belts, and several new houses on what appears to be a subdivided croft to the left. Forestry Commission plantations are visible on the skyline between Drynoch and the Cuillin Hills in the distance. (R. F. Rohde, Sco 44b).

one of the most fertile areas of the island and at the same time one of the most afflicted by clearances and the practice of rack-renting by an unscrupulous landlord (Cameron 1996). In the wake of the Crofters Holdings Act in 1886, Kilmuir crofters and cottars petitioned their landlord for the restoration of their former grazings at Monkstadt and Duntulm farms, which "were in the hands of strangers while . . . the rightful owners were huddled together on rocks and moss not fit for cultivation" (Hunter 1976: 176, quoting Crofting Files AF67, Scottish Record Office).

The estate, consisting of seven townships, was bought by the Congested Districts Board in 1904. The lands of Monkstadt farm were given to the adjoining townships of Totscore (fig. 18.6) and Bornesketaig (plate 18.4), and a new settlement of Linicro was formed in 1914 along with the creation of new croft holdings in Flodigarry (plate 18.5). All the townships in this parish were enlarged at the expense of sheep farms by the Congested Districts Board and later the Department of Agriculture.

Over the next 25 years, the area of arable land increased from 990 to 1,345 hectares and the total area of crofting tenure from 9,847 to 18,049 hectares. Ten large sheep farms were incorporated into the croft-

ing landscape, and 85 new crofts and 268 existing holdings were enlarged. "Landlessness had been eliminated and the number of thatched houses had fallen from 336 to 137 while the number of modern stone-built houses with slated roofs had increased from 20 to 304" (Hunter 1976: 206). When the Smallholders Act became law in 1912, the estate was registered in the name of the Board of Agriculture for Scotland and is now a Crofting Estate under the Secretary of State for Scotland.

Fifty years after Kilmuir was taken into state ownership there were . . . great changes in the material conditions of the people, with a consequent change in their outlook. Black house to white house; lorry for the creel and cart; tractor and plough for the *caschrom*; bus or car instead of walking or riding; daily post office van instead of thrice weekly delivery; free medical services; free school transport; etc. (Barron 1985: 477)

At the same time, a massive reduction in population occurred, from a peak of 3,625 in 1841 to a quarter this amount in 1961. In contrast to the district as a whole, this locality has seen a further fall of almost 40

A

B

Figure 18.6. Totscore township, Scotland, UK.
A. (27 September 1936). Totscore township with Monkstadt House (now derelict) visible in the distance. Adam's photo shows haymaking in various stages and unharvested potatoes late in the season. Evidence for more extensive cultivation and drainage of surrounding fields is clearly visible in 1936. (R. M. Adam, RMA-H5268).
B. (1 September 2006). Aerial photographs show that every scrap of land on the coastal plain has been turned by cas-krom (hand-plough) or horse-drawn plough at one time or another. Today, these fields have been abandoned for so long that little trace remains of this intensive land use apart from the ancient head dike and the single croft in the right midground. (R. F. Rohde, Sco 45b).

percent in the population since 1971 (Skye and Lochalsh Local Plan 1999). Accompanying these socioeconomic and demographic trends, arable cultivation (including temporary grass) fell by one half. Since 1963, cattle numbers have halved from 2,349 to 1,176 while sheep numbers have almost doubled from 16,974 to 30,220, reflecting the wider trend of extensification. In 1949, the number of part-time jobs (apart from crofting) in the nine townships of Kilmuir Parish was 16, half of which were fishing for export and the others estate work, public services and road maintenance, and textile manufacturing (Fraser Darling 1955). Today crofting continues to provide supplementary income to tourism- and service-based employment, much of it dependent on commuting to the nearby town of Portree (Skye and Lochalsh Local Plan 1999).

Discussion and Conclusions

Deagrarianization in the Northwest Highlands has much in common with trends throughout the more marginal agricultural areas of Europe, which are characterized by a diversity of cultural landscapes shaped by traditional land-use practices (Plieninger et al. 2006). These landscapes support high levels of biodiversity and ecological services, both of which are being lost due to deagrarianization, extensification, and land abandonment (Birnie and Mather 2005, Plantureux et al. 2005). The repeat photographs of crofting townships illustrate the links between socioeconomic, agrarian, and ecological systems, and by implication the relationship between cultural and biological diversity (O'Rourke 2006). The species-rich hay meadows that were developed on crofts over more than a century of stable agricultural tenancy were the result of long rotations of late cutting with after-grazing by cattle interspersed with short periods of cropping. The repeat photographs show that all but the largest and most productive fields have suffered deterioration evidenced by the invasion of *Juncus effusus* and *Pteridium aquilinum* and the loss of conservation value based on floristic composition (Tiley and Jones 2005). Land abandon-

ment and impoverishment of marginal fields have also resulted in scrub woodland regeneration, which might otherwise be welcome from a biodiversity perspective, were it not that scarce, hard-won agricultural land is being lost. Furthermore, the larger fields that are now cropped for silage are likely to be less floristically diverse due to reseeding and application of nitrogen fertilizer. Both trends are consistent with the change from "traditional" croft management to more modern capital-intensive fodder production dependent on economies of scale, extensification of sheep farming, and complete abandonment due to deagrarianization and decrofting (Dodgshon 2006).

The repeat photographs from all 19 townships of this study clearly illustrate these interrelated trends that constitute nothing short of an ecological revolution between the 1930s and the present (Merchant 1997, Rohde and Hoffman 2008). This has been accompanied by an economic transformation in the local economy due in part to the recent creation of employment in fish farming, tourism, services, and a boom in house building related to the national surge in property values. The reasons for these changes are complex and revolve around European agricultural and development policy as well as global changes in commodity production, transport, and tourism. However, agriculture now plays a minor role in an environment that was once almost wholly dependent on it to sustain the local economy.

Two very different perceptions of the crofting landscape were espoused during the decades following Adam's trips to the Northwest Highlands that echo into the present. One portrayed it as devastated and degraded "and that is the plain primary reason why there are now few people and why there is a constant economic problem" (Fraser Darling 1955: 192); the other identifies the landscape "as a wilderness of mountain flanked glens and mist shrouded islands, inhabited by a quaint, noble and once warlike race" (Lorimer 1999: 518). Fraser Darling sums up the first viewpoint throughout his influential book on crofting agriculture:

It is possible that the wilderness value of the West Highlands for the jaded townsman will still be sufficient to justify a large subsidy to maintain a sufficient population of people following practices of misuse to prevent any natural healing of the devastation. . . . Man in numbers, combative and political man, who has no place in nature, has entered this complex organism like a protozoan parasite in its blood-stream and had brought the organism to a state of debilitation. . . . The greatest value the mass of Highland land could give to the nation would be as a continuing productive wild land in which perhaps twice as many people could live than are there at present. . . . A period of a century is probably needed together with skilful management, to repair the damage. (Fraser Darling 1945: 192–193)

Thirty years later, Bryden and Houston (1976), commenting on the ability of the recently formed Highlands and Islands Development Board (HIDB) to effect socioeconomic improvements in the crofting areas, diagnosed the "problem" in somewhat different terms but reiterated Darling's sentiment:

[In] the 1960s the central problem was widely considered to be depopulation. . . . Unless special measures were taken, it was argued, many areas would soon reach a critically low level of population density, the continuation of basic public services would be threatened and, at best, only a semi-derelict economy would survive. [And] since the main cause of the Highland problem in agriculture appeared to be the main cause of the Highland problem in the 1960s, it seemed absurd to look on the development of that industry as helping towards its solution. (Bryden and Housten 1976: 132)

Development planners within the HIDB analyzed the socioeconomic "problem" as stemming from subsistence agricultural practices, protected from market forces, largely as a result of crofting—"a stultifying form of land tenure" (Highlands and Islands Development Board 1967: 4, quoted in Carter 1974: 293). However, without crofting agricultural activity

and the security of tenure associated with the institution of crofting, it is doubtful that a population large enough to engender a local economy would exist in the Northwest Highlands today (Hunter 1991). Furthermore, the modernization ideology of the 1960s has been replaced today by nature conservation as the driver of landscape change. Production subsidies are in the process of being replaced by an assortment of area-based and management activity subsidies that act as incentives for crofters to pursue environmental and conservation objectives as well as social and economic development. An equally diverse array of government landscape designations (which include Sites of Special Scientific Interest, National Nature Reserves, National Scenic Areas, Special Protection Areas, Special Areas of Conservation, Areas of Great Landscape Value, Listed Wildlife Sites, Sites of Importance for Nature Conservation, and Biodiversity Action Plans), aimed at reviving the area's "natural heritage," are an increasingly important factor in the local economy, although these are rarely directly related to the economics of "traditional" crofting. Furthermore, if Europe's new carbon management agenda is not carefully integrated into existing agricultural policy and support schemes, this is likely to have an adverse effect on crofting in the Northwest Highlands, leading to further stock reductions and a spiral of decline in crofting activities (Yuill and Cook 2007).

Crofting and its association with cultural heritage are central to the popular currency of modern tourism, the mainstay of the economy in the Northwest Highlands today (Rohde 2004). At the same time and for similar reasons, the myth of the area as "wilderness" is given increased symbolic importance in the promotion of tourism, as well as by conservation nongovernmental organizations and the defenders of deer stalking on sporting estates (MacDonald 1998). These contradictory ways of seeing the environment have become "naturalized" in the landscape, corresponding closely with Darling's prescient vision. It would seem that the contradictory ideological discourse on nature and society that places traditional crofting society in a wilderness arises out of today's neoliberal economy and the globalization of the conservation lobby.

From an ecological perspective, the lack of agricultural activity now evident in the crofting areas, especially the more marginal townships, represents a distinct hiatus in the area's long history of intense anthropogenic disturbance. The changes evident in the brief period of 75 years depicted in the repeat photographs are but the latest of a long series of transformations, stretching back several millennia, brought about by the impact of humans within this evolving landscape. Although recent, the detail and clarity of these images are a valuable source of evidence toward understanding the environmental history of the Northwest Highlands. They also help to put into perspective the discourse of political ecology related to the relative values of cultural landscapes, wilderness areas, and natural heritage, all of which have been "written on the surface of the soil" and continue to influence the way we see and respond to this landscape today.

Acknowledgments

Many Highlanders contributed indirectly to this chapter, but William and Christine MacKenzie and Flora MacLean deserve special mention for their generosity and interest. Chuck Jedrej, to whom this chapter is dedicated, made many insightful comments on an earlier draft. I am also grateful to Diane Boyer for her attention to detail and help in editing, as well as to Peter Griffiths for his generous contribution of fine maps to this chapter.

All of the R. M. Adam photographs are reproduced courtesy of the University of St. Andrews Library, St. Andrews, UK. The repeat photographs are courtesy of the author.

Literature Cited

Armit, I., and I. B. M. Ralston. 2003. The coming of iron, 1000 BC–AD 500. In *People and woods in Scotland: A history*, ed. C. Smout, 40–59. Edinburgh: Edinburgh University Press.

Barnett, T. R. 1933. *The land of Lorne and the isles of the west*. Edinburgh: W&R Chambers.

Barron, H. 1985. *The third statistical account of Scotland: The county of Inverness*. Edinburgh: Scottish Academic Press.

Birnie, R. V., and A. S. Mather. 2005. Drivers of agricultural land use change in Scotland. In *Understanding land-use and land cover change in global and regional context*, ed. E. Milanova, Y. Himiyama, and I. Bicik, 147–163. Enfield, NH: Science Publishers.

Blundell, A., and K. Barber. 2005. A 2800-year palaeoclimatic record from Tore Hill Moss, Strathspey, Scotland: The need for a multi-proxy approach to peat-based climate reconstructions. *Quaternary Science Reviews* 24:1261–1277.

Brown, K. M., and B. Slee. 2004. *Exploring the relationship between common property, natural resources and rural development: The case of crofting common grazings*. Aberdeen Research Consortium Discussion Paper Series, People, Environment and Development, No. 2004-3. Aberdeen, Scotland: Macaulay Institute.

Bryceson, D. F. 1996. Deagrarianization and rural employment in sub-Saharan Africa: A sectoral perspective. *World Development* 24:97–111.

Bryden, J., and G. Houston. 1976. *Agrarian change in the Scottish Highlands: The role of the Highlands and Islands Development Board in the agricultural economy of the crofting counties*. Glasgow Social and Economic Research Studies 4. London: Martin Robertson.

Cameron, E. A. 1996. *Land for the people: The British government and the Scottish Highlands, c. 1880–1925*. East Lothian: Tuckwell Press.

Cameron, E. A. 1998. The Highlands since 1850. In *Modern Scottish history: 1707 to the present*, Volume 2, ed. A. Cooke, I. Donnachie, A. MacSween, and C. A. Whatley, 47–72. East Linton: Tuckwell Press.

Carter, I. 1974. The Highlands of Scotland as an underdeveloped region. In *Sociology and development*, ed. E. de Kadt and G. Williams, 279–314. London: Tavistock Publications.

Chapman, P., and M. Shucksmith. 1996. The experience of poverty and disadvantage in rural Scotland. *Scottish Geographical Magazine* 112:70–75.

Dodgshon, R. A. 1993. Strategies of farming in the Western Highlands and Islands of Scotland prior to crofting and the clearances. *Economic History Review* 46:679–701.

Dodgshon, R. A. 1994. Rethinking Highland field systems. In *The history of soils and field systems*, ed. S. Foster and T. C. Smout, 64–74. Aberdeen, UK: Scottish Cultural Press.

Dodgshon, R. A. 2006. Heather moorland in the Scottish Highlands: The history of a cultural landscape, 1600–1880. *Journal of Historical Geography* 32:21–37.

English, P. R. 2000. *Arnisdale and Loch Hourn: The clachans, people, memories and the future.* Arnisdale: Arnisdale and Loch Hourn Community Association.

Fossitt, J. A. 1996. Late Quaternary vegetation history of the Western Isles of Scotland. *New Phytologist* 132:171–196.

Fraser Darling, F. 1945. *Crofting agriculture.* Edinburgh: Oliver & Boyd.

Fraser Darling, F. 1955. *West Highland survey: An essay in human ecology.* Oxford: Oxford University Press.

Highlands and Islands Development Board (HIDB). 1967. *First Annual Report.* Inverness: Highlands and Islands Development Board.

Hirons, K. R., and K. J. Edwards. 1990. Pollen and related studies at Kinloch, Isle of Rhum, Scotland, with particular reference to possible early human impacts on vegetation. *New Phytologist* 116:715–727.

Hunter, J. 1976. *The making of the crofting community.* Edinburgh: John Donald Publishers.

Hunter, J. 1991. *The claim of crofting: The Scottish Highlands and Islands, 1930–1990.* Edinburgh: Mainstream.

Lorimer, H. 1999. Ways of seeing the Scottish Highlands: Marginality, authenticity and the curious case of the Hebridean blackhouse. *Journal of Historical Geography* 25:517–533.

MacDonald, F. 1998. Viewing Highland Scotland: Ideology, representation and the "natural heritage." *Area* 30:237–244.

Mackenzie, A. F. D. 2006. A working land: Crofting communities, place and the politics of the possible in post–Land Reform Scotland. *Transactions of the Institute of British Geographers* 31:383–398.

Merchant, C. 1997. The theoretical structure of ecological revolutions. In *Out of the woods: Essays in environmental history*, ed. C. Miller and H. Rothman, 18–27. Pittsburgh, PA: University of Pittsburgh Press.

Miers, M. 2006. *Western seaboard: An illustrated architectural guide.* Edinburgh: Rutland Press.

O'Rourke, E. 2006. Changes in agriculture and the environment in an upland region of the Massif Central, France. *Environmental Science and Policy* 9:370–375.

Pacione, M. 1996. Rural problems and "Planning for Real" in Skye and Lochalsh. *Scottish Geographical Magazine* 112:29–38.

Parliamentary Papers. 1884. (Napier Report). *Report of the Commissioners of Inquiry into the condition of the crofters and cottars in the Highlands and Islands of Scotland, 1884.*

Parliamentary Papers. 1895. Royal Commission *Report and Minutes of Evidence*, Vol. 38–39.

Plantureux, S., A. Peeters, and D. McCracken. 2005. Biodiversity in intensive grasslands: Effect of management, improvement and challenges. *Agronomy Research* 3:153–164.

Plieninger, T., F. Hochtl, and T. Spek. 2006. Traditional land-use and nature conservation in European rural landscapes. *Environmental Science and Policy* 9:317–321.

Rohde, R. F. 2004. Ideology, bureaucracy and aesthetics: Landscape change and land reform in Northwest Scotland. *Environmental Values* 13:199–222.

Rohde, R. F., and M. T. Hoffman. 2008. One hundred years of separation: The historical ecology of a South African "Coloured" Reserve. *Africa, Journal of the International Africa Institute* 78:189–222.

SEERAD (Scottish Executive Environment and Rural Affairs Department). 2007. *Agricultural Statistics 1984–2005, by Parish.* Supplied by Lynda.Reid@scotland.gsi.gov.uk, 21 November 2006.

Seppä, H., K. Antonsson, and M. Heikkilä. 2003. *Holocene annual mean temperature changes in Europe: Pollen-based reconstructions.* Geological Society of America, XVI INQUA Congress, Paper Number 48–5, Reno, NV.

Short, D., and A. Stockdale. 1999. English migrants in the Scottish countryside: Opportunities for rural Scotland? *Scottish Geographical Journal* 115:177–192.

Simmons, I. G. 2001. *An environmental history of Britain: From 10,000 years ago to the present.* Edinburgh: Edinburgh University Press.

Skye and Lochalsh Local Plan. 1999. Inverness, UK: The Highland Council.

Smout, T. C. 1994. Perspectives on the Scottish identity. *Scottish Affairs* 6:101–113.

Tiley, G. E. D., and D. G. L. Jones. 2005. Is biodiversity declining in the traditional haymeadows of Skye and Lochalsh, Scotland? In *Proceedings of the XX International Grassland Congress*, June 26–July 1, Dublin, Ireland, ed. F. P. O'Mara, 628. Dublin, Ireland: University College Dublin.

Tipping, R. 2003. Living in the past: Woods and people in prehistory to 1000 BC. In *People and woodlands in Scotland: A history*, ed. T. C. Smout, 14–39. Edinburgh: Edinburgh University Press.

Tipping, R., A. Davies, and E. Tisdall. 2006. Long-term woodland dynamics in West Glen Affric, northern Scotland. *Forestry* 79:351–359.

Yuill, B., and P. Cook. 2007. Trends in agriculture and supporting infrastructure within the HIE area 2001–2006, with commentary on the North West Highlands area. Unpublished report for Highlands and Islands Enterprise.

Chapter 19

Photography and Rephotography in the Cairngorms, Scotland, UK

Peter R. Moore

Comparison between vintage images and contemporary rephotographs provides powerful evidence of change to landscape and lifestyle. The process of rephotographing a location also offers an opportunity to collect additional information and knowledge about the area depicted (Klett, chapter 4). One of the best-known hills—and the most visible from the popular tourist destinations—An Carn Gorm, lent the Anglicized form of its name to the area of hills surrounding it and, since 2003, to a much wider area: the 3,800-square-kilometer Cairngorms National Park (fig. 19.1). The park encompasses a mix of habitats, landforms, and land use, ranging from hills, moorland, forest, and agricultural land to lochs, rivers, and communities.

The study is ongoing and seeks to record documented change using rephotography. Its purpose is largely educational and interpretive, with the rephotographic technique being used as a tool to reveal change, foster a connection and understanding of the landscape, make links with human involvement in the processes of change, and inform—perhaps guide—future management practices.

By definition, rephotographers are drawn to locations and make photographic compositions determined up to around 150 years previously. Usually, the initial images were taken by another photographer, but in some notable long-term studies (Webb et al., chapter 1; McClaran et al., chapter 12) a single photographer or group of photographers may have returned to a location many times, over many years. In Scotland, photographs from the mid- to late nineteenth and early twentieth centuries tend to have been taken most abundantly in areas popular with Victorian and Edwardian tourists. These visitors to the area, mainly the English gentry and wealthy lowland Scots, came by train, arriving to stay in newly built villas in the towns and at remote shooting lodges. They followed the example of Queen Victoria, who regularly visited the Highlands beginning in 1842 and set up Balmoral Castle as a holiday retreat. It was the height of fashion: they "toured," stalked deer, shot grouse, and recorded their exploits among the picturesque and romantic wild landscapes by the new medium of photography.

Photographic Record

The photographic record of the Cairngorms includes examples of pioneering photography and early popular photography, together with a proliferation of printed media from the early twentieth century. More recent personal collections of photo-

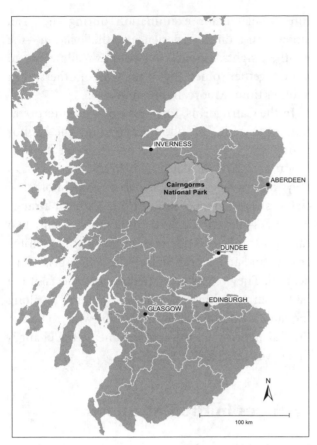

Figure 19.1. Location map showing the Cairngorms National Park boundary.

graphs, tourist promotional imagery, and photojournalistic records extend the evidence.

Original Photography

Horatio Ross was one of the founders of the Photographic Society of Scotland and took some of the earliest photographs of life in the Cairngorms. In exhibitions in 1858 and 1859, Ross included several pictures taken in the deer forest of Glen Feshie (Stubbs 2001–2008). As the photographic process evolved and saw wider use, other visitors created their own comprehensive photographic records of their visits (e.g., J. E. A. Steggall), while some commissioned professional photographers to accompany them (e.g., Walter Winans). Probably as a result of the difficulty of the photographic process and

the uncertainty of results, professionally printed album photographs were popular souvenir purchases. These were followed by booklets and postcards produced in Scotland by G. W. Wilson and Company, based in Aberdeen, and, most notably, the Dundee-based photographers James Valentine and Company, who can be credited as being the largest and longest-surviving photograph-publishing business in Britain (Lambert 1996).

During the twentieth century, a number of individual photographers made significant landscape studies in the Cairngorms, among them Seton Gordon (Gordon 1912, 1921, 1925), Robert Adam, Walter Poucher (Poucher 1947), and John Markham (in Fraser Darling 1947, Pearsall 1950). Of these, Robert Moyes Adam (1885–1967) is perhaps the most significant exponent; he is remembered for his prolific Scottish landscape work and his documentation of rural life, much of which changed dramatically or even disappeared completely within his lifetime (Rohde, chapter 18). In a 1958 interview, he revealed something of his motivation: "Suppose I catalogued (Scotland's) wildlife and its topography as a permanent record against industrial and other changes of the future. Suppose I were to preserve for my own botanical interest, the land as I see it in my lifetime" (Bruce-Watt 1958 quoted in Smart 1996). Adam's collection of negatives—some 15,000—were made from the late 1890s to the 1950s. They form a major documentary resource that has been featured in many book publications and magazines and is now archived in the University of St. Andrews Library.

Rephotography

The rephotographic process is governed by the quality, location, and content of an earlier image (Boyer et al., chapter 2; Klett, chapter 4). In selecting areas for rephotographic studies, an abundance of imagery from a variety of sources offers the greatest potential to build up a picture of the changing environment and the changing lifestyle. The images are inevitably restricted to compositions made previously, typically by another photographer, and may be further biased toward well-known or easily accessed viewpoints of popular subjects. In some situations, these viewpoints may be managed and maintained as such, thus influencing natural, discerned change.

Rephotography provides a way to explore the passage of time—both the concept and the reality of change—at a number of levels (Klett, chapter 4). To relocate the point from where an earlier picture was taken is to step into the footsteps of someone else. It is also to embark on a journey of one's own discovery and so begin the process of understanding the original photographer's choices in selecting the viewpoint and composing the picture. One also gains an understanding of the limitations imposed by the physical nature of the land and, perhaps, by the photographic process.

By faithfully replicating a view, it is possible to begin to make sense of the landscape change, particularly in terms of human intervention and influence. Rephotography allows the two dimensions of the original image to be interrogated and compared on like terms with a contemporary counterpart. From an interpretive standpoint, images speak in a universal language to young and old, and rephotographed "pairs" present a set of criteria for comparison and comment—providing the "provoke, relate, and reveal" elements of interpretation.

Yet the rephotographic process offers more: a chance to "connect" physically with common objects and features depicted in a photograph and persisting within the contemporary view: a tree, a rock, a building, or perhaps a distinctive door handle. The experience of direct personal connection can inform the future interrogation and interpretation of the earlier and the contemporary images, often providing a better understanding of the change and allowing recorded detail to be validated firsthand (Moore, in preparation).

At a personal level, the search for the locations of photographs taken by or including a relative or ancestor becomes a personal journey toward an understanding. The combination of information contained within the image and the common fea-

tures available for examination during the visit, along with a direct experience of the location—the weather, sights, smells, and sounds—allows us to piece together something of the lives of others from another time (Moore, in preparation).

In the Cairngorms, interrogation and interpretation of a series of photographs and the rephotographing of locations tell as much about the evolution and growth of tourism in the local area as they do about general social and technological changes. This chapter uses six images from a range of sources—an early private photograph album, two early postcards, two documentary landscape photographs, and an illustration from a 1960s tourist publication. They represent a selection of aspects of the Badenoch and Strathspey districts, which lie along the catchment of the river Spey in the western half of the Cairngorms, and describe some of the changes that have occurred.

Changes in the Cairngorms

Figure 19.2A was taken on the An-t-Aonach ridge by J. E. A. Steggall and looks northwest across Glenmore Forest, with Loch Morlich and the Monadhliath (the Grey Hill range) visible in the background. It is one of a number of images compiled within the albums forming the Steggall Collection held by the University of St. Andrews Library. These albums—46 in total—document the varied travels of John Steggall, and four of the albums include images taken in the Cairngorms (Lambert 1996). This 1901 view was taken at an altitude of roughly 660 meters. It depicts a group of hill walkers—a man, a woman, and a young boy—in typical period dress, on an overcast day at a small pool near a traditional route up to the summit of Cairn Gorm. The path climbs the shoulder of An-t-Aonach to the summit of Cairn Gorm via The Marquis' Well. The party, with photographer Steggall, would probably have started their walk from Glenmore Lodge, near the shores of Loch Morlich, climbing up along the river valley of the Allt Mor ("big burn"—a Scottish word

A

B

Figure 19.2. An-t-Aonach ridge, Scotland, UK.

A. (August 1901). "Cairngorm. On The Way Up." A group of hill walkers poses on the An-t-Aonach ridge next to a pool of standing water. The view is northwest across Glenmore Forest, with Loch Morlich and the Monadhliath visible in the background. Vegetation is short, probably due to a combination of wind and grazing. (J. E. A. Steggall, Record No. JEAS-21-37, Courtesy of the University of St. Andrews Library).

B. (2007). The vegetation is now taller, obscuring the boulders present in the 1901 view. A few stunted *Pinus sylvestris* (Scots pine) are visible in the foreground. The nearby trail is wide and eroded. Water is still available, but it is not considered potable by modern standards. (P. R. Moore).

for a stream) through the outpost pines of the natural treeline and onto the open hill.

In 1901, the vegetation is typically short for this altitude and aspect due to wind-clipping and probably heavy grazing by deer. Many boulders are visible along a slight ridge beyond the figures in the photograph. The 2007 rephotograph indicates that grazing pressures have changed and hints at a change in climatic conditions toward warmer temperatures at this altitude. The exposed boulders have been overgrown by vegetation and some have been removed and used in path building nearby. The location is now within the upper limits of the treeline as evidenced by the stunted *Pinus sylvestris* (Scots pine) visible.

The group is posed beside a small area of standing open water. Water remains available at this location but is peaty and certainly not potable within modern drinking water standards.

The woman wears a blouse and long skirt and is crouched, apparently sipping water from a beaker as she steadies herself on a stick. Her hair is drawn up. She has a coat and a wide-brimmed hat (possibly straw) slung on a strap over her shoulder.

The man stands behind her. He is dressed in a tweed jacket with a tweed cap and stands with a pipe in his hand. The young boy has a watch chain visible on his waistcoat and a cape or Mackintosh slung over a strap he wears over his shoulder.

The woven cotton, linen, and tweed clothing of the party provides the most obvious evidence of change. Tweed, long a staple of utilitarian dress in the Highlands and adopted by tourists as fashionable country wear, also lent itself well to outdoor pursuits through its qualities of warmth and durability. While tweed wear is still seen regularly in the Cairngorms—worn by Lairds, stalkers, gamekeepers, and some shepherds—for outdoor pursuits, waterproof or windproof outer wear, often brightly colored, made with high tech synthetics together with layered synthetic fleece is worn by the vast majority who venture to the hill. In what may be considered to be a mirror of fashion influence, this highly technical and ergonomically designed clothing is also now popularly worn as casual, high street fashion.

The original path remains in use but is wide and eroded near this location, damaged both by footfall, which has eroded the vegetation and the thin, peaty soils, and by the scouring of water runoff. In the last 45 years, the pattern of visitor use of Cairn Gorm has changed, influenced most by the building of the Ski Road in 1960, which passes a hundred meters below the vantage point of the photograph. Most visitors to Cairn Gorm now arrive by car or public transport and explore the hills along routes and paths built to service visitor access from the car park at the end of the road and, therefore, bypass the location of this image.

Figure 19.2B shows Krummholz pines—small, stunted trees, twisted by the wind in this exposed location—that have now established in the intervening years. The expansion of the treeline up the hill began between the late 1950s and early 1980s (Watson, pers. comm.) and has been encouraged more recently by a reduction in the numbers of red deer (*Cervus elaphus*) at Glenmore, together with a generally wetter and milder climate.

Figure 19.3A was taken by a James Valentine and Company photographer around 1910 (Glen 2002) and published as a postcard. It shows a view looking north on the Great North Road (now Grampian Road) in the village of Aviemore, described as Aviemore (Easter), and located away from the major hotels and railway station. It shows a woman, dressed

in dark clothing with a white-ruffed blouse, riding a bicycle toward the camera.

In the original image, the edge of the road is uncurbed. A broad verge of grass extends outside a fenced pasture on the right (east) side at this location. The fence "strainer" is of cast iron and the gate has ornate wrought ironwork. Square-milled wooden fence posts and line wire stock fencing extend down both sides of the road. St. Andrew's church, built in 1903, is on the right side of the view and may well have influenced the production of this image. The houses are large villa style and set back from the roadside.

Many of the villa houses survive in the 2007 image (fig. 19.3B), though they are partly obscured by planted *Populus* sp. (poplar) trees lining the roadside. In the distance to the north, the hillside remains afforested. The road has been widened and a tarmacadam surface has been laid. Until the early 1970s, this was a section of the Great North Road, which passed through many Highland villages along its route to link Edinburgh with John O'Groats. It remains a busy thoroughfare.

Plate 19.1A was taken of the village of Newtonmore probably just before the turn of the twentieth century. The view looks northeast along the main

Figure 19.3. Aviemore (Easter), Scotland, UK.
A. (ca. 1910–Glen, 2002) (Easter) Looking north along the Great North Road (now Grampian Road) in the village of Aviemore. Several large villa houses are visible; St. Andrew's church is on the right. (James Valentine and Co., No. 64324. Published as a postcard by S. G. MacDonald, post office, Aviemore).
B. (2007). Many of the villa houses are still present, albeit obscured by planted, nonnative *Populus* sp. (poplar) trees. The road is a busy thoroughfare carrying local traffic. (P. R. Moore).

street of the village and was published as a postcard. It includes remarkable detail featuring a number of people, walking, standing, or talking on the street. Their presence, late on this afternoon (determined by the position of the shadows), raises a number of questions that cannot be answered without further information that is unavailable from the picture alone—was there an event or occasion? Was the creation of the photograph the event in itself?

The dominant subject within the image is *Acer pseudoplatanus* (broadleaved sycamore), a nonnative tree in the area, that was in full leaf when the photographer secured this image. The tree stands at the roadside beside a pointed stone wall and has an advertising sign attached to its trunk.

On the right stands a small building clad with corrugated iron, advertising as a "Jeweller" on its fascia; beyond it, a wooden shed with a corrugated iron roof is offering a "car-hire" service—exceptionally early for such an enterprise, but quite possibly befitting of the prospective patrons frequenting the village at this time. Beyond this is a small roadside cottage. The road has a loose surface, is cambered, and without a pavement, the carriageway defined by a slight concavity to the surface, providing rudimentary drainage.

Several people, dressed formally, are walking in the streets. Three ladies walk away from the camera in the shadow of the tree at left. The ladies wear hats and long, dark dresses. Two men walk toward the camera, both carrying sticks and wearing hats. A gentleman leans on his bicycle on the right-hand side, looking toward the camera. A dog, a dark Scottish terrier, stands near him. Two girls stand in the street, wearing light pinafores.

In 2005 (plate 19.1B), the village of Newtonmore retains its linear form. Although the road remains a major thoroughfare, particularly for traffic from the west coast heading north, along with other Badenoch villages, Newtonmore was bypassed by the building of the main A9 trunk road in the 1970s. The buildings to the right foreground, including the small stone cottage and the corrugated iron–clad huts in use as retail and service premises in the early

picture, have been removed. New houses have been added to the center right of the view, infilling space that was formerly part of the grounds of one of the major hotels (just out of view to the right).

The tree, probably planted as one of several boundary trees at the time the hotel was built in the 1870s, has survived and grown considerably in the intervening years. In 2005, its lower branches were removed by the local authority and the remnant ironwork that previously supported the advertising signs still protrudes from the trunk of the tree, partly subsumed by tree growth and bark. At left, an art deco building has been constructed and its most recent incarnation was as a film set within the British Broadcasting Company (BBC) television series *Monarch of the Glen.*

The left (west) side of the street has changed little in form, and the nearest building (left) is the oldest surviving house in Newtonmore. It has, over the last 75 years, variously been used as a draper's, a chemist's, a café, and a guest house (Newtonmore Community Web site) and it is now a private dwelling. This reflects a widespread pattern of change that has taken place with the majority of these commercial premises and services such as banks, butchers, bakers, and news agents having been replaced in the village by a petrol station diversifying into selling basic requirements, a single, dedicated cooperative supermarket, and an automated cash point. Most needs are met by supermarket chain outlets at least 27 kilometers distant, and commercial premises tend to be café outlets or shops more focused on tourist sales of crafts and trinkets. A village post office remains viable, though is threatened by national policies of closure.

The location of the photograph is the main junction of two roads, where the military road from Forts Augustus and William to the west meets the Great North Road. In the late nineteenth century, at the time of the original photograph, the location would have been a busy center of the village and closest to the railway station, which at Newtonmore lies away from the village itself. The Hotel Newtonmore (more recently Main's Hotel) is a large, railway-inspired hotel, one of several built in the mid- to late nineteenth century to serve the influx of Victorians visiting Badenoch by train and from which, up until as late as 1916, a stagecoach still operated between Newtonmore and Tulloch, linking by road the Highland and the West Highland railway lines.

This photograph is one of the early examples of tourist imagery from this location, and, quite possibly, a commission by the hotel proprietor has set a trend for the next 60 years in documenting the village from this location. The tree is a central, persistent component of these images, up until the 1960s.

Rephotography captures this socioeconomic change that has occurred worldwide as once isolated towns and villages become tourist destinations. Detailed interpretation of the content of the nineteenth-century image also points toward the capture of a way of life that has been lost.

With the closure of the hotel in 2001, the gradual loss of shop premises at this end of the village, and reduced reliance on the railway, the village center, as such, has now shifted approximately half a kilometer to the vicinity of the village hall and post office. This view is no longer the choice of tourist photographers wishing to capture the village of Newtonmore in the twenty-first century, illustrating the shifting geography of the region's cultural heritage.

Figure 19.4A, taken in 1932 by Robert Moyes Adam, looks southwest and shows the river Spey about 1.6 kilometers south of the village of Newtonmore, close to its confluence with the river Calder. The vantage point for this view lies at the end of a short cart track built from the main (Great North) road, possibly formed to enable the extraction of gravel from the river for the construction of the nearby bridge, which was built of concrete and completed in 1927. The proximity of the track just a few meters away may have influenced the photographer in his selection of the vantage point. The location now lies within the ownership boundary of a private house.

The river Spey cuts a wandering route across the floodplain and its dynamic nature, at the time, is hinted at by fresh gravel deposits and a scoured, wide channel. A few scattered *Betula pendula* var. *pubescens* (birch), which appear to be of a uniform age, are established on the drier moraines above the river. They are in the range of 80 to 100 years old. Other vegetation in the 1932 view appears to be heavily

Figure 19.4. River Spey and the view to Loch Laggan, Newtonmore, Strathspey, Scotland, UK.
A. (21 May 1932). River Spey and the view southwest to Loch Laggan, Newtonmore, Strathspey. A cart track lies a short distance away, making the location easily accessible. The vegetation is heavily grazed, probably by sheep. The river pools were likely used for fishing for salmon or trout. (R. M. Adam, RMA-H2908, courtesy of the University of St. Andrews Library).
B. (2003). Sheep grazing ended in the mid-1980s, allowing for a pulse of regeneration including *Alnus glutinosa* (alder) and *Salix* spp. (willow). Enclosures have permitted recovery of *Calluna vulgaris* (heather) in the foreground. The river's course has shifted as a result of major flood events, the most recent being in 1989–1990. (P. R. Moore).

grazed. Old stock fencing still present nearby indicates that this location had long been grazed by stock, probably sheep. The topography of the land permits livestock to benefit from floodplain grazing but also to safely retreat to higher ground (left of picture) in the event of flooding. While the riverbank with its layer of alluvium is grass covered, the drier moraines support *Calluna vulgaris* (heather). The floodplain in the foreground shows evidence of what are likely to be long-established mole *Talpa europaea* (mole) hills, grassed over and only evident by their form.

At the right-hand side of this upstream view, where the river narrows, deepens, and speeds between large rocks, salmon or trout fishing has taken place, evidenced by the naming of river "pools" depicted on early maps (e.g., Ordnance Survey Maps—25 inch 1st edition, Scotland, 1855–1882). Accordingly, any emergent vegetation escaping the attentions of grazers would have been cleared by hand as part of riverbank management necessary to facilitate rod fishing.

The view was rephotographed in May 2003 (fig. 19.4B), and there have been many changes in the 71-year interval. The most obvious change is the level of grazing pressure, the resulting regeneration of the birches, and the thickening of grassy vegetation along the riverbank. The distinct age class of trees suggests a pulse of regeneration that would correspond with Second World War years (1939–1945) and its aftermath, during which a lack of management and a subsequent reduction in grazing pressure allowed the trees to establish. The area was grazed by sheep until the mid-1980s, but the sale of the land for development as a house plot, changing farm-production emphasis and a consequent removal of livestock, has resulted in a second pulse of regeneration including *Alnus glutinosa* (alder) and *Salix* spp. (willow) on the wetter margins. No riverbank management has taken place since ca. 1987. In addition, a rabbit-proof stock fence has been introduced across the view and the slope above the river enclosed, allowing recovery of the *Calluna vulgaris* (heather) vegetation cover.

Adam's picture predates the construction of a dam on the headwaters of the river Spey, built in order to transfer water into another catchment to augment hydroelectricity production. This has resulted in reduced incidence of flooding of the river. The bank at the lower right of the photograph has remained almost constant and is underpinned by resistant bedrock, which holds the course of the river at this point. On the floodplain (middle and left of the picture), the course of the river has shifted considerably. While channel change is a constant process at a minor scale, ground examination indicates at least two phases of river movement that have been preceded by major flood events, the last in 1989–1990. These changes are similar to others viewed on regulated rivers worldwide (Bierman, chapter 9).

Figure 19.5A is a more remote scene, taken of a roughly bulldozed track, cutting across a *Calluna vulgaris*–clad slope wooded with mature *P. sylvestris* above and below. Two *B. pendula* (silver birch) trees are visible downslope to the left. In the Cairngorms, mechanized access to remote locations has typically been required in support of forestry or the deployment of shooting parties and to the servicing of recreational facilities such as the ski areas. In addition, the development of tracked earth moving vehicles during the twentieth century changed the scale and speed of such an operation and allowed access to remote areas.

Robert Adam took this picture on 15 April 1946 as one of a series around this location, taken over several days. It is interesting to speculate on Adam's motivation: he may have been moved to document the recent disruption to the land and chose this location on the bulldozed road surface, in the first clear area he encountered walking up the hill.

The track was formed during the Second World War, bulldozed through ancient Caledonian pine forest to enable the commercial extraction of timber, and would have been fairly recent at the time of Adam's visit. The uphill side of the track shows signs of instability, with loose glacial rubble falling across the surface. The soils at this location are extremely thin, and the ground vegetation, predominantly of

Figure 19.5. Coire Ruadh, Glen Feshie, Scotland, UK.

A. (15 April 1946). Robert Moyes Adam captured this view of a roughly bulldozed track through a *Pinus sylvestris* wood in Coire Ruadh, Glen Feshie. There are two *Betula pendula* (silver birch) trees downslope to the left, and the slopes are clad with *Calluna vulgaris*. In 1954, this area became part of the Cairngorms National Nature Reserve. (R. M. Adam, RMA-H8307, courtesy of the University of St. Andrews Library).

B. (15 April 1996). Fifty years on, there has been little change. There is some regeneration of *P. sylvestris*, *Calluna vulgaris*, and *Cytisus scoparius* (broom), providing stability to the uphill slope. *P. sylvestris* is held in check by grazing or browsing by wild populations of red deer (*Cervus elaphus*) and roe deer (*Capreolus capreolus*). (P. R. Moore/SNH).

C. (15 April 2006). A decade later, *P. sylvestris* has increased substantially, largely due to greater control of red deer. The track has narrowed, reflecting a change in the type of vehicles being used off road, as well as a perceived increase in the number of hikers using the track. (P. R. Moore).

Calluna vulgaris, is undercut. The soil and root mat overhangs the track edge. In 1954, within a decade of Adam's original picture, the area was declared as part of the Cairngorms National Nature Reserve and has, since that time, been managed as a native pinewood habitat.

Fifty years on, in 1996 (fig. 19.5B), the scene is readily identifiable and shows remarkably little change. Most trees remain unaltered, although one of the two *B. pendula* trees has been lost. The dis-

tinctly twisted pine tree on the slope above the track—clearly a veteran tree when originally photographed—has neither grown nor regressed noticeably in the five decades between the photographs. Some regeneration is occurring with *P. sylvestris*, *Calluna vulgaris*, and *Cytisus scoparius* colonizing and providing mechanical support to the loose gravels. Through this, the uphill slope is acquiring a degree of stability.

The double-wheel track indicates continued regular use at this time, infrequently by vehicles but increasingly as a footpath (D. Duncan, pers. comm.). The wheel tracks on the undrained surface are susceptible to surface wash and water runoff, and a combination of these factors is not permitting recolonization. The *P. sylvestris* is being held in check by grazing or browsing pressure with saplings mostly multiheaded and only just growing above the height of surrounding vegetation. There are no records of domestic stock grazing this area and *Lepus timidus* (mountain hare) occurs infrequently. Browsing can therefore be attributed to wild populations of red deer and roe deer (*Capreolus capreolus*).

A decade later—April 2006 (fig. 19.5C)—the regenerating *P. sylvestris* have established and have begun to alter the scene radically. This is due to a number of factors, but principally through the cooperative control of red deer on this estate and adjacent land units. There has undoubtedly been a shift in climatic conditions during this period also, with less severe winters and wetter, warmer summers, which have permitted fast and strong growth from the young trees. The track has narrowed and a wider, single-track footpath has formed. It reflects not only a change in the type of vehicles being used for deer recovery—from four-wheel-drive vehicles such as Land Rovers, to either quad bike or all terrain vehicles with a smaller wheelbase, softer tires, and lower ground pressure—but also a perceived increase in the number of hill walkers using the track as an access route to or egress route from popular hilltops. The recovery of previously established roads and tracks and the increased woody vegetation in landscapes are principal subjects for repeat photography (McClaran et al., chapter 12), and these records are examples of what can be determined from general patterns and local cases.

These repeated images also demonstrate the limitations of remote photographic analysis and the additional information available to the rephotographer. Fieldwork to ground-truth these images across a wider area at this site in 2008 indicates that the regeneration evident within the repeated images is, in fact, very localized and associated in particular with the disturbed ground at the sides of the engineered tracks. There is some logic to the claim that the tracks themselves and their regular use by visitors help to reduce grazing pressure through displacement; however, the thick overlying surface layer of mosses and organic litter in the adjacent clearfell areas is also impeding regeneration. There is visual evidence of this lack of recovery at the center right of the images.

Plate 19.2A was taken in the early 1960s and used as an illustration within a tourist guide to the Cairngorms, published by Jarrold in a Cotman Colour Book entitled *Around the Cairngorms* (Brooks 1964). The picture was taken from the car park in Coire Cas at the end of the ski road, probably in the month of July. It shows signs of recent excavation and construction that represent the early stages of development of the Cairn Gorm ski area, which continued throughout the 1960s. The support pylons for the chairlift on the White Lady ski run are visible at the center left of the picture. A vehicle track is discernible at least as far as the mid station, and the head wall of Coire Cas holds a wreath of snow, an annually occurring feature, and one of the few snowfields in the Cairngorms to be named (Gordon 1925). It is called Cuidhe Crom, or "crooked wreath."

The repeat photograph (plate 19.2B), taken in August 2007, shows a prominent new building, built to house the workings of the funicular railway that was opened in 2001. The funicular train can be seen approaching on its raised rails. Coire Cas is entirely clear of snow at this time, and it is perhaps easy to jump to a conclusion aligning this as evidence of climate change. However, Forsyth notes, "it is counted a late season, if the Cuidh(e) Crom does not break in May and if the whole wreath has not disappeared by the middle or end of June" (Forsyth 1900) and indeed loss of the snow at this time seems to be the trend in the last two decades. Photographic evidence also exists that illustrates the wreath lasting into August in a number of years with long snow cover. It is increasingly unpredictable, compounded by large swings of temperature, resulting in rapid snowmelt.

The Cairn Gorm ski area has been the site of what may be described as quite catastrophic development during its evolution—the signs are apparent within the original image. Some restorative work in the form of reseeding, fertilizing, and drainage, together with restricted use of vehicles on site, was noted in 1974 and the observation made that the "artificial landscape" created by this work was "better than a scarred hillside . . . liable to wash away" (Nethersole-Thompson and Watson 1974). The recurring sensitivities to the impacts of the ski area and its associated infrastructure have boiled near the surface through the decades. A condition of planning permission for the funicular railway, the latest development, came with rigorous environmental specifications for the earthworks and construction, together with consideration for sympathetic coloring of buildings and infrastructure, all of which have helped to minimize the visual impacts of this significant imposition on the landscape and have enabled some work toward environmental restoration.

History and the Use of Rephotography

The human history of the Scottish Highlands in the eighteenth and nineteenth centuries has been affected by major landscape change: the felling of the old-growth pines in the Caledonian forest, largely for ship building; the clearing of the land of people; and the subsequent introduction of sheep (Rohde, chapter 18). The effects of each of these factors may be discerned within the Cairngorms, although they predate the technology of photography. From the 1850s, photography documents the development of the Highlands, from the building of transport infrastructure through the expansion of towns and the growth of tourism. A significant loss of manpower in the World Wars of the twentieth century marked a break in intensive management, which was never replaced. In the latter half of the twentieth century, a changing emphasis toward commercial forestation, a growing awareness of nature conservation, and the

ebb and flow of government-provided financial incentives have driven land management and land use on varying courses.

With the decline in sheep husbandry and programs for culling red deer in the last decade, rephotography shows the recovery of habitats at a local scale and supports the wider hypothesis of the gradual reforestation of the land, which has accelerated in recent decades.

The development of the village of Aviemore (fig. 19.3) and the Cairngorms over the last half century plots the rise in tourism and outdoor recreation. Until the 1950s, Aviemore was little more than a stop on a railway line, but since 1960, it has developed as a winter resort through excellent transport links and ready access to the hills. The development coincided with and undoubtedly benefited from a period of quite extreme weather conditions during the 1960s and 1970s, which often provided skiing (of variable quality) for six months of the year.

In 1960, a road was extended from Glenmore up to the foot of the ski area. The construction involved the movement of thousands of tons of rock. Although certainly not unique as an engineering feat, the scale and speed of the Ski Road construction demonstrated the increasing ability of humans to change the landscape and highlighted the nonexistent, or at best rudimentary, "after care" of the modified land. As a consequence, the recovery of the landscape has been slow. In the 1960s, bus tours and private car drivers made what was regarded at the time as an intrepid trip up to the Cairn Gorm car park to "the most extensive snow fields in Britain . . . only free of snow for a few weeks of the year" (Brooks 1964). In recent years, snow cover, and therefore skiing, has been unpredictable (Watson et al. 2003, 2004), and temperature swings and rapid snowmelt have presented their own challenges to the ski industry and to the infrastructure, washing out hill roads and scouring footpaths.

The more recent gradual transformation of the Cairngorms to a year-round tourist destination may increase pressure from walkers, climbers, and mountain bikers, posing additional challenges to the landscape and wildlife and requiring more innovative management techniques in the future.

Discussion and Conclusions

In the Cairngorms of Scotland, photographs are most abundant from locations that are, or have been, popular tourist destinations. The images document the then contemporary landscape, land use, fashion, and lifestyle and provide opportunities for rephotographers to identify, explore, and interpret the changes taking place in a rapidly changing landscape as tourism and forest habitat regeneration replace pastoral or other uses of the land.

The "then" and "now" images produced by rephotography may be interrogated at a number of levels: from the "spot-the-difference" interrogation by infants, to the close and exacting scrutiny of students, to a personal exploration of the place depicted. They also provide stimulus for discussion at the local level, stimulating anecdotes and drawing sometimes obscure knowledge from older members of the community responding to the photographic images (Moore, in preparation) and the local experience can be transferred to and compared at an international level.

Rephotography provides some of the tools to fill the gap between the past and the present of this unique region—the first step toward unraveling the complexities of our history—and allows reflection upon our own links, management, and use of the land, which in turn allows us to begin to make some sense of our lives and our place in the Cairngorms today.

Acknowledgments

Stuart Rae offered valuable comments on the draft of this chapter, and Adam Watson corrected and advised on several aspects. In 2005, I visited many rephotographers around the world on a Winston Churchill Memorial Trust traveling fellowship. The experiences and discussions from those meetings have influenced how I have approached rephotography. I thank the Trust for that opportunity and the various rephotographers—several of whom have contributed to this chapter—for their time and for sharing their thoughts. I thank Paul Bierman, Diane Boyer, Eric Higgs, Mark Klett, Darrell Lewis, Doug Spowart, Ray Turner, Rob Watt, Bob Webb, and Byron Wolfe.

Literature Cited

Brooks, J. 1964. *Around the Cairngorms*. The Cotman Colour Book Series. Norwich: Jarrold.

Bruce-Watt, J. 1958. The photographer of the high tops. *Scotland's Magazine* February:28–33.

Forsyth, The Rev. Dr. W. 1900. *In the shadow of Cairngorm*. Inverness: The Northern Counties Publishing Company Ltd.

Fraser Darling, F. 1947. *Natural history in the Highlands and Islands*. New Naturalist Vol. 6. London: Collins.

Glen, A. 2002. *The Cairngorm gateway*. Dalkeith: Scottish Cultural Press.

Gordon, S. 1912. *The charm of the hills*. London: Cassell & Co.

Gordon, S. 1921. *Wanderings of a naturalist*. London: Cassell & Co.

Gordon, S. 1925. *The Cairngorm hills of Scotland*. London: Cassell & Co.

Lambert, R. A. 1996. Strathspey and Reel: Photography and the Cairngorms. *Inferno: St. Andrews Journal of Art History* 3:68–81.

Moore, P. R. (in prep.). *Photography, rephotography and change, in the contexts of space, time and place*. PhD thesis, University of Aberdeen.

Nethersole-Thompson, D., and A. Watson. 1974. *The Cairngorms: Their natural history and scenery*. London: Collins.

Newtonmore Community Web site. 1998–2004. www.newtonmore.com (accessed 10 April 2008).

Ordnance Survey Maps—25 inch 1st edition, Scotland, 1855–1882.

Pearsall, W. H. 1950. *Mountains and moorlands*. New Naturalist Vol. 11. London: Collins.

Poucher, W. A. 1947. *A camera in the Cairngorms*. London: Chapman and Hall.

Smart, R. 1996. Robert Moyes Adam. Unpublished flyer to accompany The Landscape Photographs of R. M. Adam exhibition at St. Andrews University, St. Andrews.

Stubbs, P. 2001–2008. Edinphoto Web site. www.edin photo.org.uk (accessed 3 June 2010).

Watson, A., J. Pottie, and D. Duncan. 2003. Five UK snow patches last until winter 2002/03. *Weather* 58:226–229.

Watson, A., J. Pottie, and D. Duncan. 2004. No snow patches survive through the summer of 2003. *Weather* 59:125–126.

Chapter 20

Learning Landscape Change in Honduras: Repeat Photography and Discovery

J. O. Joby Bass

The complexities of humans interacting with biophysical landscapes result in cultural landscapes that represent both environmental modification and the contexts for human life. As records or reflections of human activity, cultural landscapes contain an abundance of information, both ecological and social. Being also communicative and somewhat instructive—we make them and they return the favor—landscapes also often can provide insights into an array of cultural or social phenomena.

Using repeat photography, I conducted a study of landscape change in Honduras. From the photograph collection of geographer Robert C. West, I selected for replication a group of approximately 100 photographs taken in 1957 throughout the country. The methodology revealed some of the variety and complexity of conditions and change in the Central American country, particularly in land cover. The experiences of conducting the methodology also offered chances to discover what people think and say about change, including land cover. This opportunity for discovery proves to be an invigorating and insightful experience, deriving from implementation of the method and then providing contextual information on the processes behind the patterns. This aspect deserves consideration as an application of repeat photography. After all, as people interact with and impact the biophysical world, what we do has a lot to do with how we perceive the world (Lowenthal 1967).

Looking at the Landscape

A long-standing direction of research in many disciplines, geography among them, has tried to understand how humans make and impact landscapes, how these landscapes both reflect and teach, and how humans perceive and utilize them (Tuan 1974, Lewis 1976, Cronon 1983, Norton 1989, Rowntree 1996). As few places on Earth are not impacted by human activity, most of the world can broadly be seen as a collection of cultural landscapes (see Sauer 1956 and Turner et al. 1990). These range from the obvious, such as strip malls, to the not so obvious, such as a tropical rain forest where species composition has been altered and managed over time by humans engaged in the complex, environmentally manipulative activities of hunting and gathering. Cultural landscapes are records of activity but are also, of course, seen and interpreted differently (Cosgrove 1984, Duncan 1990). They reflect the efforts of humans to survive, to manipulate resources (including other humans), and to create significance and communicate meaning. As such, they hold a bounty of information about who people are and what we are up to, though this may not be inherently obvious (Mitchell 1996). As specific landscapes are made and then understood from different perspectives, with sometimes different intentions, they are indeed communicative devices (Richardson 1994). These, then, stand to affect perception and consequently impact cultural and biophysical conditions.

Using repeat photography to study cultural landscape change in Honduras began as an exercise in assessing the method for insights it might provide into patterns of change and persistence. What would repeat photograph comparisons have to say about conditions in the places Robert C. West visited nearly a half century before me? What manifestations of the increasing extent of modernization and global–local links would be evident in landscapes and what might be derived about the character and velocity of change events and processes (see Foote 1985)? What information would the method offer about how humans have been interacting with their biophysical surroundings in recent decades (Humphrey 1987, Works and Hadley 2000, Lewis 2002)? Would the method be able to accommodate the temporal and spatial complexity inherent in environmental change (Southworth et al. 2004)? Would it provide the appropriate sort of sample that would not "emphasize change by looking for it" (Vale and Vale 1983: 184)? Most of the photographs I chose for baseline data are of landscapes that are clearly cultural while others are implicitly so.

Repeat Photography in Honduras

Until far beyond his 1980 retirement, Robert West worked at understanding the cultural, physical, and historical geography of Latin America (see Anderson 1998). The photographs, taken throughout his career during numerous field trips, were his efforts to capture evidence and examples of types of landscapes and landscape elements characteristic of places. One of his broad scholarly objectives involved under-

standing how humans, acting as members of larger cultural groups with specific, complex perspectives and endeavors, interact with the biophysical world to create particular types of landscapes and places. In seeking to catalog what he saw, West collected and preserved a historical record of places, landscapes, and regions throughout Latin America. Before he stopped working, West spent countless hours cataloging and labeling this collection of over 6,000 black-and-white photographs (Crossley 2007). Cut lines provide information about when, where, and why he took each photograph. Often, his descriptions turn out to be humbling, insightful lessons on landscape assessment, particularly while reading them in situ; rarely, some turn out to be inaccurate. He donated a copy of the collection to Louisiana State University Cartographic Information Center, housed in the Department of Geography and Anthropology, where he worked, and another to the University of Texas at Austin Benson Latin American Collection.

I chose approximately 100 photographs from the West Collection (for details on the selection process, see Bass 2004). All were taken during a 1957 summer field season during which West visited various places in Honduras, mostly in the southern and western parts of the country. Reflecting two of West's interests, many of the photographs are of places and landscapes inhabited by indigenous populations and of colonial mining centers and landscapes. I spent much of 2001 in Honduras rephotographing West's photographic views, learning much about repeat photography, Honduras, and the people who live there (fig. 20.1). Through the process of conducting repeat photography, I learned how to better see the landscape. I also learned a great deal about some of the many interesting changes taking place and, importantly, how people think about them.

Discovering the Countryside

Due to problems of scale and compositional bias, one photograph pair may simply be too specific to be of much use in general terms. Foote (1985) even

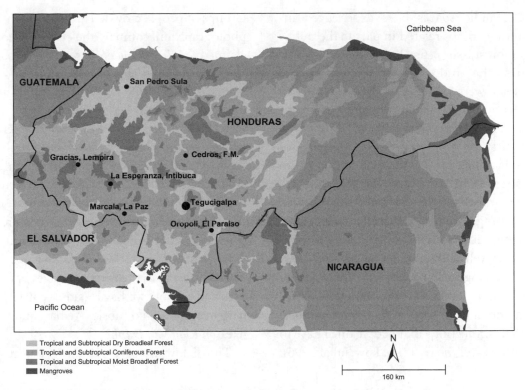

Figure 20.1. Map of Honduras showing terrestrial biomes and places discussed in the text.

pointed out how the binary "then and now comparisons" of most repeat photography miss some of the complexities of change, particularly the temporal. However, one repeat photograph pair alone may contain plenty of significance. Plate 20.1 was taken in the colonial town of Gracias, administrative center for the department of Lempira in western Honduras, and shows the market area. The 1957 photograph shows Lenca women from the nearby town of La Campa selling pottery made there, most of it carried in on people's backs. In the 2001 photo, women from La Campa still sell their pottery, though they came to town in pickup trucks and there are fewer of them. The men in the newer photograph appear more *ladino* in their dress than in 1957 and mountain bikes, a pickup truck, an electronics store, and an illuminated Pepsi sign indicate the modernization that has taken place. The change surrounding Castillo (Castle) San Cristóbal in the background is also clear. The *castillo* is now a historic attraction and a public park. In 1957, the hillsides surrounding the *castillo* were bare of visible woody vegetation while

in 2001 the same hillsides appear covered in trees, most of them planted by people.

Comparing across whole sets of photograph pairs helps minimize some of the method's problems of scale and bias. Comparing these two sets, I observed different patterns of change. Much of this involved vegetation, showing a general increase in trees (74 of 75 photograph pairs) (see also Hecht 2004 for a concise overview of patterns of vegetation increases in the region). This takes different forms in different settings for different reasons, both within settlements and across the countryside (see also Bierman, chapter 9; Lewis, chapter 15).

Areas of apparent reforestation sometimes have actually experienced an agricultural shift from corn or cattle to shade coffee (Bass 2006a). Settlement forests have grown as settled areas have grown and as landscape choices and values seem to have changed (fig. 20.2). Rural areas, often appearing as background or peripheral in many photographs, consistently show an increase in trees through apparent decreases in land cleared for agriculture (plate 20.2). Public spaces have experienced increases in trees, perhaps related in part to the public discourse on forest issues (fig. 20.3, Bass 2006b).

Sometimes the more significant observable changes at a site were in the realm of economics or commercialization. This would probably be true for much of Latin America or anywhere else. Plate 20.3 shows that, first, the coffee production center of Marcala, La Paz, has grown to include this once-peripheral street in its commercial district. Second, it shows the commercial character of that inclusion: stores, advertising, brands indicating global–local linkages, changes in resource access, changes in material culture, as well as changes in vegetation.

Tensions of change often evoke nostalgia as a sort of self-reflection. Retaking the photograph in figure 20.3 offered the chance to experience the cultural and emotional connections that people often have to both their homes and the past (see Lowenthal 1996).

Groups of people crowded around me to see the old photograph and compare the current scene. One exclaimed, "*¡Mire! El puente casa. No hay todavía. ¡Que cheque!*" ("Look! The bridge house. It's not there anymore. How cool!"). Everyone smiled at the older photo.

In the smaller nearby Lenca settlement of Guajiquiro, an old photograph of the church and grass plaza evoked similar expressions: "Ahhh! How precious, the church!" They said they preferred the way it used to look. They also asserted that, in general, things are better now. Sure, the kids are sort of a problem: "They are lazy. They don't want to work. They want to buy clothes all of the time. They won't work." Otherwise, the recent changes have brought a better life: "Now we have roads, vehicles, lights. We have televisions and music. We have movies. Do you like *Yon Klah Van Danh*?"

The fieldwork of conducting repeat photogra-

A

B

Figure 20.2. Santa Ana Cacauterique, La Paz, Honduras.
A. (July 1957). This view shows the Lenca village of Santa Ana Cacauterique, with its church (right center) and official buildings with tiled roofs and plastered, whitewashed adobe walls. Most of the dwellings are of wattle and daub with thatched roofs. (R. C. West, H1, 16-8).
B. (July 2001). The town has grown, and the residents have planted many trees, mostly *Pinus, Juniperus,* and *Eucalyptus.* Like many other towns in the region, the surrounding rural landscape shows more trees, as corn and bean fields (*milpas*) are being converted to coffee farms (*fincas*), often with shade trees of *Inga* spp. (*guama*), or have reverted to pine forest with a decrease in agricultural activity with economic diversification. The match is not exact due to the density of the foreground trees. (J. O. Bass, HOND. PP48b).

A

B

Figure 20.3. Cedros, Francisco Morazán, Honduras.

A. (June 1957). The sixteenth-century mining town of Cedros, Francisco Morazán, in which winding streets follow the contours of the hillside. Most of the mines have been abandoned for about 200 years. The town sits just upslope from the wide Siria Valley, a hillside settlement pattern characteristic of mining settlements with no indigenous predecessor. (Davidson 1985). (R. C. West, H1, 5-11).

B. (September 2001). The photograph site, from a hill overlooking town, is now in a public park that has been landscaped with many trees (a mix of *Pinus, Quercus, Mimosaceae, Laurus* with *Cocos nucifera* and *Araucaria heterophylla*), some of which obscure the view. Many of the trees are decorated with signs extolling the virtues of trees and forest and the pertinence in planting and caring for them—a landscape of conservation ideology. (J. O. Bass, HOND. PP26b).

phy offers plentiful opportunity for reflection and discovery. It offered insights into what people say about change, into values and environmental perception, and sometimes into the disjuncture between these perceptions and conditions. These encounters offered a wealth of contextual insights that help inform the photographic observations, providing "thicker descriptions" (Geertz 1973) of conditions than would the photograph pairs alone.

Field Encounters Inform Photographs

A chance encounter with a Lenca man (see West 1998 for an overview of the Lenca culture region) and others while I was en route to a camera station (see fig. 20.4) illustrates another cultural aspect of repeat photography. I heard his metal hoe chinking against rocks before I saw him. I topped a rise in the giant, dibble-planted cornfield, and saw the Lenca man hoeing up feral mustard greens. He stood up and leaned on his hoe, smiling, as I approached. I stopped near him and we exchanged greetings, briefly touching our palms together in a gesture of shaking hands. We briefly discussed his corn and how lucky the people in the area were to have been spared most of a recent drought. Then I asked him about the small oak forest patch behind him. He nodded toward it and confirmed for me that it was indeed bigger than it used to be. "It's a small forest. The trees, the forest, are good. Good for the air, for the animals, for the rain, the temperature." He told me that, owned by a local man, the forest was used partially to supply firewood and construction materials for locals. I asked him if he thought it would be all right for me to go check it

Figure 20.4. Llano Yarula, La Paz, Honduras.
A. (July 1957). This *chagüite* (swale), called Llano Yarula, near Santa Elena supported livestock grazing on aquatic grasses and sedges. *Chagüites* are found in the western Honduran highlands between 1,200 and 1,800 meters. Crops were cultivated on the surrounding slopes, with only one tiny patch of trees in the upper left and a few shrubs in upper center. (R. C. West, H1, 16-5).
B. (July 2001). The slopes still support crop raising. The tree coverage (*Quercus* and *Laurus*) at the upper left has expanded. The shrubs at upper center have become a small forest, and other smaller patches of trees also appear. The far slope vegetation has increased perhaps eight- to tenfold, probably due to an increase in the size of *Quercus* and to new plantings. A new road carries vehicle traffic. An increase in settlement is evident and is closely related to changes in tree cover, with newly settled areas planted in trees, part of the complexes that may be called dooryard gardens that produce food, flowers, and shade. (Kimber 1973, Anderson 1954). (J. O. Bass, HOND. PP42b).

out. He nodded assuredly, "Of course." Then, looking back toward it, he continued, "It's beautiful, the forest. We need forest here. It's good."

I continued my walk to the tree patch. It was a small, fenced oak–laurel forest patch with only minimal underbrush and a few birds. I stood in the shade at its edge and looked over at another hill. There at the top was the new patch of trees that appears in the top center of the new photograph (fig. 20.4B). I headed for it.

At a barbed wire fence, I looked up to see a big dog hurrying across a packed earth yard, hopping a handmade palmetto broom as he came after me. From a hole in an adobe house, a woman's voice yelled at him to stop and then at me to tell her what I wanted. I crawled through the barbed wire, watching the dog, and crossed over to her window. I told

her that I had old photographs of the area and tried to show one to her. She glanced at it, but quickly. I told her how the older photograph appeared to indicate that this area had more trees than in the 1950s and I asked her about the two new patches near her house.

She nodded across the barren earth of her yard toward the oak–laurel stand and said that the man who owned the land there had let the trees grow up. He had fenced it off to keep animals out and had essentially raised a forest. She said that he sells the wood for both firewood and construction materials "because we are poor here—*pobrecitos*. We are poor so we still have to cook with firewood." In 1995, 95 percent of rural Hondurans still used firewood as their primary source of energy, making up 60 percent of the country's total energy consumption

(UNPD 2000). The Lenca highlands are among the country's poorest areas and remain very rural in character. Many people in the area still depend on firewood for cooking energy, though the number is rapidly decreasing.

I asked about another, smaller tree patch that I had passed on top of the hill. Her face lit up. "That's mine," she smiled, "I planted it." She told me that she had planted the patch of *Perymenium strigillosum* (tatascan) for fence posts. Standing in even rows, the trees were young and not very big. She said that they need lots of fence posts because they have so much barbed wire fencing. Though fences often follow old "living fences," barbed wire is the rule now, and this requires fence posts, again, "because we are poor—*pobrecitos*. We have to make everything. This house, our meals, firewood, everything. We are poor here." Though self-deprecating, she smiled a lot. She left the dusty windowsill and came outside to hold the dog so I could leave.

Throughout my fieldwork, people were quick to offer me their insights on change and environmental perception. These ethnographic encounters, I believe, offer rich contextual information that better informs an analysis of change. The fact that I was discovering unexpected increases in trees through repeat photography made me aware of how often people brought up the issue. One afternoon in the town of Marcala, La Paz (see plate 20.3), I leaned on a steel gate talking to the lady who lived on the other side of it. She grew up in Marcala but had spent some of her adolescent years in Chicago, Illinois, where she went to high school. I asked her about the town's growth. She said that, yes, Marcala had grown a lot, especially since the 1980s. When she was a kid, she said, it was so different. "It was much smaller. It was cooler, too. Yes, it is much hotter now. This time of year ten or twenty years ago, you needed a sweater to be outside like this. Now it's so hot." She held up her bare arms to show. "Because everyone is cutting the trees," she replied, as if it were obvious. "We thought of putting a coffee *finca* (farm or plantation) back there on the hill. Everyone says, 'First

thing is you cut all of the trees. Then you plant coffee.' Hmmph. So we didn't do anything. We left the trees there. No coffee." She said that the situation is becoming dire and that they need trees. She pointed at the sky. "When I was a kid here, at this time of year you could go days without seeing the sun. It was all clouds, days at a time. Now? No. Look." She pointed to a nursery down the hill behind her house. The nursery sat on her property but was a project of a Catholic nongovernmental organization (NGO). She supports the project by letting them use the land. In exchange, she asks for some of the trees, mostly pines, casuarina, mahogany, acacia, and a few others. She believes strongly in the project, which focuses on planting trees near water sources.

"The local people don't understand. They think we are crazy. They say, 'Why would you not have a *finca* because you want some trees?' It's hard to make them understand. But at least there are people like the organization who understand, who are trying. The majority of people here, they are living for today. Well, at least we have the fortune to live differently, and to think about it. We don't have to cut our trees for firewood and stuff. We have other options. But they have to warm their food somehow. And grow their food somewhere. It is not an easy situation."

From the 1950s to the 1990s, Marcala did see a slight drop in precipitation (Zuniga Andrade 2000), and data for the country do indicate a slight warming trend for the same period of perhaps ±0.5 degrees Celsius. However, as repeat photograph pairs (both oblique and aerial) indicate, there was not an obvious corresponding drastic decrease in area tree cover; in fact, the opposite is true (Bass 2006b).

A couple of hours north of Tegucigalpa sits the colonial mining town of Cedros (fig. 20.3) (see Davidson 1985 on indigenous and colonial settlement patterns in Honduras; see West 1958 on colonial mining activity). Sitting and drinking a Coke in the open door of the building where Honduras's first congress was held in 1824, I watched a mule loaded with firewood sticks follow a small boy over uneven cobblestones. The whitewashed wall across the narrow street blindingly reflected the late morning sun. A bus horn blared from the other end of town as a pickup truck rattled slowly toward me; across its windshield, a sticker read CHICO BOSS OF THE USA. Eric Clapton's song "Cocaine" came through the open door of a house down the sidewalk.

The *señora* of the house and the store in its front room talked about her town. She was quick to say how *sano* (healthy) it is and asserted that it has barely changed. "It's the same for years. Almost nothing changes." The one thing, she offered up, that has changed is this: "It is not as cool as it was years ago. It's hotter now," she affirmed. "Look around. It's because of all of the deforestation. We have cut all of our forest and so it's hotter now."

Sometimes I had the privilege of getting a close-up look at the tree issue in a different way. In most cases, I found that areas of settlement have seen an increase in trees—settlement forest—over the past few decades. Walking through the town of Oropoli, El Paraiso, having just rephotographed a scene overlooking the town and showing an increase in both settled area and trees, I saw a man standing against the front wall of his cement block house. He greeted me as he repaired a mahogany slingshot that he uses to kill rabbits and raccoons out at his *milpa* (swidden field). He grinningly said that, of course, as God's creatures, they have to eat too but that he just can't let them have it all. I showed him the early photograph and asked him about the town and its trees. His face lit up and he pointed to the side of his house to his fenced yard. "Want to see my trees?" he asked.

We stepped through the house, my appearance surprising his wife, and into the backyard, where I got a tour of the yard. The dark, fenced area was filled with trees, an arborescent kitchen garden. As he had only moved into town from his rural land 12 years before, many of the trees were still small. We waded through the brushy shade as he grabbed a limb on each tree and identified it. All were fruit trees. The soil beneath us was dark and damp, an anomaly in the area. This small kitchen forest was made up of *Citrus* spp. (orange and lime), two *Mangifera* spp. (mango), *Carica papaya* (papaya), *Musa* sp. (banana), *Cocos nucifera* (coconut), *Tamarindus* sp. (tamarind), and *Crescentia cujete* (calabash), in addition to several ornamental bushes and plants. Many other houses had similar plantings.

Other field encounters saw conversations about trees and change turn to other issues, different but compelling. One day near Lepaterique, a man, Francisco Morazan, came pushing a long, wooden wheelbarrow up a steep embankment toward the road where I was standing to rephotograph a scene overlooking the town and surroundings. He stopped to see what I was up to. I showed him the photograph and told him that I was looking out to see what had changed in his town. I said that it seemed to have changed a bit. "Yes. Lots of change."

"The place has grown."

"Oh yes. Much bigger."

I said it looked like there might be more trees than in the 1950s. "Yes. We have lots of projects here now. Lots of organizations, lots of forest projects, like COHDEFOR [Corporación Hondureña de Desarollo Forestál]. Lots of other projects too. Especially since [Hurricane] Mitch. There have been lots of organizations and lots of donations. It's better now here. The town and life are better, *gracias al Señor*. Things have improved a lot. We have offices now. And trees."

The conversation shifted a bit. It was late September 2001. "Did you know the twin towers?" he asked. We agreed that the attacks were a shame, *que bárbaro* (how barbaric). Then he held forth on the future in a way that told me not so much about the future but about another change happening in Latin America. "There are big problems in the world. What happened with the towers was a prophecy. It was in the Bible, *La Palabra* (The Word). There are more things to come, too. It is in *La Palabra*. I know. I know these things. It is because I am a Christian, an Evangelical. I know. The Word says so." As we know that the relationships that people have with the biophysical world can be linked to larger cultural philosophies, including the theological or cosmological (Glacken 1956, 1967; Doughty 1981), to what other changes might this change contribute?

Discussion: Change

The Honduras through which Robert West traveled in 1957 was different than it is today. The country was far less developed, particularly in rural areas. Travel in many places was limited to the feet of people or mules; resource consumption was far more restricted; communications, for many places and most people, depended on the actual human transport of information; and education in the contemporary sense was very limited. This is not to say that economic diversification was not also occurring. It was, slowly bringing a growing middle class and organized labor. The hacienda was in decline and the elite position in society was being appropriated by bankers, many of them the offspring of Middle Easterners who had begun to migrate to Honduras in the 1890s (Argueta 2001, pers. comm.).

Frank Tannenbaum (1966) offered salient and general descriptions of where and how most people in Latin America lived during the time West took his photographs. The typical community was a plaza town of 100 to 200 families with a church, municipal building, jail, and a few stores. Houses were closed in on themselves and contacts between the town and the outside world, including the capital city, were limited; the town bought and sold little. Most goods were locally produced, including construction materials, clothing, and food. Most people, he said, had little connection to the modern Western world. Illiteracy was the norm, books and newspapers rare. Tools were basic and many people still carried burdens on their own backs. In essence, he described a near total lack of communication between "country folk and city dwellers."

A few years before West's visit, only 1.5 percent of Hondurans read a newspaper (Tannenbaum 1966). The population of Honduras was only approximately one-quarter urban, and most everyone practiced agriculture for a living; this is far from the Honduras, and the Latin America, that exists today in most places (Argueta 1999). Unfortunately, though, this is still the image held by many in the developed world (Aparicio and Chavez-Silverman 1997).

Honduras in 2001 was more urban, more connected, more modern, and more economically diverse than in 1957. Buses, retired from US school districts, were blasting down highways and crawling through mountains all over the country, carrying people, goods, and information to the farthest reaches of the country. Radios, televisions, and Internet connections were carrying information into homes and heads daily. In contrast to the earlier 1.5 percent, a newspaper source reported that over 60 percent of the population had access to a daily newspaper. The modernization that characterizes conditions includes increased communication and contact between the countryside and urban areas, macroeconomic, social integration, commercialization, and the consolidation of political sectors that are connected to agriculture, which is connected to a lot of the population (UNPD 1998: 110–111; UNPD 2000). In some ways, recovery from Hurricane Mitch's destruction accelerated much of this change (UNPD 1999).

Tannenbaum noted some of this coming, as did Wagner (1961: 249) in Costa Rica. Tannenbaum noted what he called the "tugging at [Latin Americans'] way of life, language, amusements, and their social and personal standards" by North America (Tannenbaum 1966: 174). As he put it, "[t]he greatest change of our time [in Latin America] is the consumers' revolution propelled by the United States" (ibid.: 202). He even noted an increase in American advertisements on television as a significant part of this cultural change (ibid.: 174). Wagner (1961: 249) called it a "commercial revolution." The development of municipal water and electric services has affected settlement patterns that also appear coincidental with changes in vegetation. Many smaller towns in rural areas show little change in population and in rural/urban ratios between the 1961 census and the 1988 census. However, since 1988 and significant infrastructure expansion, many of them have grown significantly, as have their urban settlement forests.

Larger changes in economic patterns, urbanization, and land use associated with them can often be linked to environmental conditions (Stonich and DeWalt 1996; Rudel, Bates, and Machinguiashi 2002; Aide and Grau 2004). Observations of change, such as through repeat photography, may offer more insight by considering potential larger cultural processes within which changes take place. For example, institutions (Klooster 2003), accessibility (Southworth and Tucker 2001), laws (Sunderlin and Rodriguez 1996), and social contexts (Klooster 2000) can all be clearly linked to environmental conditions. As ethnographic encounters can provide insights into how people perceive their landscapes—perception being an important consideration in understanding how people use and impact landscapes—understanding the contexts of change may help explain why it happens. In this way, understanding landscape changes or conditions may be improved by considering the larger contexts for them and how people see this.

Since the time of West's and Tannenbaum's visits, changes—economic diversification and growth, increased health care access, explosions in transportation and communication infrastructures and opportunities, and educational expansion, among others—mean that life for many is very different now. Some no longer farm, some no longer cut firewood to cook, many still do these but do them less. Perhaps this leads to reforestation: trees that once might have been seen as unhealthy when grown near houses are, perhaps with changes in improved health care and insect control, more common in towns, leading to more domestic settlement forest. Rural-to-urban migration, often encouraged by municipal development of utility services, brings farming families to towns, who may plant kitchen orchards that contribute to settlement forests. With the development of national and international environmental concerns came the landscaping of many public spaces with trees. These trees are very often adorned with signs announcing virtues of forests and the environment, announcing their virtue as symbolic forests, landscape manifestations of contemporary public sentiment regarding environmental conditions in the area.

Discussion and Conclusions

The complexities of landscape interpretation are perhaps no better explained than by geographer Karl Raitz:

> [O]ne must attempt to interpret an often overwhelming complexity of objects and relationships, layer upon fragmentary layer, derived from nature and culture, in and out of temporal chronology; a seemingly endless flux of clues within a montage where each is potentially causally related to everything or nothing. . . . The field interpreter's difficulties are further magnified by uncertainties inherent in attempting to understand relationships, cause and effect, from the testimony of local informants. . . . Understanding and explanation are at risk if the field researcher naively attempts to reconstruct a link between the construction of certain landscape relationships and an informant's explanation based on oral tradition and the vagaries of perception and memory. (Raitz 2001: 121–122)

Explaining the general patterns of and reasons for the increases in trees I found is difficult for a variety of reasons. The sites and contexts of the vegetation increases vary, as do the reasons for them. Some is explicitly cultural; most, if not all, is related to human activity (see Koukal and Schneider 2001). More complex is rectifying the disjuncture between these increases and the consistent assertions regarding its perceived decrease.

Nearly all the landscapes I observed showed increases in vegetation from 1957 to 2001 (though see Sandoval Corea 2000 for an overview of local forest conditions in the country). Some of the changes I observed subsequently led to efforts to better understand the complexities of these changes, from understanding land cover change to understanding the part that symbolic, communicative landscapes may play in perception and cultural discourse. Some of the increases in trees that I observed can be linked to the coffee economy (global, regional, and local), local adaptive strategies, and global conservation discourse. Others have shown how some of the land-cover increases in the region can be complexly related to socioeconomic processes (Painter and Durham 1995, Hecht 2004, Southworth et al. 2004).

Using repeat photography allowed me to see various types of physical and cultural change that reflect a variety of larger forces at work. The fieldwork offered the chance to interact with and learn from residents. Conversations that turn to change often offer insights into how people both experience and perceive change and how they perceive contemporary conditions.

This provides context for the photograph pairs; for example, as I was observing consistent increases in trees in photograph pairs, locals were consistently teaching me about decreasing forest cover and its negative impacts. Assessing changes through photograph pairs necessitates a contextualized understanding of the many processes, agents, and events that help shape landscapes, and this includes perception. Most people with whom I spoke throughout Honduras both expected and asserted that I would find deforestation when comparing 2001 with the past (1957). This disjuncture seemed to be generally pervasive, as long-time locals and recent arrivals such as Peace Corps volunteers expressed similar perceptions. Understanding this disjuncture may prove to be key in other ongoing efforts to address human use of environmental resources (see also Nyssen et al. 2009). As memory is indissolubly linked to perception and local conditions (Hoelscher and Alderman 2004), addressing memory and perception is key to understanding how people see and think about the environments in which they live. How people see and think about the world is as-

suredly tied to how they act (Dansereau 1975). As well, learning what people say about all of this may provide some insight into the complexities of change, how people see it, and the interrelationships between conditions, perception, and information flows.

Repeat photography can offer both specific empirical information as well as a holistic or synthetic perspective (see Smith 2007). Again, changes in trees and changes in road infrastructures, education, policy, desires, economic options, and perceptions may from time to time be related. Seeing changes together shows specific characteristics or phenomena as changing together. Through the "discovery process"—the knowledge gained by becoming familiar with a place, by wandering through it, by listening to people who live there, and by learning how to see the landscape in the process of locating sites, assessing conditions, and talking to people—repeat photography can also provide a synthetic, contextual perspective on specific changes, on how people see them and why, and, even at times, on other changes less visible through the method but likely as significant. This useful and insight-inspiring method will hopefully continue to develop so that interpretations developed through it can be better contextualized and assessed and help us understand some of the many complex factors at play in our ongoing constructions of Earth as the home of humans.

Acknowledgments

I thank William Davidson, William Doolittle, and the late Terry G. Jordan for their support and guidance; the late Robert C. West for inspiration; the Tinker Foundation and the Peninsula Community Foundation for funding support; Mario Argueta at La Universidad Nacional Autónoma de Honduras for research assistance and insight; and the many Hondurans who helped me locate photograph sites, who opened their homes to me, and who helped me try to understand the complex landscapes in which they live. Special thanks to John Anderson at

Louisiana State University's Cartographic Information Center for help with the West Photograph Collection. Robert West's photographs are from the Dr. Robert C. West Latin American Photograph Collection, Cartographic Information Center, Department of Geography and Anthropology, Louisiana State University, Baton Rouge.

Literature Cited

Aide, T. M., and H. R. Grau. 2004. Globalization, migration, and Latin American ecosystems. *Science* 305:1915–1916.

Alcerro-Castro, T. 1989. *Mis dos mundos*. Tegucigalpa: Graficentro Editores.

Anderson, E. 1954. Reflections on certain Honduran gardens. *Landscape* 4:21–23.

Anderson, K. 1998. Bob West, geographer. In *Latin American geography: Historical-geographical essays, 1941–1998*, ed. R. C. West, 1–18. Baton Rouge, LA: Geoscience Publications.

Aparicio, F., and S. Chavez-Silverman. 1997. *Tropicalizations: Transcultural representations of Latinidad*. Hanover, NH: University Press of New England.

Argueta, M. 1999. Del burro a Internet: Desarollo del transporte en la Honduras del siglo XX. *El Heraldo* 17 December:27.

Bass, J. O. 2004. More trees in the tropics. *Area* 36:19–32.

Bass, J. O. 2006a. Message in the plaza: Landscape, landscaping, and a forest discourse. *Geographical Review* 95:556–577.

Bass, J. O. 2006b. Forty years and more trees: Land cover change and coffee production in Honduras. *Southeastern Geographer* 46:51–65.

Cosgrove, D. 1984. *Social formation and symbolic landscape*. Totowa, NJ: Barnes and Noble Books.

Cronon, W. 1983. *Changes in the land: Indians, colonists, and the ecology of New England*. New York: Hill and Wang.

Crossley, P. 2007. The Robert West Photograph Collection. Paper presented to Conference of Latin Americanist Geographers, 1–2 June, Colorado Springs, CO.

Dansereau, P. 1975. *Inscape and landscape: The human perception of environment*. New York: Columbia University Press.

Davidson, W. V. 1985. Etnografía histórica y la arqueología de Honduras: Un avance preliminar de la investigación. *Yaxkin* 8:215–226.

Doughty, R. 1981. Environmental theology: Trends and prospects in Christian thought. *Progress in Human Geography* 5:234–248.

Duncan, J. S. 1990. *The city as text: The politics of landscape interpretation in the Kandyan kingdom*. New York: Cambridge University Press.

Foote, K. 1985. Velocities of change of a built environment, 1880–1980: Evidence from the photoarchives of Austin, Texas. *Urban Geography* 6:220–245.

Geertz, C. 1973. *The interpretation of cultures*. New York: Basic Books.

Glacken, C. J. 1956. Changing ideas of the habitable world. In *Man's role in changing the face of the Earth*, ed. W. L. Thomas, 70–89. Chicago: University of Chicago Press.

Glacken, C. J. 1967. *Traces on the Rhodian shore: Nature and culture in Western thought from ancient times to the end of the eighteenth century*. Berkeley: University of California Press.

Hecht, S. 2004. Invisible forests: The political ecology of forest resurgence in El Salvador. In *Liberation ecologies, second edition: Environment, development, social movements*, ed. R. Peet and M. Watts, 64–103. New York: Routledge.

Hoelscher, S., and D. Alderman. 2004. Memory and place: Geographies of a critical relationship. *Social and Cultural Geography* 5:347–355.

Humphrey, R. R. 1987. *90 years and 535 miles: Vegetation changes along the Mexican border*. Albuquerque: University of New Mexico Press.

Kimber, C. 1973. Spatial patterning in the dooryard gardens of Puerto Rico. *Geographical Review* 63:6–26.

Klooster, D. 2000. Beyond deforestation: The social context of forest change in two indigenous communities in highland Mexico. *Yearbook, Conference of Latin Americanist Geographers* 26:47–59.

Klooster, D. 2003. Forest transitions in Mexico: Institutions and forests in a globalized countryside. *Professional Geographer* 55:227–237.

Koukal, T., and W. Schneider. 2001. *Tree resources outside the forest in Central America. Development of methods for assessment and monitoring of natural resources to support regional planning*. Final Report, funded by the European Commission (INCO-DC); im Eigenverlag des Instituts für Vermessung, Fernerkundung und Landinformation, S. 21.

Lewis, D. 2002. *Slower than the eye can see: Environmental change in northern Australia's cattle lands.* Darwin, Northern Territory: Cooperative Research Centre for the Sustainable Development of Tropical Savannas.

Lewis, P. 1976. Axioms of the landscape. *Journal of Architectural Education* 30:1.

Lowenthal, D. 1967. *Environmental perception and behavior.* Research Paper No. 109. Chicago: University of Chicago, Department of Geography.

Lowenthal, D. 1996. *Possessed by the past.* New York: Free Press.

Mitchell, D. 1996. *The lie of the land.* Minneapolis: University of Minnesota Press.

Nagendra, H., D. Munroe, and J. Southworth. 2004. From pattern to process: Landscape fragmentation and the analysis of land use/land cover change. *Agriculture, Ecosystems, and Environment* 101:111–115.

Norton, W. 1989. *Explorations in the understanding of landscape: A cultural geography.* New York: Greenwood Press.

Nyssen, J., Mitiku Haile, J. Naudts, R. N. Munro, J. Poesen, J. Moeyersons, A. Frankl, J. Deckers, R. Pankhurst. 2009. Desertification? Northern Ethiopia rephotographed after 140 years. *Science of the Total Environment* 407(8):2749–2755.

Painter, M., and W. H. Durham. 1995. *The social causes of environmental destruction in Latin America.* Ann Arbor: University of Michigan Press.

Raitz, K. 2001. Field observation, archives, and explanation. *Geographical Review* 91:121–131.

Richardson, M. 1994. Looking at a world that speaks. In *Re-reading cultural geography,* ed. K. E. Foote, P. J. Hugill, K. Mathewson, and J. M. Smith, 156–163. Austin: University of Texas Press.

Rowntree, L. 1996. The cultural landscape in American human geography. In *Concepts in human geography,* ed. C. Earle, K. Mathewson, and M. S. Kenzer, 127–160. Lanham, MD: Rowman and Littlefield.

Rudel, T., D. Bates, and R. Machinguiashi. 2002. A tropical forest transition? Agricultural change, out-migration, and secondary forests in the Ecuadorian Amazon. *Annals of the Association of American Geographers* 92:87–102.

Sandoval Corea, R. 2000. *Honduras: Su gente, su tierra y su bosque.* Tegucigalpa: Graficentro Editores.

Sauer, C. O. 1956. The agency of man on Earth. In *Man's role in changing the face of the Earth,* ed. W. L. Thomas Jr., 49–69. Chicago: University of Chicago Press.

Smith, T. 2007. Repeat photography as a method in visual anthropology. *Visual Anthropology* 20:179–200.

Southworth, J., and C. Tucker. 2001. The influence of accessibility, local institutions, and socioeconomic factors on forest cover change in the mountains of western Honduras. *Mountain Research and Development* 21:276–283.

Southworth, J., C. Tucker, and D. Munroe. 2004. Modeling spatiality and temporally complex land-cover change: The case of western Honduras. *Professional Geographer* 56:544–559.

Stonich, S., and B. DeWalt. 1996. The political ecology of deforestation in Honduras. In *Tropical deforestation: The human dimension,* ed. L. Sponsel, T. N. Headland, and R. C. Bailey, 187–215. New York: Columbia University Press.

Sunderlin, W., and J. Rodriguez. 1996. *Cattle, broadleaf forests and the agricultural modernization law of Honduras.* Occasional Paper No. 7. Jakarta, Indonesia: Center for International Forestry Research.

Tannenbaum, F. 1966. *Ten keys to Latin America.* New York: Random House.

Tuan, Y. 1974. *Topophilia: A study of environmental perception, attitudes, and values.* Englewood Cliffs, NJ: Prentice-Hall.

Turner, B. L., III, W. C. Clark, R. W. Kates, J. F. Richards, J. T. Mathews, and W. B. Meyer, eds. 1990. *The Earth as transformed by human action: Global and regional changes in the biosphere over the past 300 years.* New York: Cambridge University Press.

United Nations Program for Development (UNPD). 1998. Informe sobre desarollo humano, Honduras: 1998. Programa de las Naciones Unidas para el Desarollo, Tegucigalpa, Honduras.

United Nations Program for Development (UNPD). 1999. Informe sobre desarollo humano, Honduras: 1999. Programa de las Naciones Unidas para el Desarollo, Tegucigalpa, Honduras.

United Nations Program for Development (UNPD). 2000. Informe sobre desarollo humano, Honduras: 2000. Programa de las Naciones Unidas para el Desarollo, Tegucigalpa, Honduras.

Vale, T., and G. Vale. 1983. *US 40 today: Thirty years of landscape change in America.* Madison: University of Wisconsin Press.

Wagner, P. 1961. *Nicoya: A cultural geography.* University of California, Publications in Geography 12. Berkeley: University of California.

West, R. C. 1958. The mining economy of Honduras during the colonial period. Paper presented at XXXIII Congreso Internacional de Americanistas. July 20–27, San Jose, Costa Rica.

West, R. C. 1998. The Lenca Indians of Honduras: A study in ethnogeography. In *Latin American geography: Historical-geographical essays, 1941–1998*, ed. R. C. West, 67–76. Baton Rouge, LA: Geoscience Publications.

Works, M., and K. Hadley. 2000. Hace cincuenta años: Repeat photography and landscape change in the Sierra Purépecha of Michoacán, Mexico. *Yearbook of the Conference of Latin American Geographers* 26:139–156.

Zuniga Andrade, E. 2000. Es real el cambio climático en Honduras? *Revista Geográfica* 1:51–56.

Chapter 21

Using Rephotography of Artwork to Find Historic Trails and Campsites in the Southwestern United States

Tom Jonas

The mists of history sometimes obscure the details of how expeditions moved across once trackless areas, such as the southwestern United States. Knowledge of the exact route of an expedition is desirable for several reasons. First, environmental reconstruction based on historic textual or visual materials requires the ability to re-create the route, reoccupy positions where notes and sketches were taken, and document changes (Lewis, chapter 15). In other cases, the route of an expedition might be of importance for historical interpretation of the trip itself. Finally, some routes might be important because of family history of participation (Hanks et al., chapter 3). Vintage images are invaluable in determining specific locations visited by expeditions, and reoccupying the sites where these images were made provides tangible evidence of roads once traveled.

Historical reconstruction of routes can be inhibited for any number of reasons, including changes in geographic names between the time of the expedition and the present, subsequent construction or land-use changes that obliterate or alter a locality, or merely the lack of geographic precision of whomever recorded an expedition's progress. One example is the Colorado River expeditions of John Wesley Powell in 1869 and 1871–1872; attempts to replicate some of the photographic images from the second

expedition, and thereby document aspects of the first, met with obstacles ranging from freeway construction to the recent inundation of canyons by large reservoirs behind dams (Stephens and Shoemaker 1987). In the case of the Powell expeditions, extant photography was available, but most expeditions in the region before that time either lacked any visual documentation or relied on artists' renditions of the scene (e.g., Huseman 1995).

I am a trail researcher. To me, historical research, whether it is documentary analysis or a hike in the back country, is an adventure. I enjoy learning about the history of an exploration but I'm even more interested in its geography, and I want to find the actual trails as precisely as possible (fig. 21.1). Reading a 150-year-old account of the impressions of one of the first European Americans to view the western landscape is even more thrilling to me when I'm standing right in the spot that the author is describing. In this chapter, I relate some of my experiences tracking the routes of early expeditions in Arizona with an emphasis on the use of artwork created either during the expedition or afterward to establish better details on how these expeditions moved across the landscape. Additional information and details are posted on my Southwest Explorations Web site (www.southwestexplorations.com).

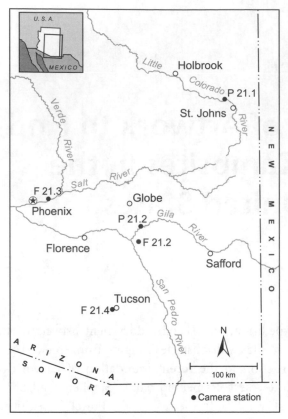

Figure 21.1. Map of locations mentioned in the text.

Background

The study of historic trails begins with documentary research. US government expeditions during the mid-nineteenth century usually wrote an official report to Congress. These reports were published by one or both branches of Congress as executive documents, usually within a year or two of the completion of the expedition. Some were printed in fairly large quantities since they were in great demand as trail guides by citizens who planned to settle or travel through the new western territories and possessions. Original or reprint copies of these reports can be found in libraries and antiquarian bookstores.

The official report contains the diary or itinerary and various other useful features, often including listings of geographic coordinates, geological and zoological studies, mileage charts, maps, and the artist's illustrations. In addition to the official report,

other diaries and journals of the same trip may exist as published books or be hidden in a historical repository somewhere, waiting to be discovered by a diligent researcher. All of these are called primary documents because they were written by the original participants at or close to the time of the event.

Some of the early exploration routes were used only once, and many were not practical for wagon or railroad travel. The mule pack trains that carried the expedition's personnel and baggage left no roads or wagon ruts behind for us to discover. Later roads were built on paths that were more appropriate for wheeled vehicles, and the tracks of the original explorers faded with the passage of time. Now we have only the archival documents to help us rediscover their paths.

Expedition photographs and the artist's illustrations from the official report can provide us with very precise location information about the landmarks and campsites they depict. During the time period that I study (1840–1860), photography was still in its infancy and very difficult to use in the wilderness. Early photography was also unreliable; Joseph C. Ives's 1857–1858 expedition along the lower Colorado River attempted to take daguerreotypes but met with failure (Huseman 1995). John C. Frémont engaged professional daguerreotypist Solomon N. Carvalho (Carvalho 1857) to photograph his fifth expedition in 1853, but the collection was later lost in a warehouse fire (Newhall 1976). Rather than deal with the manifold difficulties of making photographs in the field, most expeditions in that period employed artists, rather than photographers, to illustrate their travels.

Using photography to replicate a drawing was a natural for me. My first exposure to the concept of rephotography came from a 1984 book by my former college photography professor, Allen A. Dutton (Dutton 1984). This book fascinated me, in part because it contained images of my hometown accompanied by old photographs of the same places taken decades earlier. I had occasionally seen some of the old photos, but to see them paired with modern photos of the same place was magical.

Many photographers have been able to achieve a near-perfect correspondence between old and new photographs by carefully relocating the original vantage point and using a camera and lens combination that captures the same field of view. The work of Fielder (1999) is a notable example, and there are many others (see Boyer et al., chapter 2; Klett, chapter 4; and other chapters in this book). Because of the inherent inaccuracies of hand-drawn scenes, it is extremely difficult to get this same image correspondence between an artist's drawing and a photographic replication. Still, an artist's illustrations can help to determine campsite locations with a high level of precision.

More often than not, the artists probably did their drawings near camp. This is undoubtedly because there was not much time to do a careful drawing while trying to keep up with a pack or wagon train on the move. Discovering precise campsite locations is important because mileages in the expedition reports are given between camps, and an error in one affects the plots for all the ones that follow. We can ascertain the general location of a campsite by using diary accounts but a picture of the campsite gives us an exact position. Many drawings depict tents, wagons, men, or mules somewhere in the scene, further helping to pinpoint the exact position of the campers relative to the surrounding scenery.

An Example of Campsite Relocation

Finding a Sitgreaves Expedition Campsite

Several years ago a study of the 1851 Expedition of Lorenzo Sitgreaves led me to a low sandstone hill on the bank of the Little Colorado River near St. Johns, Arizona. As I walked up to the crest of the hill, I could see my objective straight ahead. In the distance, beyond the dry river bed, was Leroux Island, just as it appeared in Richard H. Kern's plate 8 in the official report of the Lorenzo Sitgreaves Expedition that explored northern Arizona in 1851 (Sitgreaves

1853). On the right were the pointed hills, and nearby was the distinctive rock on which two men stood in the 150-year-old illustration (plate 21.1A). I carefully positioned myself as close as possible to where the artist had drawn his sketch and I secured a matching photograph (plate 21.1B). But my primary objective wasn't to duplicate Kern's image. I was there to locate the expedition's Camp Number 4, and this was it.

The search for Leroux Island began with a reading of Sitgreaves's official report (Sitgreaves 1853). A rough plot of the route Sitgreaves described was then made on a modern map. The diary has no mention of Leroux Island, but this name appears on the map in his report, and there is also a lithographic illustration of it in the report (plate 8), which describes it as "near camp 4." The illustration was intriguing. It was not apparent which feature in the picture was the island, but if the scene could be located the approximate position of Camp 4 would be revealed.

In June 2000, I went to the area to check the trail plot against the actual terrain. As I crossed the dry Little Colorado River some of the picture elements began to line up toward the west. Near the area of the presumed campsite, I climbed a hill and spotted Kern's rock and Leroux Island straight ahead. I had guessed correctly about the campsite location but I would never have known this without the lithograph from the Sitgreaves report (plate 21.1A).

The Value and Limitations of Explorer Maps

Historical maps are another important source of data used to reconstruct the routes of historic expeditions. Original maps published with the expedition's official report are obviously a critical source of information but are lacking in detail and are typically inaccurate geographically. Early survey and navigation methods were very crude by today's standards, and maps were drawn on a scale that did not provide much detail. Other old maps, such as the township and range maps created by the General

Land Office (GLO), can provide valuable supporting data with greater detail.

Mid-nineteenth-century surveyors usually employed a sextant to measure the altitude of the sun or a star above the horizon. The resultant reading could then be used to determine the local latitude. These measurements were surprisingly accurate, usually within 2 kilometers or so of the actual position. Longitude is a different story because it was determined by comparing a precision clock set to a standard time such as that of the observatory in Greenwich, England, with a local time reading calculated from solar observations. The difference between the two time readings gave the longitude of the observation point. The main problem was that the timepieces often stopped or became irregular because of the rough handling they received in a wagon or on a mule. Even with frequent astronomical observations to keep the clocks calibrated correctly, small errors in the clock's time translated to rather large errors in the calculated longitude; as a consequence, it is not unusual to find these readings off by 15–30 kilometers.

The scale of nineteenth-century exploration maps is also a limiting factor. The recording of a 2,400-kilometer journey on a single map imposes significant constraints on the amount of detail that can be shown.

Because of these limitations, the original map can show only the approximate route taken by the travelers. A good example of this is the map made by Lieutenant William H. Emory in 1848 (Emory 1848). Emory was mapping the route of the United States Army under General Stephen Watts Kearny as it marched west to seize New Mexico and California from the Mexicans in 1846–1847. Emory's map shows the army's trail following close to the Gila River between Dripping Springs Wash and the San Pedro River near the future site of Winkelman, Arizona. If we read his itinerary, however, we find that the Indian trail they were following hugged the river for 10 kilometers and then turned eastward along a tributary for a few kilometers before heading south again to the San Pedro. Since the map is of little help in this area, how do we determine the precise route of their little detour? Which tributary did they follow away from the river, and where did they cut across the rugged country to the San Pedro?

A lithograph by John Mix Stanley titled "A Tributary of the Gila" (fig. 21.2A) in Emory (1848) gives

Figure 21.2. Saddle Mountain, Arizona, USA.
A. (5 November 1846). John Mix Stanley's "A Tributary of the Gila" plate is from Emory (1848). The view shows a pack train at lower center, some *Carnegiea gigantea* (saguaro), and some apparently idealized rock outcrops and trees in the foreground and midground. (J. M. Stanley, courtesy of the author).
B. (15 February 2003). This view of Saddle Mountain, taken from a vantage point east of Winkelman, Arizona, shows that the original plate has exaggerated topography. At the time of my match, *Carnegiea* was still common in the view, but *Parkinsonia microphylla* (littleleaf paloverde) is the most common tree, and it does not resemble the trees depicted in the original drawing. (T. Jonas).

us a snapshot of the detour and helps to answer both questions. Emory called the peaks in the lithograph "Saddle-Back Mountain," which has been shortened on modern maps to "Saddle Mountain." On a field trip to the area, I first found the spot that gave me a view of the peaks that best matched the lithograph. I then drew a pencil line on a topographical map intersecting that spot and the peaks. The only tributary of the Gila River crossed by this line is Ash Creek, which answered the question of where the expedition departed the course of the Gila River.

Stanley's lithograph (fig. 21.2A) shows men on mules riding through the canyon. After searching the first 5 or 6 kilometers of Ash Creek on foot, I found that there is no place on Ash Creek that matches the elements in Stanley's view. This suggests that the view must depict the army's route through a tributary canyon after they left Ash Creek and headed south again toward the San Pedro River. My companions and I explored all possible side canyons along my pencil line to find the likely exit point from the canyon. After a thorough reconnaissance, we found a view on a southward side canyon that corresponds reasonably well with the artist's depiction, except for some of the details in the foreground rocks. We found little rocky spires like these in nearby parts of the canyon and finally concluded that the artist had added these features to the foreground (compare the scenes in figs. 21.2A and 21.2B). The army had followed a side canyon south to a low ridge and then crossed into another drainage that led them to the San Pedro River about 2 kilometers from its mouth on the Gila River. The Stanley picture had given us valuable clues about the detour that were not shown on Emory's map.

I cannot overstate the value of computer mapping software and Global Positioning System (GPS) receivers for this type of work (also see Hanks et al., chapter 3; Hoffman and Todd, chapter 5). Where we used to have to plot trails on paper maps and measure mileages with a ruler, now we can draw a line on a computer screen and instantly see the virtual distance traveled. The GPS receiver records our travels and waypoints and automatically plots the coordi-

nates on the computer map. After confirming a trail location we can record its coordinates in a digital file that can be shared over the Internet and used by others with similar software. The USGS 7.5-minute topographic quadrangles are still the usual maps of choice, even with computer software, but more and more maps and aerial photos are becoming available on the Internet daily.

Other Challenges to Route Finding Using Artistic Images

Problems with Artists' Illustrations

Several prominent artists were employed by government expeditions during the mid-nineteenth-century explorations of the southwestern United States. Among them were John Russell Bartlett, Richard and Edward Kern, John Mix Stanley, and German artists Frederick Von Egloffstein and Heinrich Balduin Möllhausen. Often there were other members of the party who had some artistic talent and drew sketches to accompany their personal journals of the trip. Each sketch, professional or amateur, shows us a vignette of the journey and provides a bit of data that can help us more precisely plot the course.

The artists drew the field sketches in their diaries or sketchbooks, sometimes noting the coloring of the various elements of the scene. When the trip was over, the journals, sketchbooks, and field maps were taken to Washington, DC, where the final report was prepared using the field diary, final maps were drawn for printing, and field sketches were turned into engravings, lithographs, or woodcuts by craftsmen at the printing establishment. The artists sometimes used these field sketches to create more careful works, such as oil paintings and watercolors, after they returned home. The pictures that we see reproduced in the reports were usually made by a Washington, Baltimore, or New York craftsman who had never been to the West. Because of this, I often see errors and inaccuracies in the final picture where the lithographer or engraver has filled in detail that was unclear or missing in the field sketch.

It is interesting to observe how faithfully the artists rendered their scenes. It is not unusual to find elements added to a picture later, like the two men on the rock in the Leroux Island lithograph in plate 21.1A. Comparing the illustration to the modern photograph of the same scene, it becomes obvious that the lithographer drew the two men much smaller than they should be, likely for dramatic effect. Human figures add interest and scale to a scene, and these artists wanted to create appealing pictures for their viewers, so people were often added to the scene in the artist's studio or the lithography shop. This lithographer also seems to have misinterpreted the shape of the left side of the rock in Kern's field sketch, drawing it as a bush instead of the edge of the rock (plate 21.1A).

Another common inaccuracy seen in artists' illustrations is the exaggeration of perspective (e.g., fig. 21.3). This is present in most artists' views I've worked with. The artist naturally wanted to communicate the drama and majesty of the landscape to the viewers back home so he would often increase the apparent height and steepness of mountains and other picture elements or compress a wide-angle view down to sketchbook-page format. Huseman's (1995) juxtapositions of modern photographs with the artists' illustrations from the Ives Expedition show numerous examples of these exaggerations, as does John Russell Bartlett's watercolor "On the Salinas" (fig. 21.3A). This view was made at a day camp on the Salt (Salinas) River near modern Mesa, Arizona. They resorted there to get some relief from the Arizona midday summer sun in the shade of the *Populus fremontii* (cottonwood) trees at the river. While there, Bartlett sketched the view to the east.

In rephotographing Bartlett's view (fig. 21.3A), I found that the mountains in the drawing had been sketched much larger and closer together than they actually are. This certainly makes the landscape appear more dramatic, as you can see by comparing it with the modern photograph taken from approximately the same viewpoint (fig. 21.3B). It is also surprising to see the veritable forest of *Salix* sp. (willow) and *P. fremontii* trees that the explorers found in a riverbed that is now treeless and home to sand and gravel quarries and a landfill.

A

B

Figure 21.3. Red Mountain, Arizona, USA.
A. (4 July 1852). This view, entitled "On the Salinas—North of the Gila, New Mexico," is a pencil-and-wash drawing by John Russell Bartlett. It depicts Bartlett's day camp on the Salt (Salinas) River, near the modern city of Mesa, Arizona. The prominent peak at left center is Red Mountain, and Four Peaks is visible to its right. Note the dense stands of *Salix* sp. (willow) and *Populus fremontii* (cottonwood) trees around the river. (J. R. Bartlett, courtesy of the John Carter Brown Library at Brown University). B. (5 January 2005). My match is a now-rare image of the usually dry Salt River in flood. Despite the distortion in the original drawing, Red Mountain can clearly be identified at center and a snowcapped Four Peaks is to its right. The photograph was taken near the Country Club Road bridge at the northern edge of Mesa, Arizona, and *Populus* and *Salix* are now relatively rare in this location. (T. Jonas).

The Difficulty of Finding the Correct Viewpoint

Due to the subjective nature of nonphotographic illustrations, the artist's original viewpoint is somewhat more difficult to locate than it is with a photograph. Exaggerated perspective, height distortions, artistic license, lithographic additions, vague foregrounds, and subsequent changes in the terrain all contribute to the problem. Usually, the approximate position of the artist can be located after carefully comparing the details of the original work with the actual landscape, but sometimes we find that the original viewpoint is no longer available or will no longer provide us with a workable scene. This appears to be the case with John Russell Bartlett's watercolor of Tucson (fig. 21.4A), from Sentinel Peak (today nicknamed "A" Mountain for the large letter near its summit representing the University of Arizona).

Boundary Commissioner Bartlett's party left the Pima villages of central Arizona in July 1852 and followed the southern road toward El Paso, Texas, that passed through the northernmost Mexican outpost of Tucson (Bartlett 1854). During their three-day layover there, Bartlett drew his sketch from the eastern slopes of Sentinel Peak. When I tried to match this view, I had some difficulty finding the perfect viewpoint. I climbed to what appeared to be the right altitude above the desert floor and then began walking around the mountain to find the place where the foreground, middle ground, and distant picture elements assume the same relationship as they do in the 1852 watercolor. Unfortunately, the Bartlett viewpoint I wanted appeared to be a little beyond the edge of a cliff; subsequent research suggests that road crews removed that part of the mountain to relocate Mission Road, which now passes the eastern base of Sentinel Peak. The repeat photograph (fig. 21.4B) is from the closest location I could get to the original vantage point. Bartlett's watercolor shows the earliest known view of the old San Agustín Mission complex (large white buildings at right center), which was built by Spanish Jesuit missionaries in the late 1700s. After the mission was largely abandoned in the 1820s, the adobe structures gradually deteriorated into the desert, accelerated by

Figure 21.4. Tucson, Arizona, USA.

A. (1852). John Russell Bartlett's watercolor of Tucson shows two prominent white buildings at right center, which were the San Agustín Mission convento and chapel that were built in the eighteenth century and in disrepair when Bartlett visited (Thiel and Mabry 2006). The Mexican town of Tucson is visible above it on the other side of the Santa Cruz River, and the Santa Catalina Mountains appear in the background. (J. R. Bartlett, courtesy of the John Carter Brown Library at Brown University).

B. (29 April 2006). Photograph of Tucson, taken from the lower slopes of Sentinel Peak. The last vestiges of the adobe mission complex, limited to a few foundations, graves, and trash middens, are not visible in the photograph. The author has concluded that the viewpoint used by Bartlett no longer exists due to removal of the edge of the hill during road construction, so the photograph was made about 100 feet uphill from the estimated original viewpoint. (T. Jonas).

the activities of building-material scavengers and treasure hunters. The last standing wall of the mission's convento fell victim to an adjacent landfill project in the 1950s (Thiel and Mabry 2006).

The degree of realism in an illustration varies from artist to artist. Many nineteenth-century artists tended to romanticize their images to evoke stronger emotions in the viewer. Expeditionary artists, viewing a dramatic scene for the first time, were no different. Some of the scenes rendered fairly accurately by Balduin Möllhausen on the Ives Expedition were so dramatically exaggerated by other artists in the group that they hardly appear to be depicting the same place (Huseman 1995).

Although many artists exaggerated to some degree, there were also a few who preferred a more realistic style. Boundary Commissioner Bartlett was one of those; some of his scenes are so accurate in perspective that if they were laid over a photograph of the scene, the elements would nearly match. The inventory of instruments carried by the Boundary Survey party lists two camera lucidas, optical devices that allow the user to make faithful tracings of the scene in front of him. Bartlett may have used one of these to make some of his original field sketches, including the Tucson image. Since the scene doesn't correspond perfectly to reality (compare the silhouette of the distant mountains), I believe the finished work was done freehand based on the tracing. In fact, Bartlett's finished watercolor could have been made on-site since he mentions that he returned to the location the next morning and "completed my sketch of the valley and town."

Finding a Detour around the "Needle's Eye"

Perhaps the most exciting rephotography find I've been involved in is locating the scene of Seth Eastman's 1853 watercolor "Great Cañon, River Gila" (plate 21.2A). The location of this beautiful and well-known landscape image has been somewhat of a mystery to modern historians due to the inaccessibility of the canyons of the Gila River. Another misleading factor is a sketch of the same scene in the di-

ary of boundary surveyor Amiel Whipple with the caption, "Cañon at the mouth of the Rio San Francisco." The reason for the confusion is that the mouth of today's San Francisco River is located near the New Mexico border in the Gila Box Riparian National Conservation Area, a difficult-to-access reach southwest of Clifton, Arizona. Casual viewers had accepted that this scene was in that canyon but no one had actually seen it there.

In 2002, Professor David H. Miller, who was researching the boundary survey, wrote to ask if I was familiar with the location of the scene. As we studied the documents, it quickly became clear that the view was not in the Gila Box but in a canyon that is downstream from Coolidge Dam, some 180 river kilometers to the west. The river known in 1851 as the San Francisco was what we know today as the San Carlos River. We plotted possible photograph locations, and the following year Dr. Miller organized a trip into the canyon along with Dr. Jerry E. Mueller, Dr. Harry Hewitt, and myself. We had scheduled the trip during November because all water releases are stopped at that time for the maintenance of Coolidge Dam and its irrigation system, but we were doubly fortunate in our purpose because Arizona was in the middle of a drought and there had been no significant flow in the river for several months. This allowed us the rare opportunity of hiking down the usually wet riverbed to our destination on dry ground.

We were not sure how far we would need to walk (it turned out to be 6 kilometers one way) but as we turned a corner in the canyon we began to see the upper cliffs from Eastman's image (plate 21.2A) looming ahead. When we reached the artist's viewpoint the scene was awesome. The cliffs here are 180–300 meters high and the two prominent boulders in the river bottom are about 5–8 meters high. It was amazing to see how faithfully the artist rendered this scene—very little had changed in the intervening years.

When the river is flowing here, it would be nearly impossible to pass this point on the river either on foot, by mule, or by boat, hence the name "Needle's Eye." A damaged and flattened inflatable raft nearby testified to the navigational difficulties. While there, we discovered that Eastman had shown the water flowing toward the viewer but the view actually looks downstream; Eastman had never been to this place and apparently based his watercolor on a drawing by Bartlett, who had not been there either. Bartlett's sketch was, in turn, based on a field sketch by someone on the survey crew. Dr. Mueller, who is an authority on the artwork of the boundary survey, suggests it may have been drawn by surveyor Amiel Whipple or assistant Frank Wheaton. Whoever the original artist was, it is apparent that he made a very accurate drawing since this third-generation watercolor image very closely matches the detail of the current view (plate 21.2B); he likely used the camera lucida to sketch this scene. Finding this view was an exciting discovery, but it also showed us the limit of the river survey and the "jumping off place" where the team left the Gila River to detour around this canyon.

Discussion and Conclusions

Artistic renderings of landscapes can be used to find historical trails and campsites and thereby answer questions concerning the routes of expeditions or certain details of their experiences. Artists' illustrations may not afford the same level of interpretive value as photographs, but they can still be successfully used to add to the body of historical knowledge by helping us locate and map how expeditions traversed the landscape. While in many cases artwork was deliberately or accidentally altered or exaggerated, some illustrations are surprisingly and significantly true to life (plate 21.2) and capable of yielding a remarkable amount of information, in terms of both historical details of an expedition as well as landscape change that followed their visit. In conducting our investigations, we never know quite what we will encounter by way of image accuracy until we have located the original vantage point, which tends to be a moment as rewarding as it is insightful.

Literature Cited

Bartlett, J. R. 1854. *Personal narrative of explorations and incidents in Texas, New Mexico, California, Sonora, and Chihuahua, connected with the United States and Mexican Boundary Commission, during the years 1850, '51, '52, and '53.* New York: D. Appleton. Repr., Chicago: Rio Grande Press, 1965.

Carvalho, S. N. 1857. *Incidents of travel in the far West; with Col. Fremont's last expedition.* New York: Derby and Jackson. Repr., New York: Arno Press, 1973.

Dutton, A. A. 1984. *Phoenix then and now.* Phoenix: First Interstate Bank of Arizona.

Emory, W. H. 1848. *Notes of a military reconnaissance from Fort Leavenworth, in Missouri, to San Diego, in California, including parts of the Arkansas, Del Norte, and Gila rivers.* House Executive Document 41. Washington, DC: Wendell and Van Benthuysen.

Fielder, J. 1999. *Colorado 1870–2000.* Englewood, CO: Westcliffe Publishers.

Huseman, B. W. 1995. *Wild river, timeless canyons: Balduin Möllhausen's watercolors of the Colorado.* Tucson: University of Arizona Press.

Newhall, B. 1976. *The daguerreotype in America.* New York: Dover Publications.

Sitgreaves, L. 1853. *Report of an expedition down the Zuni and Colorado rivers.* Senate Executive Document 59. Washington, DC: Robert Armstrong. Repr., Chicago: Rio Grande Press, 1962.

Stephens, H. G., and E. M. Shoemaker. 1987. *In the footsteps of John Wesley Powell.* Boulder, CO: Johnson Books.

Thiel, J. H., and J. B. Mabry, eds. 2006. *Rio Nuevo Archaeology Program, 2000–2003: Investigations at the San Agustín Mission and Mission Gardens, Tucson Presidio, Tucson Pressed Brick Company, and Clearwater Site.* Technical Report No. 2004–11. Tucson, AZ: Desert Archaeology. www.cdarc.org/pages/library/rio_nuevo/ (accessed 14 February 2008).

Chapter 22

Persistence and Change at Mesa Verde Archaeological Sites, Southwestern Colorado, USA

William G. Howard, Douglas J. Hamilton, and Kathleen L. Howard

Preservation of archaeological sites is a national priority in the United States; Mesa Verde National Park is perhaps the best example of the way in which this is carried out. In the late nineteenth century, before the park was established, local residents in southern Colorado recognized the need for preservation of archaeological sites that were being degraded by collectors seeking artifacts. Persistent efforts by these residents and other concerned citizens over the next few years finally yielded results: in 1906 the United States Congress passed the American Antiquities Act (Fiero 2006: 23). This legislation makes it illegal to "appropriate, excavate, injure, or destroy any historic or prehistoric ruin or monument, or any object of antiquity, situated on lands owned or controlled by the Government of the United States, without the permission of the Secretary of the Department of the Government having jurisdiction over the lands on which said antiquities are situated." Three weeks later, on 29 June 1906, Congress created Mesa Verde National Park.

In addition to preservation of sites at Mesa Verde and other national parks and monuments, the National Park Service is committed to making representative examples accessible to the public for visitation and interpretation. A delicate balance must be achieved to ensure that visitors can safely tour sites, and that sites are not damaged by visitation. The present preservation policy of the National Park Service is to minimize restoration (also called reconstruction or rebuilding) of structures or parts of structures to their original appearance. Restoration has two obvious problems: the original appearance may not be known, and after restoration, viewers would not be able to distinguish original from restored elements. Indeed, as early as 1908, when Jesse W. Fewkes was the first archaeologist to excavate sites at Mesa Verde National Park, he was well aware of these problems and only made those repairs necessary to make sites safe for visitation (Fiero 2006: 27–28).

The preferred procedure for preservation is stabilization, or repair, which is "to retain and make stable the standing architecture of archaeological sites, [including] documenting that architecture prior to any modifications, and then documenting the changes made in the intervention process" (Fiero 2006: xvii). Stabilization materials are continually being tested and methods are being developed to make sites safe and to preserve them from deterioration caused by both visitation and natural elements.

Repeat photography can be used to facilitate preservation at archaeological sites in several ways: (1) for short-term "before and after" documentation of stabilization work (Fiero 2006: 47–48, 73); (2) for long-term evaluation of the effectiveness of stabilization materials and methods (Fiero 2006: 48); (3) to document the changes of appearance of sites over many years, as a result of excavation, restoration, and stabilization (Howard et al. 2006); (4) to document long-term changes of flora and soil erosion at and near sites (Malde 2000); and (5) to document the effects on structures and environment of rapidly occurring catastrophic events such as wildfires, rock falls, earthquakes, and other "great hazards" (Howard et al. 2006).

In this chapter, we describe the use of repeat photography to document both gradual changes, over nearly a century, and rapid changes of archaeological sites and environment at Mesa Verde National Park (Howard et al. 2006). Our work contrasts with other applications of repeat photography described in this book and shows different aspects of the technique and its usefulness in the natural and cultural sciences.

Mesa Verde National Park

Located in the southwest corner of Colorado on the vast Colorado Plateau, Mesa Verde National Park encompasses some 210 square kilometers of rugged terrain (fig. 22.1). To visualize the topography, imagine a set of long, narrow tablelands or mesas, joined side-by-side at the north, extending southward like the slender fingers of outstretched hands,

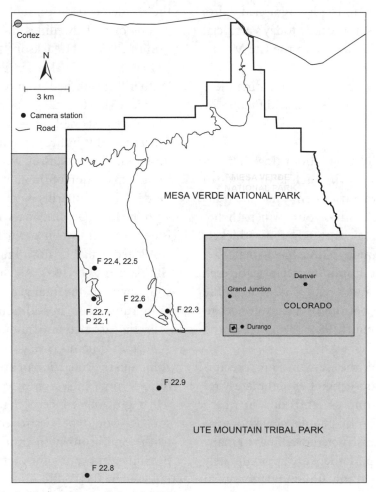

Figure 22.1. Map of Mesa Verde National Park, Colorado, USA.

and separated from each other by deep, narrow canyons. This arrangement of mesas and canyons, comprising about 1,330 square kilometers, is known collectively as the Mesa Verde, of which Mesa Verde National Park is a small part.

At their north end, the mesa tops have an elevation of about 2,425 meters; they slope gently southward to an elevation of about 1,820 meters. There is dense growth of *Pinus edulis* (pinyon pine), *Juniperus osteosperma* (Utah juniper), and *Quercus gambelii* (Gambel oak); the variegated green of their foliage led early Spanish explorers to use the name Mesa Verde (green table) to describe what they saw. The mesas and canyons extend into the adjacent Ute Mountain Tribal Park, a part of the Ute Mountain Indian Reservation. As there was no Mesa Verde National Park in 1891, Gustaf Nordenskiöld and his companions explored and photographed sites throughout an area that transcends today's artificial boundaries. Some of his sites are in the Ute Mountain Tribal Park.

Nordenskiöld describes how the alternating mesa tops and canyons made travel very difficult for explorers:

> We have been wandering about together on horseback for several days in tracts known only to the Indians, where we must often, when the trails made by Indians end, make our own path between cliffs and stones, through dense thickets and bushes. We have many a time had to make detours of miles to get around a precipice, or seek for several hours to find a path leading from the plateau down to the bottom of a canyon over to the other side. (Arrhenius 1984: 15)

The upper levels of the canyons are precipitous cliffs of the Cliff House member of the late-Cretaceous Mesa Verde Group of sandstone, in which are many alcoves. Prehistoric people, now known as Ancestral Puebloans, constructed dwellings in many of these alcoves. In the park there are approximately 600 cliff dwellings; 80 percent contain no more than 10 rooms (Fiero 2006: 14). In some larger alcoves there are villages of more than 30 rooms; the largest

and best known is called Cliff Palace, which has more than 150 rooms and 22 kivas (underground ceremonial rooms). Because of the number and state of preservation of cliff dwellings and mesa-top ruins, Mesa Verde National Park is considered to be the nation's premier archaeological park. The United Nations Educational, Scientific, and Cultural Organization (UNESCO) has also designated it a World Cultural Heritage Site (Fiero 2006: 21).

By AD 1300, the Ancestral Puebloans no longer continuously lived in the cliff dwellings. Remarkably, the existence of those structures remained unknown to any but indigenous people for more than 500 years. The first written record of the sighting of ruins in southwestern Colorado was in 1765 by the Spanish expedition of Juan Maria Antonio Rivera, and it was not until 1874 that William Henry Jackson of the Hayden expedition took the first known photographs of cliff dwellings in the Mesa Verde region (Smith 1988: 8–11, Jackson 1876).

In 1880, the Quaker family of Benjamin K. (BK) Wetherill established Alamo Ranch at Mancos, Colorado, a short distance northeast of Mesa Verde. There were six siblings—Anna, Richard, Al, John, Clayton, and Winthrop; the five brothers and Anna's husband, Charlie Mason, were to figure prominently in the exploration of the canyons and mesas of Mesa Verde. The Wetherills had a friendly relationship with the local Ute Indians, who told them of the existence of cliff dwellings. In 1882, Al found and investigated what is now known as Sandal House (Blackburn 2006: 69). As the brothers hunted for stray cows in the many canyons, they discovered more ruins; soon, in addition to ranching they were collecting artifacts and guiding visitors to sites.

At that time no laws governed the excavation of ruins and the collection of artifacts. As various expeditions came in search of collectible materials, BK felt a growing concern that valuable cultural resources would be destroyed. He therefore wrote to the Smithsonian Institution describing the cliff dwellings, indicating the methods of exploration being used by the Wetherills, and attaching notes from Richard's journal as an example. Also, he asked for guidance. His letter found its way to archaeologist

W. H. Holmes, Curator of Collections; Holmes rebuffed him (Blackburn 2006: 21). Fortunately, there arrived in Mancos, in 1891, a young man who had scientific training, was a skilled photographer, and could provide some of the guidance the Wetherills desired.

Nordenskiöld at Mesa Verde

On 2 July 1891, 23-year-old Gustaf Nordenskiöld (fig. 22.2) from Stockholm, Sweden, arrived in a horse-drawn buggy at Alamo Ranch. He was on a tour of the southwestern United States to facilitate his recovery from tuberculosis (Arrhenius 1984: 4). In Denver he learned of interesting cliff dwellings near Mancos and decided to spend a few days there. After proper introductions, he hired Richard and Al, two of the five Wetherill brothers, as guides for a week-long trip to get an overview of Mesa Verde and

Figure 22.2. (ca. 1894). Gustaf Nils Adolph Nordenskiöld. (Unknown photographer).

the surrounding area. What he saw astounded him. Al Wetherill later noted that, "When he returned to the ranch his enthusiasm had increased almost beyond his control" (Fletcher 1977: 215).

Nordenskiöld realized that no comprehensive study of this remarkable place, with excavations and photographs, had been published. He was eager to make a collection of artifacts that could be exhibited in a museum in Sweden, and to assemble a definitive collection of photographs. It was fortunate that he did not lack funds, for his one-week tour became a four-month expedition. At the outset, his only camera equipment was what he referred to as his "instantaneous" Kodak, which he had brought with him from Europe. He knew this was inadequate for the task ahead, so he wrote to his parents asking them to send his large camera. While awaiting its arrival, he concentrated on excavating ruins.

Nordenskiöld was not an archaeologist, but he had broad scientific training and was skilled in photography. He kept careful records of his excavations, including detailed sketches of layouts, notes on stratigraphy, and locations of artifacts. Digging and heavy work were done by the Wetherills and other hired hands, as were cooking, packing, and other camp work. The logistics for all of this were difficult: all supplies, camp gear, tools, and equipment had to be packed by horses some 15 miles from the ranch to the sites, then carried into cliff dwellings. Sometimes the workers had to use local trees to construct scaffolding in order to gain access to ruins. A lot of time-consuming hard work was involved.

It soon became apparent that his large camera was not going to arrive when he needed it, so Gustaf ordered a new one from Denver (Reynolds and Reynolds 2006: 65, 69). Unable to proceed with photography, he continued excavating. Then on 17 September, he journeyed to Durango to send a second shipment of his collected relics by rail, and to pick up the new camera. Unfortunately, the express agent refused to accept the shipment, and Gustaf learned that his first shipment on 8 September had been impounded. To further complicate matters, he was arrested for trespassing on Indian lands (Howard et al. 2006: 105). After the US attorney general and other high-level officials in Washington, DC, interceded, he and his shipments were finally released on 5 October with an apology. One can imagine Gustaf's frustration at having to spend more than two weeks of valuable time languishing under house arrest in Durango while waiting for his case to come to trial!

The new camera was a large-format view camera with glass-plate negatives, probably 8- by 10-inch. With it were two lenses, a Rapid Rectilinear and a wide-angle lens (Reynolds and Reynolds 2006: 101). Gustaf now ceased excavations and began a frenzy of photography, as is shown by excerpts from his diary (Arrhenius 1984: 33–34):

> Monday, October 12. 14 exposures of Cliff Palace . . .
> Wednesday, October 14. 12 exposures of Cliff Palace, . . . several small cliff dwellings . . .
> Friday, October 16. 15 exposures of ruin 9 . . .

Between 10 October and 1 November, Gustaf took about 150 photographs with the view camera. These included not only cliff dwellings but also mesa-top ruins and scenes of the environment.

By 1 November, he had completed his work at Mesa Verde. He then hired Al Wetherill and Roe Ethridge to guide him on a trip to Grand Canyon, a trip fraught with hardship and privations that further degraded his health. In January 1892, he returned to Sweden, where he began a prodigious amount of work. He wrote his best-known book, *Cliff Dwellers of the Mesa Verde* (Nordenskiöld 1893), which received international acclaim and is still a classic of American archaeology. He also assembled an exhibit of artifacts and photographs for the Columbus Jubilee in Madrid, Spain. For this, in 1894, the Queen of Spain awarded him the Order of Isabella Catolica. He continued his previous work on photography of snow crystals, wrote nine technical papers, and translated 11 of Jules Verne's novels from French to Swedish (Reynolds and Reynolds 2006: 134, 170–171). All of this took a heavy physical toll. Gustaf's health continued to worsen, and he died on 5 June 1895, three weeks before his 27th birthday.

When Mesa Verde National Park was created in 1906, Nordenskiöld's work established standards for its future archaeology. In 1984, the Mesa Verde National Park superintendent Robert Heyder said of him, "I shudder to think what Mesa Verde would be today had there been no Gustaf Nordenskiöld. It is through his book that the notoriety of the cliff dwellings became established, and the volume might well be called the harbinger of Mesa Verde National Park as we know it today" (Arrhenius 1984: i).

Rephotographing Nordenskiöld's Sites

In 1986, Bill Howard was examining some of Nordenskiöld's 1891 photographs on display at Mesa Verde National Park. He recognized that nearly a century had passed, and repeat photography could yield much valuable information. He discussed this with Don Fiero, at that time chief of interpretation. Don suggested that it would make a good Volunteer-In-Park (VIP) project, and that it would fit in well with the Nordenskiöld Centennial Celebration that the park was planning for 1991.

Howard assembled a team consisting of himself, Kathleen Howard, Douglas Hamilton, and Shirley Hamilton. The team prepared a proposal with the objective of using repeat photography to document (1) evidence of natural weathering and erosion; (2) effects of human intervention, including reconstruction and stabilization; and (3) environmental changes, both botanical and geological. Robert Heyder, then superintendent, approved the project and made it possible for the team to have access to ruins and to use facilities available at the park. Because Nordenskiöld photographed some sites in the Ute Mountain Tribal Park, the team obtained a permit and hired a guide for access to those places.

The team conducted archival research at the Mesa Verde Research Library, where copies of negatives and prints of Nordenskiöld's photographs are available. In order to facilitate identification of sites, copy prints were made in the park darkroom. Considerable fieldwork was then necessary to find locations, determine camera sites, and do repeat photography (for additional information on techniques, see Boyer et al., chapter 2; Klett, chapter 4; Hoffman and Todd, chapter 5). While photographs of some of the more popular ruins were easily identified, Nordenskiöld's captions for many were brief and not sufficiently specific for identification, so much exploring was necessary.

The team used a 4- by 5-inch Calumet view camera equipped with a Kodak Ektar 127-millimeter lens, and Kodak T-Max 100 black-and-white sheet film. We also used a 35-millimeter camera. While we attempted to use Nordenskiöld's camera locations, often this was not possible because of trees or other shrubs, removal of locations by subsequent excavation, or obstruction of view by reconstruction. Also, we were not able to repeat photographs at the same month, day, and hour as the originals as is often desirable in repeat photographic studies (Klett, chapter 4; Hoffman and Todd, chapter 5). Altogether we rephotographed 85 of the photographs Nordenskiöld took with his large-format view camera.

For four years the team made numerous visits to Mesa Verde National Park to locate and photograph sites. At the end of each trip, negatives that had been exposed were developed in the park darkroom. In 1990 Vic Goodwin of Vic's Photo Shop in Cortez, Colorado, made enlarged prints of photographs selected by the team, and the corresponding Nordenskiöld photographs. Park personnel fabricated seven display panels, and the team mounted pairs of photographs for an exhibit. One room of the Mesa Verde Museum was designated for the exhibit, which remained in place for several years (Howard et al. 1991).

Repeat Photography Examples

Cliff Palace

Cliff Palace is the signature cliff dwelling of Mesa Verde National Park. Al Wetherill reported sighting it in 1885, when, at the end of a long day of hard scrambling, he viewed it from the west side of what is now called Cliff Canyon. He had no time to investigate, and it was not until 1888 that Richard Wetherill and his brother-in-law, Charlie Mason, returned to explore the ruin (Blackburn 2006: 73).

Figure 22.3A shows Nordenskiöld's 1891 photograph of the southern part of Cliff Palace. Upon first seeing this cliff dwelling, he wrote:

In a long but not very deep branch of Cliff Cañon, a wild and gloomy gorge named Cliff Palace Cañon, lies the largest of the sites on the Mesa Verde, the Cliff Palace. Strange and indescribable is the impression on the traveller [*sic*], when, after a long and tiring ride through the boundless, monotonous piñon forest, he suddenly halts on the brink of the precipice, and in the opposite cliff beholds the ruins of Cliff Palace, framed in the massive vault of rock above and in a bed of sunlit cedar and piñon trees below. (Nordenskiöld 1893: 59)

Figure 22.3B is the 1989 photograph of the same scene. Cliff Palace was excavated in 1909 by Jesse Walter Fewkes, who reported that it was in good enough condition to be opened to the public, but that it had been thoroughly looted for commercial purposes and many of the relics were gone forever (Smith 1988: 71).

Spring House

Nordenskiöld gives a very picturesque description of his first view of Spring House:

What a striking view these ruins present at a distance! The explorer pictures to himself a whole town in miniature under the lofty vault of rock in the cliff before him. But the town is a deserted one, not a sound breaks the silence, and not a movement meets the eye, among those gloomy, half-ruined walls, whose contours stand off sharply from the darkness of the inner cave. (Nordenskiöld 1893: 43)

A

B

Figure 22.3. Cliff Palace, Mesa Verde National Park, Colorado, USA.
A. (1891). This view of the south part of Cliff Palace shows some fairly well preserved structures. Built in a large alcove, this is the largest cliff dwelling in the park, with 22 kivas and more than 150 rooms. Closer to the opening there is more damage, notably to the four-story tower. The structures in front of this tower are covered by rubble. (G. Nordenskiöld, N087B).
B. (1989). The kivas, which were unexcavated by Nordenskiöld, are now exposed. Fewkes did considerable repair of the tower, kivas, and other structures when he excavated the ruin in 1909. Paths for visitors have been added to facilitate access. (B. Howard, B087B).

Spring House (figs. 22.4 and 22.5) is so named because of a spring in the back of the alcove. Nordenskiöld first inspected this ruin on 25 July 1891 but did no excavating. He returned to excavate on 12 September and to take photographs on 20 October (Arrhenius 1984: 19, 27, 36). Spring House has not been excavated after Nordenskiöld, nor has it been opened to the public. It was stabilized in 1935.

Navajo Watch Tower

This ruin was originally called Navajo Canyon Watch Tower because of its location overlooking Navajo Canyon. One of several watch towers at Mesa Verde, this is of particular interest because it is completely exposed to the elements (fig. 22.6). The Park Service has never excavated it, but they stabilized it in 1966. Nordenskiöld's photograph shows that, despite its exposure, it has survived remarkably well. The repeat photography shows the effectiveness of stabilization methods.

Kodak House

Nordenskiöld named Kodak House (fig. 22.7) because he stored his Kodak camera there while mak-

ing a trip back to the Alamo Ranch (Nordenskiöld 1893: 21). On 24 July 1891, he excavated a grave and noted: "Photographed the grave unopened and then with some soil removed" (Arrhenius 1984: 19). This is interesting because at that time he had only his Kodak camera; we do not know if these Kodak photographs still exist.

This cliff dwelling has elements that are well preserved even though many structures are very exposed. Structures are visible in alcoves on two levels; the lower alcove is higher and deeper than the upper one. The Kodak House check dam (plate 22.1) is on the cliff-top above the dark vertical stripes.

Navajo Canyon

In addition to his photographs of cliff dwellings and mesa-top ruins, Nordenskiöld took many photographs of Mesa Verde's natural environment. These images enable us to document differences of flora that have occurred in the intervening years. One example is the sweeping view of the mouth of Navajo Canyon (fig. 22.8). The increase of the density of woody plants exemplifies areas of Mesa Verde National Park that have not been subjected to cata-

A  B

Figure 22.4. Spring House, Mesa Verde National Park, Colorado, USA.
A. (1891). Spring House is a five-story cliff dwelling. Nordenskiöld noted: "The rooms of the building lie on two ledges within the same cave. . . . I carried out only quite superficial excavations in a few of the rooms on the lower ledge" (Nordenskiöld 1893: 43). His photograph shows that, as would be expected, structures closer to the opening of the alcove have suffered significant damage from weathering. (G. Nordenskiöld, N144).
B. (1988). No further excavation has been conducted, nor has the site been opened to regular visitation. This photograph shows that stabilization has prevented significant further deterioration. Some rubble was removed, and several walls were reinforced, but overall the ruin appears almost the same. (B. Howard, B144).

strophic changes such as wildfires, chaining (uprooting shrubs and trees by dragging a large chain between bulldozers), or overgrazing.

Pool Canyon Ruin

This small cliff dwelling in Pool Canyon provides an example of the consequences of an instantaneous change resulting from the unstable nature of the Mesa Verde sandstone (fig. 22.9). At an unknown time between Nordenskiöld's photograph and 1988, a large slab fell from the roof of the alcove onto a structure, crushing a wall.

Kodak House Check Dam

Check dams were widely used by the occupants of Mesa Verde in order to trap water in drainages for irrigation. Dams were made by laying rock walls across drainages at convenient locations. The area behind a dam was filled in with soil in which crops were planted. Longer drainages have several dams distributed along their courses.

The photographs of this check dam above Kodak House (plate 22.1) are perhaps the most interesting ones in our collection because they show the remarkable state of preservation of an exposed and untouched site over almost a century. They also show the effects of both gradual and instantaneous change, the latter during a wildfire. In 2000, a fire started in Pony Canyon on the Ute Mountain Tribal Park and quickly swept over Wetherill Mesa. Kathleen Fiero, who was at that time supervisor of stabilization, describes the aftermath:

August 14, 2000. I view Wetherill Mesa for the first time after the devastating Pony fire. The burned area is like a moonscape—not a blade of

A

B

Figure 22.5. Spring House, Mesa Verde National Park, Colorado, USA.

A. (1891). Nordenskiöld's photograph shows that even though the cliff dwelling had been unoccupied for 600 years, the interior was in good condition. He noted that it was "remarkable on account of two stone pillars put up to support a large floor" (Arrhenius 1984: 21). Such use of pillars is not common in Mesa Verde cliff dwellings. (G. Nordenskiöld, N143).

B. (1988). This photograph shows that this part of the ruin is still unchanged in almost every detail. (B. Howard, B143).

A

B

Figure 22.6. Navajo Watch Tower, Mesa Verde National Park, Colorado, USA.

A. (1891). Originally called Navajo Canyon Watch Tower, this was constructed on a naturally occurring rock point that seems made to order for such a structure. It is accessible from the nearby cliff that is partially visible in the lower right corner of the photograph. (G. Nordenskiöld, N110).

B. (1989). Some of the top of the right-hand wall is missing in this photograph, and there is evidence of stabilization; otherwise the structure is mostly the same as in 1891. This degree of preservation is quite remarkable since the structure is totally unprotected. (B. Howard, B110).

Figure 22.7. Kodak House, Mesa Verde National Park, Colorado, USA.

A. (1891). The alcove sheltering Kodak House has two levels, with structures on both. At the far right of the lower level, the high structure is barely within the alcove, and is thus almost totally exposed to the elements, as are those on the second level. (G. Nordenskiöld, N147).

B. (1987). On the left side of the second level, there is some deterioration of walls; otherwise, even though the alcoves are shallow and some structures are exposed, the entire ruin appears similar to its condition in 1891. The Park Service did stabilization work in Kodak House beginning in 1963 (L. Martin, pers. comm., 31 July 2007). (B. Howard, B147).

Figure 22.8. Navajo Canyon, Ute Mountain Tribal Park, Colorado, USA.

A. (1891). Nordenskiöld photographed Al and John Wetherill herding cattle at the mouth of Navajo Canyon on the Ute Mountain Indian Reservation. In this rather open meadow there are few shrubs. (G. Nordenskiöld).

B. (1990). This view shows much denser growth of shrubs in the canyon bottom. The trees on the slopes, probably *Juniperus* (juniper) and *Pinus edulis* (pinyon pine), have also increased. Grazing has not been permitted here for many years. In general, denser vegetation was observed at Mesa Verde in 1991 than in 1891; other factors contributing to this are climate change and suppression of fires. (B. Howard).

Figure 22.9. Pool Canyon cliff dwelling, Ute Mountain Tribal Park, Colorado, USA.
A. (1891). This site in Pool Canyon is a typical small cliff dwelling of only a few rooms. When Nordenskiöld photographed it, the structures near the mouth of the alcove were in ruins. (G. Nordenskiöld, N172).
B. (1988). Sometime in the intervening century, a large slab fell from the roof of the alcove, destroying some of the dwelling. Living in an alcove in Mesa Verde sandstone could be perilous: the Park Service has found other cases of rockfall that occurred while the dwellings were occupied (L. Martin, pers. comm., 31 July 2007). (B. Howard, B172).

green grass, not a live tree. Ash several inches deep swirls around and fills my nose and lungs at the least little breeze. It is truly unbelievable, even compared to the other three devastating fires I've lived through in the park. (Fiero 2006: 149)

Six years later there has been considerable regrowth of vegetation surrounding the Kodak House check dam, but only a small charred piece of the fallen tree remains (plate 22.1C).

Discussion and Conclusions

We were able to observe, through repeat photography, both gradual and rapid changes in archaeological sites and the surrounding landscape at Mesa Verde National Park in southwestern Colorado. Examples of gradual changes include an increase of vegetation density, and geological erosion. Examples of rapid changes include the fallen slab in the Pool Canyon Ruin (fig. 22.9B) and a disastrous fire whose effects are documented by the 2006 photograph of the check dam (plate 22.1C). These emphasize that

photography needs to be repeated over many years in order to document the effects of both gradual change and instantaneous changes resulting from "great hazards."

In general, we found that the vegetation density at Mesa Verde National Park has increased in the century following Nordenskiöld's photographs. This is true of both mesa-tops and canyon bottoms and is exemplified by the photographs of Navajo Canyon and the Kodak House check dam. Clearly, gains in vegetation can be quickly reversed by wildfire, a frequent occurrence in the southwestern United States where years of fire suppression contributed to dense and fire-susceptible stands of forest vegetation.

The state of preservation of the ruins is impressive and attests to the benign climate of Mesa Verde. Six hundred years elapsed between the exodus of Ancestral Puebloans and Nordenskiöld's arrival; the condition of structures shown in his photographs is quite remarkable. When we began our project, we naturally expected to find that since Nordenskiöld's visit the more-exposed ruins would show some deterioration, while those protected by alcoves would not. The latter is certainly true, as shown by the photographs of Spring House in figure 22.4. However, photographs of such sites as Navajo Watch Tower, Kodak House, and the fallen tree at the Kodak House check dam show that many completely exposed features are well preserved. The survival of the fallen tree before the 2000 fire is an outstanding example.

Comparison of our photographs of exposed parts of ruins that have been stabilized with the same structures in Nordenskiöld's photographs shows that the stabilization methods developed and implemented by the Park Service have been very effective.

Acknowledgments

Permission to use the Nordenskiöld portrait and his photographs has been granted by Mesa Verde National Park. The authors are grateful to Linda Martin for providing information, and to Kathleen Fiero for valuable suggestions.

Literature Cited

Arrhenius, O. W. 1984. *Stones speak and waters sing: The life and works of Gustaf Nordenskiöld*, ed. R. H. Lister and F. C. Lister. Mesa Verde National Park, CO: Mesa Verde Museum Association, Inc.

Blackburn, F. M. 2006. *The Wetherills: Friends of Mesa Verde.* Durango, CO: Durango Herald Small Press.

Fiero, K. 2006. *Dirt, water, stone: A century of preserving Mesa Verde.* Durango, CO: Durango Herald Small Press.

Fletcher, M. S. 1977. *The Wetherills of the Mesa Verde: Autobiography of Benjamin Alfred Wetherill.* Lincoln: University of Nebraska Press.

Howard, W. G., D. J. Hamilton, and K. L. Howard. 2006. *Photographing Mesa Verde: Nordenskiöld and now.* Durango, CO: Durango Herald Small Press.

Howard, W. G., K. L. Howard, D. J. Hamilton, and S. M. Hamilton. 1991. *A century's perspective: Gustaf Nordenskiöld's Mesa Verde today.* Mesa Verde National Park, CO: Mesa Verde Museum.

Jackson, W. H. 1876. Ancient ruins in southwestern Colorado. *American Naturalist* 10:31–37.

Malde, H. E. 2000. *Repeat photography at Chaco Canyon based on photographs made during the 1896–1899 Hyde expedition and in the 1970s.* Chaco Culture National Historic Park, NM: US National Park Service.

Nordenskiöld, G. 1893. *Cliff dwellers of the Mesa Verde, southwestern Colorado: Their pottery and implements.* Stockholm: P. A. Norstedt and Söner. Repr., Mesa Verde National Park, CO: Mesa Verde Museum Association, 1984.

Reynolds, J. L., and D. B. Reynolds. 2006. *Nordenskiöld of Mesa Verde: A biography.* Bloomington, IN: Xlibris.

Smith, D. A. 1988. *Mesa Verde National Park: Shadows of the centuries.* Lawrence: University Press of Kansas.

Chapter 23

The Future of Repeat Photography

Robert H. Webb, Raymond M. Turner, and Diane E. Boyer

In its essential form, repeat photography is a decidedly low-technology tool that yields high information content of value to many disciplines in the natural and social sciences. In an age of multispectral scanners on satellite platforms and digital orthophotography acquired from aircraft, this simple, inexpensive technique has proven to be compatible with and complementary to the most sophisticated systems for spatial data acquisition, in fact exceeding those systems if site-specific information or a long-term perspective is desired. As the contents of this book indicate, repeat photography is broadly applied across disciplines and research areas ranging from glaciology and global change to anthropology, sociology, and even fine art.

One might be inclined to conclude that the use of a simple tripod-mounted or hand-held camera to capture oblique views of landscapes might become antiquated in the twenty-first century, falling into disfavor with the rapid technological advancements of image acquisition and analysis. While technology undoubtedly will alter the camera, rendering film completely obsolete at some point early in the twenty-first century, repeat photography has a unique place in landscape monitoring and hypothesis testing. It is likely that this elegantly simple technique, when applied to social history or landscape change or art, may evolve simply through the ongoing transformation of photography from analog to digital. In the sciences, however, the future will likely entail more profound changes as suggested or stated in many of the chapters in this book, mostly involving the evolving technology of image analysis.

Remote Sensing, Repeat Photography, and Monitoring of Landscape Change

In the simplest definition of the term, repeat photography is a remote-sensing technique. However, remote sensing as commonly understood is defined as image acquisition using satellites with multispectral scanners (e.g., Landsat Thematic Mapper) or aircraft equipped with digital cameras or multispectral scanners or both. Repeat photography typically acquires data only from the visual spectrum and, in many cases, only in black-and-white. Although this comparative framework within the visual spectrum may seem limited when compared with the range of multispectral scanners, other advantages of repeat photography make it the technique of choice for certain types of landscape-change evaluations.

The first applications of repeat photography involved monitoring of landscape change, whether in the advances or retreats of glaciers or rangeland changes attributed to climate or grazing effects (Webb et al., chapter 1). At present, the state of the art in landscape change detection may be best illustrated by the work of Hoffman and Todd (chapter 5), who analyzed changes in the spectral signatures of repeat images to quantify changes in grass cover in response to changing land-use practices. Another example of what is possible appears in Webb and Hereford (chapter 8), who analyzed aerial mapping photography, taken more frequently than repeat photography, to quantify channel changes readily apparent in oblique ground imagery. Finally, McClaran et al. (chapter 12) compared the results of repeat photography, aerial photography, and satellite remote sensing to evaluate changes at specific points in rangelands subject to controlled livestock grazing and climatic fluctuations.

Ultimately, the differences among remote-sensing techniques involve resolution, both spatial and spectral; the beginning of the monitoring interval; and the frequency of repeat imagery. Although some remotely sensed data acquired from satellite platforms have resolutions of less than 1 meter, more typically available imagery, particularly outside of North America and Europe, has a resolution from 30 to 250 meters. Replication of oblique ground photography has far greater resolution in the foreground and midground of the view, which may yield data that, while not spatially rectified, may provide detailed information on specific types of changes, particularly those in certain plant species or landscape details too subtle to be detected from high-altitude platforms with a downward perspective.

Online GIS Image Viewers and Repeat Photography

Development of online image viewers has created a revolution in landscape ecology and other disciplines involved with quantitative assessment of land-scape change. Hanks et al. (chapter 3) portend this evolving research direction of repeat photography as applied to landscape change. Google Earth (http://earth.google.com/) and TerraServer (www.terraserver.com/) are currently the most widely used GIS image viewers available. These viewers can be used for tasks ranging from the location of camera stations prior to fieldwork to quantitative estimation of landscape changes.

Geographic information system (GIS) image viewers commonly use variations on the same types of geospatial information. When this book was prepared, the standard resolution of topographic data used in these viewers (digital-elevation models, or DEMs) was 10-meter data for the United States and 90-meter data for most of the rest of the world (obtained from the Consortium for Spatial Information Web site at srtm.csi.cgiar.org/). As discussed in Hanks et al. (chapter 3), state-of-the-art GIS technology, in the form of ArcGIS, provides methodology for analyzing extensive topographic and spectrographic data of use to repeat photography. Online image viewers are rapidly advancing to perform the same capability, only much faster and without the need for expensive software or computers. Satellite imagery is already available at high resolution for most of the world; what remains is an improvement in DEM resolution before this type of technique can be applied anywhere on Earth's surface.

At present, online GIS image viewers are being used to document large-scale landscape changes. One of many examples concerns the impact of drought from 2000 through 2005 on Lake Powell in southern Utah, as well as dramatic land-use impacts on various parts of the world (Buchen 2009). This type of analysis is necessarily limited by available imagery, but historical aerial photography dating back to its inception in the 1920s and 1930s is becoming available as part of the "historical analysis" options of online image viewers, particularly in Google Earth 5.0. Full implementation of historical aerial photography with contemporary satellite imagery will open an array of potential applications for quantifying landscape changes, but the temporal range will still be less than what is typically available using repeat photography.

Plant Demographics and Population Changes

The use of repeat photography to quantitatively assess changes in valued species or plant assemblages is one rapidly evolving application of this technique. Bullock and Turner (chapter 10) discussed the use of repeat photography in evaluating species-specific changes in plant communities, offering both qualitative and quantitative techniques for estimating the amount of change that has occurred in a specified time interval. Hoffman et al. (chapter 11) apply this concept to specific succulent plant species with well-defined trunks and estimate the changing demographics of threatened or endangered species. This application will be particularly appropriate when assessing broad-scale changes in rare plant populations beyond what can be accomplished with permanent ecological study plots or mark-remeasure plots that are commonly established for the purpose of documenting change in charismatic or threatened species.

Quantitative assessment of plant demographics will move the potential applications of repeat photography beyond documentation of long-term change, as illustrated by a number of chapters of this book, and into the potential for future predictions based on extrapolation from past trajectories. One particularly good way of bridging the once-yawning gap between qualitative and quantitative assessments is repeat photography of permanent ecological study plots; while measurement of the plots provides quantitative data on phytoecological changes, repeat photography can assess changes in the larger setting or in low-density species, such as columnar cacti or tree aloes, that are important aspects of habitat or ecosystems.

Some new techniques may be required to reach the full potential of repeat photography in quantitatively documenting plant demographics. Many re-

peated views show small numbers of plants, and the count statistics of change (i.e., number that died, number that are still alive, and number of new plants in the interval between views) can be small and can greatly vary among different views. The question of statistical significance of count statistics within a population or within treatments of a population is a difficult one to address. As statistical analyses of this type of data are developed, the potential for robust estimations of species demographics increases, thereby increasing the potential applications of repeat photography to population assessments of plant assemblages.

Global-Change Monitoring

Many hypotheses have been posed concerning the long-term impacts of global change, particularly driven by increases in temperatures or changes in the amount or seasonality of precipitation. Future testing of certain hypotheses, such as increasing aridity, reduction in water resources, or shifts from grasslands to shrublands, may be accomplished more efficiently with other techniques, particularly satellite- or aircraft-based remote sensing. Repeat photography has a large role in global-change monitoring, however, because many expected changes in plants and habitat are too subtle to be readily monitored from high-altitude or space-based platforms.

Several examples are given in this book of landscape changes possibly related to global change that would be difficult to detect with vertical-format remote sensing. Long-term changes in tree aloes, for example (Hoffman et al., chapter 11), can only be accurately determined using oblique ground imagery. Documentation of arroyo filling (Webb and Hereford, chapter 8) requires information on vertical changes in floodplains, and the lower resolution of data from satellite remote sensing, in particular, makes that technique a poor choice for this aspect of global-change monitoring. While overall changes in forests affected by changing fire regimes (Veblen, chapter 13; Lewis, chapter 15), landscapes affected by changing land-use practices (Nyssen et al., chapter 14; Rohde, chapter 18; Bass, chapter 20), or even the amount of ice stored in glaciers (Molnia, chapter 6; Fagre and McKeon, chapter 7) can be monitored using satellite-based or aircraft-mounted platforms into the future. Subtle details of future changes can be obtained only through repeat photography; for example, Molnia (chapter 6) notes that, while the largest changes in many of his repeated views involve melting ice, whole new ecosystems are becoming established on the landscapes emerging from beneath the glaciers. The most efficient technique for documenting what species become established in the newly exposed areas is repeat photography.

Repeat Photography and Education

Repeat photography is difficult to exceed as an educational tool for illustrating the extent and timing of landscape changes. The visual appeal of repeat photographs invites interpretation of change from any observer, whether from the general public or from trained scientists. Repeat photography plays a large role in educating every interested party in the extent of past landscape changes as well as the potential for future changes. Because of its temporal reach into nineteenth-century conditions, repeat photography provides the most useful information for determining habitat baseline conditions for many ecosystems. This information is extremely valuable if ecosystem restoration projects have an underpinning of science instead of conjecture as to the extent of ecological changes.

In our experience, the interpretative staff at national parks or other valued land areas highly value repeat photography as a tool for communicating complex issues to the general public. Repeat photography is commonly used at national parks in the western United States, for example, to convey landscape or ecosystem changes. Howard et al. (chapter 22) use repeat photography to convey a sense of change in archaeological sites, in terms of both re-

construction and stabilization and natural deterioration, which is something visitors do not generally learn from readily available information. Repeat photography can be used to bridge history, particularly of explorations, settlers, or early visitors, and landscape change that may have been brought about by land-use changes that followed those explorations and settlers.

Repeat photography is often used in the process of making land-use decisions as well. One example is the slow recovery of desert ecosystems from fire (Turner et al., chapter 17), which can justify large expenditures for active restoration of burned landscapes. Other examples include the impacts of wildlife management (e.g., elephants, Western, chapter 16), fire suppression (Lewis, chapter 15), grazing management (Hoffman and Todd, chapter 5; Mc-Claran et al., chapter 12), and forestry management (Bierman, chapter 9). In an era where the focus is on global change, repeat photography is extremely valuable for reminding all observers that landscapes are seldom stable, and instead landscapes should be expected to change in response to shifts in climate and land-use practices.

Cultural Applications

Repeat photography is a popular technique for documenting changes to urban and rural cultural landscapes, in both historic and prehistoric times. Sets of such images inherently provide social commentary and insight, ranging from changing architectural styles to automobiles to apparel, as noted by Moore (chapter 19) and Bass (chapter 20), and on iconic but fading lifestyles (Rohde, chapter 18). Klett (chapter 4) and others have experimented with and are redefining the manner in which photographs are documented and displayed, both as information and as fine art. Social geographers (e.g., Bass, chapter 20) increasingly use repeat photography to examine perceptions of change and the responses that such changes evoke. Repeat photography can be used with both photographs (Hanks et al., chapter 3) and artwork (Jonas, chapter 21) to document the routes used by early explorers and expeditions. It also serves to document long-term change in archaeological sites (Howard et al., chapter 22).

The Future of Repeat Photography

Numerous examples in this book show the utility of repeat photography in capturing the details of a changing world, and it is clear to the authors of this book that repeat photography has a prominent role to play in documenting landscape changes well into the twenty-first century. Most of the examples given show how this technique is invaluable and unique in documenting landscape changes, subtle and major. In an era of rapidly changing technology, however, some aspects of repeat photography bear closer examination, particularly if new projects using this technique are planned. The largest single change we envision is the switch from analog photography using film to digital photography requiring digital media for storage and processing.

Repeat photography is inherently linked to archival photography, both for the purpose of acquiring original images and for storing repeat images for future use (Boyer et al., chapter 2; Bierman, chapter 9). Before the invention of cost-effective digital imagery, archival storage of film was relatively straightforward with well-established guidelines; the questions for repeat photographers concerned how to efficiently create and run small archives or what larger archives were available to take film and repeated imagery for long-term accessible storage. Particularly because of its usefulness in education, and also because most original images came from larger archives, the choice of whether to create a small archive or collaborate with a larger one was relatively easy for most practitioners of this technique.

With the probable elimination of film as a medium within the next few decades, perhaps the most severe problem facing the future of repeat photogra-

phy is the issue of storage and archiving of digital imagery and derivative products from that imagery. At the time of this writing, few digital media (e.g., compact discs) have sufficient archival properties to guarantee long-term persistence of digital imagery; even worse, the technology may change, rendering even archivally stable media obsolete and potentially unreadable in the future. The issue of archival storage of landscape data is perhaps more acute with other types of remote sensing, and the answer may lie in the use of ever-more cost-efficient and cost-effective hard disks to store digital repeat photography, as well as archivally processed and stored prints.

Ultimately, repeat photography is a technique rooted in the passage of time. If collections of repeat photographs are preserved in perpetuity through proper cataloging and storage, their information spectrum will expand from periods spanning multiple decades to multiple centuries. As societies worldwide face the many challenges presented by both cultural and natural change, repeat photography will continue to play a key role in illuminating, understanding, and responding to these changes.

Acknowledgments

The authors thank Peter Griffiths and Helen Raichle for their work on the maps and graphs throughout this book, and the many researchers who have contributed to chapter and other works that use repeat photography. Use of trade names in this publication does not imply endorsement by the US Geological Survey.

Literature Cited

Buchen, L. 2009. Time-lapse videos of massive change on earth. *Wired Science* (May 29), www.wired.com/wiredscience/2009/05/earthobservatoryvideos/ (accessed 9 June 2009).

About the Editors

Dr. Robert H. Webb is a research hydrologist with the USGS in Tucson, Arizona, and has worked on long-term changes in natural ecosystems of the southwestern United States since 1976. He has degrees in engineering (BS, University of Redlands, 1978), environmental earth sciences (MS, Stanford University, 1980), and geosciences (PhD, University of Arizona, 1985). **Diane E. Boyer** is an archivist with the USGS in Tucson, Arizona. She has a degree in animal health science (BS, University of Arizona, 1983). **Dr. Raymond M. Turner** is retired from the USGS and has degrees in botany (BS, University of Utah, 1948; PhD, Washington State University, 1954). Webb, Boyer, and Turner have authored, coauthored, or edited more than a dozen books, including *Grand Canyon: A Century of Change* (Webb); *The Changing Mile Revisited* (Turner, Webb, and others); *Cataract Canyon: A Human and Environmental History of the Rivers in Canyonlands* (Webb and others); *The Ribbon of Green: Long-Term Change in Woody Riparian Vegetation in the Southwestern United States* (Webb, Stanley Leake, and Turner); and *Damming Grand Canyon: The 1923 USGS Expedition on the Colorado River* (Boyer and Webb).

Contributors

J. O. Joby Bass is a cultural geographer at the University of Southern Mississippi in Hattiesburg, Mississippi. He received a PhD from the University of Texas at Austin in 2003.

Paul Bierman is a professor of geology at the University of Vermont. He received his PhD in geology from the University of Washington in 1993.

J. Luke Blair is the GIS Manager for the USGS Earthquake Hazards team in Menlo Park, California. He received his MS in geology in 1996.

Dawn M. Browning is a research physical scientist with the US Department of Agriculture, Agriculture Research Service, based in Las Cruces, New Mexico. She received her PhD in natural resource studies with a minor in remote sensing from the University of Arizona in 2008.

Stephen H. Bullock is a research professor with the Centro de Investigación Científica y de Educación Superior de Ensenada, in Ensenada, Baja California, México. He received his PhD in biology from the University of California, Los Angeles, in 1975.

Jozef Deckers presently lectures in soil geography, soils in the tropics, and land evaluation at the University of Leuven, Belgium. His research is on integrated natural resources management in developing countries, with special focus on East Africa.

John Duncan is an environmental consultant and received an MSc in environmental policy and management from the University of Manchester, United Kingdom, in 2008.

Todd C. Esque is a research ecologist with USGS–Western Ecological Research, Henderson, Nevada. He received a PhD in ecology, evolution, and conservation biology from the University of Nevada, Reno, in 2004.

Daniel B. Fagre is a research ecologist with the USGS Northern Rocky Mountain Science Center. He is director of the Climate Change in Mountain Ecosystems Project and is stationed at Glacier National Park, Montana. He received his PhD in ecology from the University of California, Davis, in 1981.

A. T. (Dick) Grove is a retired geography lecturer at Cambridge University (UK). He started his career in Nigeria as a rural development specialist and has written extensively on the physical and social geography of Africa.

George E. Gruell is a wildlife biologist, now retired from the US Forest Service. He received his BS from Humboldt State University in 1953.

Douglas J. Hamilton is an independent researcher. For 30 years he was professor of electrical and computer engineering at the University of Arizona. He received his PhD in electrical engineering from Stanford University in 1959.

Thomas C. Hanks is a research geophysicist with the USGS in Menlo Park, California. He received a PhD in geophysics from the California Institute of Technology in 1972.

Richard Hereford is an emeritus geologist with the USGS in Flagstaff, Arizona. He received his MS in geology at Northern Arizona University in 1974.

M. Timm Hoffman is a plant ecologist with the University of Cape Town, South Africa. He received a PhD in botany from the University of Cape Town in 1989.

Kathleen L. Howard is a consulting historian in Arizona and New Mexico. She received her PhD from Arizona State University in 2002.

William G. Howard is a consulting engineer in Arizona. He received his PhD in engineering from the University of California at Berkeley in 1967.

Cho-ying Huang is an assistant professor in the Department of Geography, National Taiwan University in Taipei, Taiwan. He received his PhD in the School of Natural Resources from the University of Arizona in 2006.

Tom Jonas is a trail researcher, photographer, cartographer, and business owner in Phoenix, Arizona. He studied commercial art at Phoenix College and Arizona State University in Tempe, Arizona.

Prince Kaleme is a PhD student at Stellenbosch University and received his MSc in conservation biology from the University of Cape Town, South Africa, in 2003.

Mark Klett is Regents' Professor of Art at Arizona State University in Tempe, Arizona. He received his MFA in photography from the State University of New York at Buffalo in 1977.

Darrell Lewis is a research fellow at the History Research Centre, National Museum of Australia, Canberra. He received his PhD in Australian history at the Australian National University, Canberra, in 2005.

Mitchel P. McClaran is a professor in the School of Natural Resources, University of Arizona in Tucson, Arizona. He received his PhD in wildland resource science from the University of California at Berkeley in 1986.

Lisa A. McKeon is a physical scientist with the USGS Northern Rocky Mountain Science Center in West Glacier, Montana. She received her BA from the University of Washington in 1991.

Mitiku Haile is professor of sustainable land management and president of Mekelle University, Ethiopia. He received a PhD in soil science from Ghent University in 1988.

Bruce F. Molnia is a research geologist with the USGS in Reston, Virginia. He received his PhD in geology from the University of South Carolina in 1972.

Peter R. Moore works as an area officer with Scottish Natural Heritage based in the Highlands. He is a Churchill Fellow and a part-time postgraduate student at the University of Aberdeen, studying rephotography and its uses in exploring change in the contexts of space, time, and place.

R. Neil Munro trained as a geologist–physical geographer and has been a consulting soil scientist/land-use planner since 1972. In 2006 he returned to Ethiopia to make repeat photography of 1975 landscapes of Tigray.

Jan Nyssen is professor of physical geography with the University of Ghent, Belgium. He received a PhD in geography from Katholieke Universiteit Leuven in 2001; he carries out research and resided nearly 10 years in Ethiopia.

Jean Poesen is professor of geography with the Catholic University of Leuven, Belgium. He received a PhD in geography from Katholieke Universiteit Leuven in 1983.

Garry F. Rogers is a former academic and business executive with a PhD in geography from the University of Utah.

Richard F. Rohde is a research fellow at the Centre of African Studies, University of Edinburgh, and the Plant Conservation Unit, University of Cape Town. He completed his PhD in anthropology at the University of Edinburgh in 1997.

Simon W. Todd is a freelance plant ecologist. He received an MSc in conservation biology from the University of Cape Town, South Africa, in 1997.

Thomas T. Veblen is a Professor of Distinction in the Geography Department of the University of Colorado at Boulder. He received his PhD in geography from the University of California at Berkeley in 1975.

David Western is a conservationist and ecologist. He is chairman of the African Conservation Centre in Nairobi and director of the Amboseli Research and Conservation Program, Kenya. He received his PhD from the University of Nairobi in 1973.

Index

Island Press | Board of Directors

ALEXIS G. SANT *(Chair)*
Managing Director
Persimmon Tree Capital

KATIE DOLAN *(Vice-Chair)*
Conservationist

HENRY REATH *(Treasurer)*
Nesbit-Reath Consulting

CAROLYN PEACHEY *(Secretary)*
President
Campbell, Peachey & Associates

DECKER ANSTROM
Board of Directors
Comcast Corporation

STEPHEN BADGER
Board Member
Mars, Inc.

MERLOYD LUDINGTON LAWRENCE
Merloyd Lawrence, Inc.
   and Perseus Books

WILLIAM H. MEADOWS
President
The Wilderness Society

PAMELA B. MURPHY

DRUMMOND PIKE
President
The Tides Foundation

CHARLES C. SAVITT
President
Island Press

SUSAN E. SECHLER

VICTOR M. SHER, ESQ.
Principal
Sher Leff LLP

PETER R. STEIN
General Partner
LTC Conservation Advisory
   Services
The Lyme Timber Company

DIANA WALL, PH.D.
Director, School of Global
Environmental Sustainability
   and Professor of Biology
Colorado State University

WREN WIRTH
President
Winslow Foundation